CAMBRIDGE LIBRARY COLLECTION

Books of enduring scholarly value

Physical Sciences

From ancient times, humans have tried to understand the workings of the world around them. The roots of modern physical science go back to the very earliest mechanical devices such as levers and rollers, the mixing of paints and dyes, and the importance of the heavenly bodies in early religious observance and navigation. The physical sciences as we know them today began to emerge as independent academic subjects during the early modern period, in the work of Newton and other 'natural philosophers', and numerous sub-disciplines developed during the centuries that followed. This part of the Cambridge Library Collection is devoted to landmark publications in this area which will be of interest to historians of science concerned with individual scientists, particular discoveries, and advances in scientific method, or with the establishment and development of scientific institutions around the world.

Joint Scientific Papers of James Prescott Joule

Sir James Prescott Joule (1818–1889) became one of the most significant physicists of the nineteenth century, although his original interest in science was as a hobby and for practical business purposes. The son of a brewer, he began studying heat while investigating how to increase the efficiency of electric motors. His discovery of the relationship between heat and energy contributed to the discovery of the conservation of energy and the first law of thermodynamics. Volume 2 of his collected papers, published in 1887, contains those which he co-authored with other noted physicists, such as Scoresby, Playfair and William Thomson, later Lord Kelvin. Because he was based in Manchester, and was not an academic, Joule's work was at first ignored by the scientific establishment, but Thomson's approval helped him gain acceptance. His joint work with Thomson on thermodynamics was fundamental to the development of significant areas of twentieth-century physics.

Cambridge University Press has long been a pioneer in the reissuing of out-of-print titles from its own backlist, producing digital reprints of books that are still sought after by scholars and students but could not be reprinted economically using traditional technology. The Cambridge Library Collection extends this activity to a wider range of books which are still of importance to researchers and professionals, either for the source material they contain, or as landmarks in the history of their academic discipline.

Drawing from the world-renowned collections in the Cambridge University Library, and guided by the advice of experts in each subject area, Cambridge University Press is using state-of-the-art scanning machines in its own Printing House to capture the content of each book selected for inclusion. The files are processed to give a consistently clear, crisp image, and the books finished to the high quality standard for which the Press is recognised around the world. The latest print-on-demand technology ensures that the books will remain available indefinitely, and that orders for single or multiple copies can quickly be supplied.

The Cambridge Library Collection will bring back to life books of enduring scholarly value (including out-of-copyright works originally issued by other publishers) across a wide range of disciplines in the humanities and social sciences and in science and technology.

Joint Scientific
Papers of
James Prescott Joule

Volume 2

James Prescott Joule

CAMBRIDGE
UNIVERSITY PRESS

CAMBRIDGE UNIVERSITY PRESS

Cambridge, New York, Melbourne, Madrid, Cape Town,
Singapore, São Paolo, Delhi, Tokyo, Mexico City

Published in the United States of America by Cambridge University Press, New York

www.cambridge.org
Information on this title: www.cambridge.org/9781108028837

© in this compilation Cambridge University Press 2011

This edition first published 1887
This digitally printed version 2011

ISBN 978-1-108-02883-7 Paperback

JOINT

SCIENTIFIC PAPERS

OF

JAMES PRESCOTT JOULE,

D.C.L. (Oxon.), LL.D. (Dubl. et Edin.), F.R.S., Hon. F.R.S.E.,

PRES. SOC. LIT. PHIL. MANC., F.C.S., DOC. NAT. PHIL. LUGD. BAT., SOC. PHIL. CANTAB.,
SOC. PHIL. GLASC., INST. MACH. ET NAUP. SCOT. ET SOC. ANTIQ. PERTH,
SOC. HONOR. INST. FR. (ACAD. SCI.),
CORRESP. SOC. REG. DAN. HAFN. BOLON., SOC. PHIL. NAT. BASIL.,
SOC. PHYS. FR. PAR. ET HAL., ET ACAD. AMER. SCI. ET ARTIB.,
ACAD. REG. SCI. BELG. ET TAURIN. ADSOC. HONOR.,
HON. MEM. ASIAT. SOC. BENG., MEM. PHYS. SOC. LOND.

PUBLISHED BY
THE PHYSICAL SOCIETY OF LONDON.

LONDON:
TAYLOR AND FRANCIS, RED LION COURT, FLEET STREET.
1887.

ALERE FLAMMAM.

PRINTED BY TAYLOR AND FRANCIS,
RED LION COURT, FLEET STREET.

TABLE OF CONTENTS.

LIST OF PLATES AND ILLUSTRATIONS.

JOINT SCIENTIFIC PAPERS.

Experiments and Observations on the Mechanical Powers of Electro-Magnetism, Steam, and Horses. By the Rev. WILLIAM SCORESBY, *D.D., F.R.SS.L. and E., Corr. Memb. Inst. Fr., &c., and* JAMES P. JOULE, *Secretary of the Literary and Philosophical Society of Manchester, Mem. Chem. Soc., &c.* *

[NOTE, 1885.—On the occasion of the Meeting of the British Association for the Advancement of Science at Manchester in the year 1842, I had the happiness of forming the acquaintance of Dr. Scoresby, eminent for qualities seldom united in one man. At once an experienced seaman, a successful geographical discoverer, a hard-working and eloquent clergyman, he was also a zealous student of nature and a scientific investigator. Dr. Scoresby became greatly interested in the view I was at that time beginning to take of the relation between heat and other forms of force, and in response to my express wish to work with a powerful arrangement of magnets, he kindly invited me to Bradford, of which town he was at that time the Vicar, in order to pursue an inquiry along with him. The duties of the parish

---

* Phil. Mag. xxviii. (3rd Series) p. 448. The Experiments on the Mechanical Powers of Electro-Magnetism were made in the Vicarage, Bradford. Those on the Combustion of Hay and Corn were made at Whalley Range, Manchester.

were, however, so onerous and pressing, that the production
of our paper devolved almost entirely upon myself, so that it
was not without great objection on his part that Dr. Scoresby
allowed his name to appear with mine. Inasmuch, however,
as the facilities for the experiments were afforded by him as
well as the great magnetic battery, I felt that I could not in
justice allow it to appear other than as a joint paper.]

At the last meeting of the British Association, Dr. Scoresby
described a magnetic apparatus of very great power, and
gave an account of some experiments he had made with a
view to test its capabilities for exciting electrical currents.
The coils employed in those experiments were hastily constructed, and by no means calculated to produce a maximum
effect. We agreed, therefore, to construct and try more
efficient ones on the first opportunity.

Two kinds of revolving armature occurred to us as worthy
of trial. One of them consisted of a hollow tube of drawn
iron, 24 inches long, $1\frac{5}{8}$ths inch in diameter, and $\frac{3}{16}$ths of an
inch thick in the metal, bent into the shape of the letter U.
It had a saw-cut along its entire length, in order to prevent
the circulation of electrical currents in the substance of the
iron. Each of the legs of this armature was wound with
274 feet of covered copper wire, $\frac{1}{10}$th of an inch in diameter.
The other armature consisted of two bars of iron, each 20
inches long, 4 inches broad, and $\frac{3}{8}$ths of an inch thick. These
bars were bent edgeways into the form of a semicircle, and
then fastened together with the interposition of a piece of
calico in order to prevent currents in the iron as much as
possible. Each leg of this armature was furnished with two
coils of covered copper wire $\frac{1}{10}$th of an inch thick. The two
coils that were nearest the iron were each 276 feet long;
and each of the other two coils was 296 feet long.

Having placed the two straight steel magnets (each of
which was 4 feet 4 inches long, 4 to 5 inches square, and
had poles of $7\frac{1}{2}$ square inches surface) side by side, in a
horizontal position, and with two of their poles connected by

a suitable armature, we placed the *hollow electro-magnetic armature* on the axis of a revolving apparatus, in such a position that the poles of the armature could revolve at the distance of about ¼th of an inch from the poles of the steel magnets. The coils were arranged for quantity, and connected by means of a proper "commutator" with platinum plates (each exposing an active surface of 5 or 6 square inches) immersed in a dilute solution of sulphuric acid. The maximum amount of decomposition was effected when the armature revolved 500 times per minute. At this velocity ¾ths of a cubic inch of the mixed gases were collected per minute.

Having removed the hollow armature, we now fastened the *flat semicircular armature* upon the axis. When this armature, with its four coils arranged for quantity, was rotated at the rate of 500 revolutions per minute, we collected as much as 1·4 cubic inch of the mixed gases per minute. With the same velocity of rotation, two inches of steel wire, $\frac{1}{90}$th of an inch thick, were raised to a bright red heat; and one inch of the same kind of wire was fused.

Great as the above effects undoubtedly are in comparison with previously recorded results, we expect to be able to augment them very much by causing the armatures to revolve opposite the *true poles* of the magnets, and not, as heretofore, opposite their ends. It is proper also to observe, that on account of the imperfect hardness of many of the steel bars *, the magnets did not possess one quarter of the power due to Dr. Scoresby's principal of construction. We have not, however, hitherto cared to reconstruct the apparatus, because our principal object in the present research was to

* The bars of which the magnetic apparatus was constructed were of various lengths, but of otherwise uniform dimensions, viz. 1½ inch broad and ¼th of an inch thick. The thickness and mass were found too great for effective hardening, at least for obtaining a degree of hardness capable of sustaining the severity of the magnetic test. Œconomy and facility of arrangement were the reasons for adopting this construction, rather than the more certain and effective one of *hard thin plates*, described by Dr. Scoresby in his "Magnetical Investigations."

make experiments with the machine working as an engine, for which purpose the magnets were quite powerful enough.

The battery employed for working the machine as an engine consisted of three cells of Daniell's constant arrangement. In each cell the copper element exposed an active surface of two square feet, and the amalgamated zinc plate a surface of ⅔rds of a square foot. A pretty correct galvanometer, consisting of a circle of thick copper wire and a magnetic needle 3 inches long, was employed for measuring the currents of electricity which were transmitted by the battery through the revolving armatures. The tangents of the deflections of the magnetic needle, corrected by a small equation, indicated the absolute quantities of transmitted electricity. The quantity of zinc consumed in the battery was deduced from the deflections of the needle; the data of the calculation being derived from previous experiments on the quantity of mixed gases evolved from acidulated water by a current capable of producing a given deflection of the needle.

Our first experiments were made with the flat semicircular revolving armature, its four coils being arranged for quantity. The deflection of the needle before the engine was allowed to start amounted to 64°, which indicated a current of 2232, calling the current corresponding to 45°, 1000. The engine being then allowed to start, presently attained a velocity of 140 revolutions per minute. The needle was then observed to stand steadily at 43°, indicating a current of 920. The consumption of zinc in the battery was estimated to be at the rate of 205 grs. per hour.

Although we were not able to apply as exact a dynamometer as we could have wished, we were nevertheless enabled to arrive at a pretty correct estimation of the power developed, by ascertaining the weight which, when thrown over a wheel connected with the engine, was sufficient to keep it in uniform motion. In this way we found that the force developed in the above experiment was equal to raise 21,100 lb. to the height of a foot per hour.

On making a second experiment with the same revolving armature and battery, we obtained the following results :— Current before the engine was allowed to start, 2232; current when the armature was rotating at the rate of 180 revolutions per minute, 850; consumption of zinc per hour, 190 grains ; force given out per hour, 17,820 lb. raised a foot.

Mr. J. P. Joule has already proved that the heat evolved by voltaic and magneto-electrical currents is, *cæteris paribus*, proportional to the square of their intensity * ; and that the power of the electro-magnetic engine is obtained at the expense of the heat due to the chemical reactions of the voltaic battery by which it is worked. He has also shown, that if the whole of the heat developed by the consumption of a grain of zinc in a Daniell's battery could be converted into useful mechanical power, it would be equal to raise a weight of 158 lb. to the height of a foot †. Hence, if we designate the current when the engine is *at rest* by a, and the current when the engine is *in motion* by b, the heat evolved by the circuit in a given time, will, in the two instances, be as a^2 to b^2. But the quantities of zinc consumed being as a to b, the heat, per a given consumption of zinc, will be as a to b, or directly as the currents; $a-b$ will therefore represent the quantity of heat converted by the engine into useful mechanical effect. Therefore, putting x for the mechanical effect in lbs. raised a foot high per the consumption of a grain of zinc, we have

$$x = \frac{158\ (a-b)}{a}.$$

From the above equation it is evident that the œconomical duty will be a maximum when b vanishes or becomes infinitely small in comparison with a. In this case $x = 158$, while the power of the engine will become infinitely small with regard to work performed in a given time. We must, however, ob-

* Phil. Mag. vol. xviii. p. 308, and vol. xix. p. 260.
† Phil. Mag. vol. xxiii. p. 441.

serve that the equation can only be strictly correct when the current b is uniform, which it never can be exactly, in consequence of the resistance of the magnetic induction against the voltaic current varying in the different positions of the revolving electro-magnetic armature. Hence the current b is always, to a certain extent, of a *pulsatory* character, which has the effect of causing it to develope more heat than a *uniform* current of the same quantity. From this circumstance, as well as from the unavoidable existence of some slight currents in the substance of the iron of the revolving armature, the actual œconomical effect will always be somewhat below the duty indicated by our formula.

Applying the formula to our first experiment, we have for the *theoretical* œconomical effect,

$$\frac{158\ (2232-920)}{2232} = 92 \cdot 9,$$

while the *actual* œconomical effect was

$$\frac{21100}{205} = 102 \cdot 9.$$

In our second experiment, the theoretical œconomical effect will be

$$\frac{158\ (2232-850)}{2232} = 97 \cdot 8,$$

and the actual duty,

$$\frac{17820}{190} = 93 \cdot 8.$$

Taking the mean of the two experiments, we have for the theoretical duty 95·3, and for the actual performance 98·3. Here, therefore, in apparent contradiction to what we have just said, the actual exceeds the theoretical duty. This circumstance is, however, partly explained by the fact that the solution of sulphuric acid employed in charging the battery had been mixed immediately before the experiments were made, and was in consequence considerably heated; for

Daniell has shown that the intensity of his battery increases with its temperature, and it is evident that an increase of the intensity or electromotive force of the cells of the battery must be productive of an increased œconomical effect.

The next two experiments were made with the hollow revolving armature, its two coils being arranged for quantity. In these and the subsequent experiments, the battery was charged with a *cold* solution.

Experiment 3.—Current when the engine was kept at rest, 1381 ; current when the armature was revolving 80 times per minute, 850 ; consumption of zinc, 190 grains per hour; power developed, 8800 lb. raised a foot high per hour.　From these data, the theoretical duty will be

$$\frac{158\ (1381 - 850)}{1381} = 60 \cdot 7,$$

and the actual duty will be

$$\frac{8800}{190} = 46 \cdot 3.$$

Experiment 4.—Current before the engine was allowed to start, 1381 ; current when the engine was revolving 102 times per minute, 678 ; consumption of zinc, 151 grains per hour; power developed, 9000 lb. raised a foot per hour.　Hence for the theoretical duty we have,

$$\frac{158\ (1381 - 678)}{1381} = 80 \cdot 4,$$

and for the actual duty,

$$\frac{9000}{151} = 59 \cdot 6.$$

Lastly, we made two experiments in which the engine was fitted up with two straight electro-magnets fastened parallel to the axis.　Each of these straight electro-magnets consisted of a piece of drawn iron tube, 12 inches long, 1⅝ths inch in diameter, and $\frac{3}{16}$ths of an inch thick, cut longitudinally to prevent the circulation of electrical currents in the iron, and furnished with a coil of 210 feet of covered copper wire $\frac{1}{10}$th of an inch thick. A steel magnet consisting of a considerable number of bars was fitted up in order to excite those ends of the straight electro-magnets which were distant from the large steel magnets. The coils were arranged for quantity.

Experiment 5.—Current when the engine was kept still, 2081; current when the armature was revolving 114 times per minute, 1300; consumption of zinc, 291 grains per hour; power developed, 10030 lb. raised a foot per hour. Hence the theoretical duty will be

$$\frac{158\,(2081 - 1300)}{2081} = 59\cdot3,$$

and the actual duty,

$$\frac{10030}{291} = 34\cdot5.$$

Experiment 6.—Current before starting, 2035; current when revolving 192 times per minute, 1000; consumption of zinc, 223 grains per hour; power developed, 12,672 lb. raised a foot per hour. In this case the theoretical duty will be

$$\frac{158\,(2035 - 1000)}{2035} = 80\cdot3 ;$$

the actual performance will be

$$\frac{12672}{223} = 56\cdot8.$$

The mean of the six experiments gives a theoretical duty of 78·5, and an actual duty of 65·6. But, making allowance for the hot solution employed in the first two experiments, we may state that the actual was in general about $\frac{4}{5}$ths of the theoretical duty.

Upon the whole we feel ourselves justified in fixing the maximum available duty of an electro-magnetic engine worked by a Daniell's battery at 80 lb. raised a foot high for each grain of zinc consumed*, or, in other words, at about half the theoretical maximum of duty.

Before we leave this part of the subject, we may state that the above experiments fully bear out the idea expressed by Dr. Scoresby in his " Magnetical Investigations," that steel magnets on his construction may be employed in the stationary part of the electro-magnetic engine with much greater advantage than electro-magnets.   We have already adverted to the imperfect construction of the magnetic apparatus employed in the above experiments; had we employed one of equal weight, but constructed of thin plates of hardened steel, and furnished with armatures and batteries in proportion, we think it highly probable that a power equal to that of one horse might have been attained, the whole weight of the apparatus being considerably under half a ton.

Having thus determined the capabilities of electro-magnetism as a first mover of machinery, it will be interesting and instructive to compare it with two other sources of power, viz. steam and horses.

1. A grain of coal produces, by combustion, sufficient heat to raise the temperature of a lb. of water $1°·634$.   In other words, we may say that the *vis viva* developed by the combustion of a grain of coal is equal to raise a weight of 1335 lb. to the height of one foot. Now the best Cornish steam-engines raise 143 lb. per grain of coal; whence it appears that the steam-engine in its most improved state is not able to develope much more than $\frac{1}{10}$th of the *vis viva* due to the combustion of coal into useful power, the remaining $\frac{9}{10}$ths being given off in the form of heat.

2. A horse, when its power is advantageously applied, is

* Dr. Botto states that 45 lb. of zinc consumed in a Grove's battery are sufficient to work a one-horse power electro-magnetic engine for 24 hours. The intensity of Daniell's battery being $\frac{3}{5}$ths of that of Grove, it follows that 75 lb. of zinc would have been consumed had Dr. Botto employed a Daniell's battery,—a result not widely different from our own.

able to raise a weight of 24,000,000 lb. to the height of one foot per day. In the same time (24 hours) he will consume 12 lb. of hay and 12 lb. of corn*. He is therefore able to raise 143 lb. by the consumption of one grain of the mixed food. From our own experiments on the combustion of a mixture of hay and corn in oxygen gas, we find that each grain of food, consisting of equal parts of undried hay and corn, is able to give $0°·682$ to a lb. of water, a quantity of heat equivalent to the raising of a weight of 557 lb. to the height of a foot. Whence it appears, that one quarter of the whole amount of *vis viva* generated by the combustion of food in the animal frame is capable of being applied in producing a useful mechanical effect,—the remaining three quarters being required in order to keep up the animal heat &c.

Prof. Magnus, of Berlin, has endeavoured to prove that the oxygen which an animal inspires does not combine chemically with the blood, but is merely *absorbed* by it†. The blood thus charged with oxygen arrives in the capillary vessels, where the oxygen effects a chemical combination with *certain substances*, converting them into carbonic acid and water. The carbonic acid, instead of oxygen, is then absorbed by the blood, and thus reaches the lungs to be removed by contact with the atmosphere. Adopting this view, it becomes exceedingly probable that the *whole* of the *vis viva* due to the oxidation or combustion of the " certain substances " mentioned by Magnus is developed by the muscles. The muscles, by their motion, can communicate *vis viva* to external objects ; and, by their friction within the body, can develope heat in various quantities according to circumstances, so as to maintain the

---

    * We have been kindly informed by Mr. J. V. Gibson of Manchester, an eminent veterinary surgeon, that 14 lb. of hay and 10 lb. of corn is the average provender requisite to support a horse of average size, so as to enable him to work daily without any depreciation of his physical condition. We have, however, equalized the quantities of hay and corn, on account of the experiments on combustion having been made with a mixture containing equal portions.

    † See Phil. Mag. ser. 3. vol. xxvii. p. 561.

animal at a uniform temperature. If these theoretic views
be correct, they would lead to the interesting conclusion (which
is the same as that announced by Matteucci from other considerations) that the animal frame, though destined to fulfil so
many other ends, is, as an engine, more perfect in the œconomy of *vis viva* than the best of human contrivances.

On Atomic Volume and Specific Gravity.   By Lyon Playfair, *Esq.*, *Ph.D.*, and J. P. Joule, *Esq.**

[' Memoirs of the Chemical Society,' vol. ii. p. 401, 1845.]

Section I.

The discovery of Gay-Lussac, that gaseous bodies combine in
equal or in multiple volumes, and that the resulting compounds
stand in a similar simple relation to their constituents, is one
of the most important discoveries ever made in physical
science. Its utility has been diminished by its supposed
inapplicability to liquid and solid bodies; as its own ex-
actitude at different temperatures is entirely owing to the
equal expansibility of the same volumes of different gases
by equal increments of heat.

In its most simple form, therefore, it was *à priori* improbable that the law of Gay-Lussac should apply to the
liquid and solid forms of matter. But, as the larger number
of substances are either liquid or solid, and incapable of
passing into the gaseous state, even at very high tem-

* My work with Dr., now the Right Hon. Sir, Lyon Playfair,
K.C.B., was commenced at about the time when he occupied the post of
Chemist to the Royal Manchester Institution. It is only just to observe
that the important theoretical results arrived at with regard to atomic
volumes are almost entirely due to him, while I took the principal part
in the experiments on the expansion of salts, the maximum density of
water, &c. My own individual work was done at Oak Field House,
Upper Chorlton Road, Manchester.—Note, 1885, J. P. J.

peratures, the importance of discovering the law which governs the volumes of these forms of matter, has long been recognized, and for some time past has much engaged the attention of philosophers.

The first chemist who drew attention to this subject was Dr. Thomson, who published a Table *, in the year 1831, of the specific volumes of certain of the metals, obtained by dividing their atomic weights by their specific gravities. In this Table a remarkable coincidence of volume is observed in several of the metals most nearly allied in chemical characters.

More recently the subject has been examined in detail by Kopp, Schröder, and Persoz, whose researches have thrown considerable light on this obscure department of Physics.

Kopp † drew attention to the circumstance, that in many cases isomorphous bodies possess the same atomic volume, the law being correct when the isomorphism is strictly accurate, but approximately only when this is not the case. He admits also that perfect equality of the volume exists only at particular temperatures, on account of the unequal expansion of isomorphous crystals.

Schröder ‡ made the interesting observation, that the remainder is the same when the primitive volume of the corresponding member of a series of analogous compounds is subtracted from them : thus AO, BO, and CO leave a constant remainder when the known volumes of A, B, and C are subtracted respectively from the known atomic volumes of the compounds.

Kopp § confirms this discovery to a certain extent, believing, however, that the primitive volumes A, B and C must be assumed in certain classes of salts to be different when in combination with O from their volumes when isolated.

---

* ' Chemistry of Inorganic Bodies,' vol. i. p. 14.

† Poggendorff Annalen, Band xlvii.; and Annalen der Chemie, Band xxxvi. S. 1.

‡ Poggendorff Annalen, Band l. S. 554.

§ *Ueber das Specifische Gewicht der Chemischen Verbindungen.* Frankfort, 1841.

He also announces the discovery * of a great regularity in the physical properties of analogous organic compounds, so much so that the study of the physical characters of the compounds of one body enables us to predicate those of the corresponding compounds of another substance.

The discoveries of Schröder and Kopp, with regard to the atomic volumes of liquid and solid bodies, do not, except in a very few instances, indicate an approach to a simple mutiple ratio of volumes, and are therefore only in a small degree connected with the law of gaseous volume. We therefore thought it desirable to enter into a series of inquiries on this most important subject, and we have now the honour to lay before the Society the first part of these researches.

Hitherto the inquiry has been principally confined to solid bodies, on the just ground that their diminished rate of expansion offers less difficulty to the discovery of the law regulating volumes. But there is an objection to the use of solids, which to a certain extent counterbalances this advantage, viz. that they do not present matter in a perfectly uniform condition, free from cohesion. On consideration, therefore, we were led to believe this objection to be so powerful, that we conceived it to be preferable so to separate the particles of the body under examination as to destroy their cohesion, without at the same time altering their chemical properties. Solution in water was the obvious means of effecting this purpose, according to the notions generally entertained of solution ; and it was therefore resolved to experiment principally upon soluble bodies of well-known and defined constitution. At the same time, it was necessary to examine the relation of the solid volume to, the volume of the body when in solution, so as to indicate the connection between the solid and the liquid atom.

The specific gravities of salts are little known, and even when recorded are described so differently by different observers, that it was necessary to determine the specific gravity

* Annalen der Chemie, Band xli. S. 79.

in each of the cases upon which the experiments were
instituted. Hitherto the volumes of solids had always been
referred to an equal volume of water; in other words, the
solid form of matter had been referred to its *liquid* form.
This difference of conditions was no small impediment to
the discovery of a law which might be modified for each
form of matter *. By determining the volume of the substance in solution, we compared it in its liquid state to the
liquid form of matter in which it was dissolved; and by
contrasting the volume of the solids with each other, and
also with their volume when rendered liquid by water,
we conceived that we might be placed in more favourable
conditions for elucidating a law.

Bishop Watson † was the first chemist who endeavoured
to estimate the increase of volume when salts dissolve in
water; for, although both Gassendus ‡ and the Abbé Nollet §
had written, and Ellis ‖ had experimented upon the same
subject, they had arrived at conclusions entirely erroneous,
which were removed by Watson's more accurate experiments. Watson's apparatus was rude enough, being a matrass
capable of holding 67 ounces of water, into which he
projected 24 pennyweights of each of the salts upon which
he experimented, and noted the rise in the neck of the
matrass. He completely exploded, however, the idea that

* Before leaving the notice of the labours of those who have preceded
us in inquiring into the nature of specific gravity, we must not omit to
notice the speculations of the ingenious Persoz, who (in vol. xl. of the
Ann. de Ch. et de Phys. p. 119) drew attention to the equality in
volume of isomorphous bodies, and even of some which were not
isomorphous. Persoz also believes that the volumes of all bodies are
multiples of 56, or half the atomic weight of water; but this idea does
not agree with recorded observations, and is directly contradicted by
accurate estimations of specific gravities.—See the work of M. Persoz,
Introduction à l'Etude de la Chimie Moléculaire, p. 834 *et seq.*

† Philosophical Transactions, 1770.
‡ Gass. *Phys.* lib. i. Sect. i. cap. 3.
§ Leçons de Physique, vol. iv.
‖ Berlin Memoirs, 1750.

saline substances dissolve in water without increasing its bulk.

Between the time of Bishop Watson, whose investigations on this subject are most profound, when we consider the period at which he wrote, and that of Dalton, there were no labourers in this field to whom we need draw especial attention. In the year 1840, Dalton * made the interesting discovery, that sugar and certain salts on being dissolved in water increase its bulk only by the amount of water pre-existing in them. He generalized this observation by asserting that all hydrated salts dissolve in water, increasing its bulk merely by their amount of water of hydration, while anhydrous salts do not at all increase the bulk of the water in which they are dissolved.

But it must not be forgotten, that when Dalton published this paper, he was much enfeebled by illness, and on this account it does not derogate from the acuteness of the philosopher that Mr. Holker was unable to confirm Dalton's results in repeating the experiments in 1843 †. He did so, however, in the case of sulphate of magnesia, and approximately in that of one or two other salts. As Mr. Holker's paper has not been published, we are unable to state his claims in the progress of this subject; but we believe that an attempt was made to show a multiple relation in the increments of *isomorphous* salts, although his experiments were conducted without reference to the density or temperature of the solution on which he operated.

In the experiments about to be described, the apparatus for estimating the volume of bodies when dissolved consisted of a glass bulb to which a stem was attached. The bulbs varied in capacity from 1000 to 4000 grains of water, and the diameter of the stem was from one eighth to one sixteenth

---

* "On the Quantity of Acids, Bases, and Waters in Salts, and a new mode of measuring them." Read to the Manchester Literary and Philosophical Society, 6th October, 1840, and published as a pamphlet.

† Paper read to the Manchester Literary and Philosophical Society, but not published.

of an inch, according to the character of the experiment. In the bulbs employed for ordinary purposes, each grain of

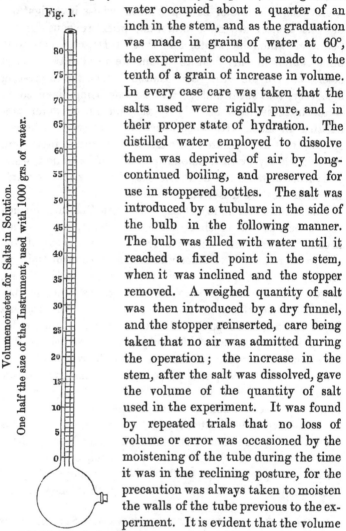

Fig. 1.

Volumenometer for Salts in Solution.

One half the size of the Instrument, used with 1000 grs. of water.

water occupied about a quarter of an inch in the stem, and as the graduation was made in grains of water at 60°, the experiment could be made to the tenth of a grain of increase in volume. In every case care was taken that the salts used were rigidly pure, and in their proper state of hydration. The distilled water employed to dissolve them was deprived of air by long-continued boiling, and preserved for use in stoppered bottles. The salt was introduced by a tubulure in the side of the bulb in the following manner. The bulb was filled with water until it reached a fixed point in the stem, when it was inclined and the stopper removed. A weighed quantity of salt was then introduced by a dry funnel, and the stopper reinserted, care being taken that no air was admitted during the operation; the increase in the stem, after the salt was dissolved, gave the volume of the quantity of salt used in the experiment. It was found by repeated trials that no loss of volume or error was occasioned by the moistening of the tube during the time it was in the reclining posture, for the precaution was always taken to moisten the walls of the tube previous to the experiment. It is evident that the volume occupied by a salt in solution must be modified by the position of the point of its maximum density. Despretz has shown *

* Annales de Chimie, tome lxx. an. 1839, p. 81.

that the temperature at which solutions are most dense becomes lower in proportion to the quantity of matter held in solution. It is also known, from the experiments of Dalton and others, that from the point of maximum density to about 30° above or below it, water and dilute solutions expand according to the square of the temperature from that of greatest density. From Despretz's table of the expansion of water, it appears that the law is not true as far as 212° As far, however, as it does hold good, it is evident, from the properties of the parabola, that the volume occupied by a salt in solution will increase in arithmetical progression with the temperature at which the experiment is made.

For instance, let *a a* in the following diagram be a parabola representing the expansion of water, and let *b b* be a similar parabola representing the expansion of a solution. Let the latter parabola have its vertex or point of greatest density opposite 30°, while the former parabola has its vertex at 40°; then $x x$, $x' x'$, $x'' x''$, &c., quantities which increase in

Fig. 2.

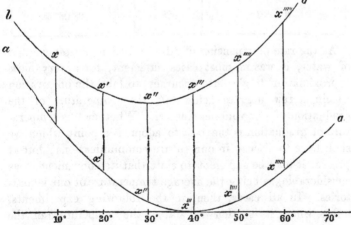

arithmetical progression, will represent the volumes occupied by the salt in solution at the temperature of 10°, 20°, 30°, &c. In a similar manner it may be shown that the volume occupied by each equivalent of a salt in solution at any given temperature will increase with the density of the solution. In order to ascertain the amount of influence exercised by a change in the position of the point of maximum density, we have made a series of experiments on the expansion of water and of solutions by heat, which we propose to lay before the Society in a succeeding memoir. In order, however, to render evident the augmentation of volume caused by increased density, we have constructed the following table of the volumes occupied by 172 grains, or one equivalent of sugar, in solutions of different degrees of density.

TABLE I.

Ratio of the quantity of sugar to the quantity of water in which it was dissolved.	Temperature.	Volume in grain-measures of water.
1 : 120	60°	99·00
1 : 10	52°	105·09
1 : 1	52°	107·01
3 : 1	52°	108·06

As the rate of expansion of dilute solutions is so near that of water, it was in most cases sufficient, for a very close approximation to absolute accuracy, to take the observation within a few degrees below 60°, the temperature of the graduation of our volumenometers. Whether this temperature of graduation is the best to adopt, is a point which we shall have to discuss in our future communications; but at present it may be sufficient to state that its convenience was considerable, as being the average temperature of our laboratories. In all cases, then, in the following experiments, unless where it is otherwise stated, the temperature of the solution was about 60°, which was also, of course, the temperature of the water before the salt was introduced. In the case of the sulphates of the magnesian class of metals, the temperature chosen was higher than 60°, in order to make up for a diminished rate of expansion, owing to a greater degree of dilution in the solution.

The specific gravity of the salts was determined in an equally simple manner. A saturated solution of the salt about to be experimented upon (made by dissolving an excess of the salt by heat and allowing the solution to cool) was placed in the apparatus already described, and a weighed portion of the salt was then introduced, care being taken that the temperature did not vary during the experiment. As the new portion of salt could not dissolve, the increase in the stem indicated the volume due to the quantity of salt introduced, and afforded data for calculating the specific gravity. In many cases oil of turpentine was used instead of the saline solution.

It was frequently desirable, especially in the case of hydrated salts rendered anhydrous, to avoid the use of water, and, in the case of organic compounds, also of turpentine; and to meet such cases we constructed the following simple apparatus, which we believe to possess various advantages.

A, in fig. 3, is the receiver of an air-pump, furnished at the top with a collar and sliding-rod, B. C is a small graduated tube filled with the substance, the volume of which has to be determined; it is closed with a stopper E, perforated with a hole of dimensions so small as to prevent any of the salt from falling out. D is a cup of mercury placed immediately below the graduated tube C. The sketch indicates the position of the apparatus on an air-pump when the experiment is about to be performed. The receiver is then exhausted as thoroughly as possible, and the indication of the siphon-gauge is accurately noted. The graduated tube is then lowered by means of the sliding-rod until it touches the bottom of the cup containing the mercury, which, after the admission of air, flows into the tube until it is filled. The whole contents of the tube are then thrown into water, and the salt is washed away by decantation. The mercury is dried by bibulous paper, and restored to the tube. If the temperature be different from that which it possessed in the first part of the

c 2

experiment, it is restored to the original temperature, or a correction is made for the difference. It is now obvious that the space in the tube unoccupied by the mercury is that which was formerly filled with the salt. To this, however, must be added a slight correction for the imperfect nature

Fig. 3.

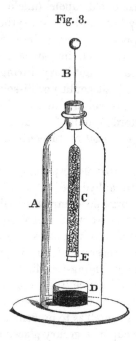

of the vacuum, which is not Torricellian—a correction which need not exceed $\frac{1}{200}$th of the volume observed. With these preliminary descriptions and observations, we now proceed to describe the details of our experiments, throwing them into various classified groups of salts for the purpose of easy reference.

The first group described is remarkable for containing a large amount of water of hydration.

Sulphate of Copper, CuO, $SO_3 + 5HO = 124\cdot88$.—The third part of an equivalent of this salt (41·62 grains) dissolved in 3140 grains of water at 32° with an increase of 13·15, but dissolved in water at 90° with an increase of 15·0.

CuO, $SO_3 + 5HO$, vol. in solution 45·0.

Half an equivalent of this salt (62·44 grains), being immersed in a saturated solution, occupied the volume of 27·7 water grain-measures.

	Sp. vol.	Sp. gr.
CuO, $SO_3 + 5HO$, vol. of salt . .	55·4	2·254.

Kopp found for the specific gravity of this salt the number 2·274.

Sulphate of Alumina, Al_2O_3, $3SO_3 + 18HO = 333·7$.—The salt used in the experiments was carefully prepared, and obtained in tolerably good crystals. The eighth part of an equivalent (41·7 grains), dissolved in 1000 of water, with an increase of 20 in one experiment and 19·9 in another, the temperature of observation being 51°.

I. Sulphate of Alumina, vol. in solution . . 160
II. „ „ „ „ 159·2
                                           ———
                              Mean . . 159·6

The same quantity of salt thrown into turpentine caused in two experiments an increase of 25·0, and in a third of 24·9.

		Sp. gr.
I. Sulphate of Alumina, vol. of salt . . .	200·0	1·668
II. „ „ „ „	199·2	1·675
	———	———
Mean . .	199·6	1·671

Sulphate of Soda, NaO, $SO_3 + 10HO = 161·48$ —Sulphate of soda, crystallized out of a strong warm solution, carries down 10 atoms of water. Of this salt, about one fourth of an equivalent (40·4 grains), on being dissolved in 1000 grains of water, caused in two experiments an increase of 23·0, at a temperature of 59°; and, in a third experiment, of 22·8 at the same temperature.

I., II. NaO, $SO_3 + 10HO$, vol. in solution. . 91·8
III. „ „ „ 91·2
                                        ———
                              Mean . . 91·5

The same quantity of the salt being immersed in a saturated solution occasioned an increase of 27·8; and on a second experiment, of 27·2 at a temperature of 62°.

		Sp. gr.
I. NaO, SO_3 + 10HO, vol. of salt . . . 111·1	1·453
II. „ „ „ 108·7	1·485
Mean . . 109·9	1·469

When sulphate of soda crystallizes from a weak cold solution, it carries down a quantity of water, corresponding to eleven equivalents. In two experiments, the volume in solution of salt procured in this way was 98; but we apprehend that the water is merely mechanical, for reasons which will be seen hereafter, as the volume of the salt itself, by a mean of several experiments, came out to 119·5, whereas had this eleventh atom of water been combined it should have been 121.

Biborate of Soda, NaO, $2BO_3$ + 10HO = 191·23.—On dissolving 40 grains of this salt in 1000 of water, the increase was 19·2 at a temperature of 55°.

NaO, $2BO_3$ + 10HO, vol. in solution 91·7.

Half an equivalent, or 95·61 grains, on being placed in a saturated solution, occasioned an increase of 55·5, and 47·8 grains caused an increase of 27·5; both experiments being made at a temperature of 55°.

		Sp. gr.
I. NaO, $2BO_3$ + 10HO, vol. of salt . . 111	1·722
II. „ „ „ 110	1·738
Mean . . 110·5	1·730

Chloride of Strontium, SrCl + 6 HO = 133·32.—There are two hydrates of chloride of strontium, the one with nine, and the other with six equivalents of water. To determine which of these hydrates was under examination, 4·324 grammes were heated to redness with a loss of 1·75 gramme = 40·47 per cent., showing that the hydrate was that with six equivalents of water, which gives by calculation 40·50 per cent.

On dissolving 40 grains of this salt in 1000 of water, the increase occasioned at a temperature of 56° was 16·0; a second experiment, in which the same quantities were used, gave exactly the same result.

I., II. SrCl + 6 HO, vol. in solution 53·3.

The same quantity of salt (40 grains), immersed in a saturated solution, caused an increase of 20·0 at a temperature of 57°; and on a second experiment, of 19·7.

					Sp. gr.
I. SrCl + 6 HO, vol. of salt	. . .	66·6	2·000
II. „ „ „		65·6	2·030
		Mean . . 66·1	2·015

Chloride of Calcium, CaCl + 6HO = 109·92.—On dissolving 55 grains, or the half of an equivalent of this salt, in 1000 grains of water, an increase of 28·0 was obtained at the temperature of 70°; and in a second experiment 27·6 at 60°.

I.   CaCl + 6 HO, vol. in solution . . 56·0
II.   „   „   „   55·2

Mean . . 55·6

The same quantity of salt thrown into turpentine caused an increase of 32·7 and 32·8 in two experiments.

				Sp. gr.
I. CaCl + 6 HO, vol. of salt	. .	65·4	1·682
II. „ „ „		65·6	1·677
	Mean . . 65·5	1·680

Chloride of Magnesium, MgCl + 6 HO = 102·16.—Millon has lately described this salt as containing $6\frac{1}{2}$ atoms of water; but as we have not been successful enough to obtain this hydrate, we retain the old formula. 25·54 grains, or the fourth of an equivalent, dissolved in 1000 grains of water at 53°, with an increase of 14·0.

MgCl + 6 HO, vol. in solution 56·0.

The same quantity, 25·54 grains, gave, in four experiments, respectively 16·5, 16·0, 16·4, 16·5.

					Sp. gr.
I.	MgCl + 6HO, vol. of salt	. . .		66·0	1·548
II.	„	„	„	64·0	1·595
III.	„	„	„	65·6	1·557
IV.	„	„	„	66·0	1·548
			Mean . .	65·4	1·562

The salts now examined are not calculated, on account of the deliquescent character of several of them, to produce absolutely accurate experimental results; but notwithstanding this circumstance, the determination of their volumes is sufficiently uniform to indicate the theory. The actual volume observed for each of the salts in solution, when divided by 9, the atomic volume of water, yields as the quotient the same number as that representing the atoms of water in the salt. Hence it is quite certain that the salts now described dissolve in water without adding to its bulk more than is due to the liquefaction of the water in chemical combination with them.

The volumes of the salts in their solid state possess a number considerably higher than that representing the liquid volume, but affect a divisor which is the same for all the salts, allowing for errors of experiments, or for alterations caused by incidental circumstances. This divisor is a number either equal or approximating to 11. When the volumes of the salts in the solid state are divided by this number, the quotient represents the number of atoms of water attached to the salt. The most natural view of this circumstance is to suppose that water in combination as a solid with a salt possesses a higher volume than liquid water, just as in the case of ice. If this view be correct, the atomic volume of the salts described is the same in the state of a solid as when in solution, the only difference being, that in the one case the volume is expressed by liquid, in the other by solid water. In this case, however, water in combination with a salt does not possess the same volume as ice, which, according to our experiments, detailed in another part of this paper, has a volume of 9·8, and not of a number approaching to 11.

Yet there is nothing extravagant in the idea that water combined with a salt may have a volume different from that of ice. Indeed, we are inclined to be of the opinion that ice represents nearly the mean of the volume of water uncombined and that of combined water. Be this as it may, it will be observed as we proceed that the number 11 is the best exponent of one class of our experiments on specific gravity; and therefore, without resting its claims to acceptance entirely on the present experiments, we assume it in the following tables as the theoretical result for each class of salts. With these views we tabulate the experiments which have been already detailed (Table II.).

There are some salts which do not take up any space in solution, except that due to their water, but which assume a volume due to one of their constituents on becoming solid; the potash- and ammonia-alums are examples of this class.

Sulphate of Alumina and Potash, Al_2O_3, $3SO_3 + KO$, $SO_3 + 24HO = 474·95$.—59 grains of alum dissolved in 1000 grains of water gave the increase of 27·0 in one experiment and 27·1 in another, both at the temperature of 60°.

> I. Alum, volume in solution . . 217·3
> II.   „   „   „   218·1
>   ———
>   Mean . . 217·7

On throwing 59·37 grains into a saturated solution, an increase of 34·4 was obtained in the first experiment, of 34·7 in a second, of 34·3 in a third, and of 34·2 in a fourth, all at a temperature about 60°.

				Sp. gr.
I. Alum, vol. of salt . . .	275·2			1·726
II. „	„	277·6		1·711
III. „	„	274·4		1·730
IV. „	„	273·6		1·735
	Mean . .	275·2		1·726

TABLE II.—Showing the Volumes occupied by certain Salts containing a large amount of Hydrate Water.

Name.	Formula.	Atomic weight.	Volume in Solution.			Volume as Salt.				
			Volume in solution.	θ, or volume of water, as unity.	Volume by calculation.	Volume of salt by experiment.	11, an assumed number, as unity.	Volume by calculation.	Specific gravity by calculation.	Specific gravity by experiment.
Sulphate of Copper	CuO, SO_3+5HO	124·88	45·0	5	45	55·4	5	55	2·270	2·254
Sulphate of Alumina	$Al_2O_3, 3SO_3+18HO$	333·7	159·6	18	162	199·6	18	198	1·685	1·671
Sulphate of Soda	NaO, SO_3+10HO	161·48	91·6	10	90	109·9	10	110	1·468	1·469
Biborate of Soda	$NaO, 2BO_3+10HO$	191·23	91·7	10	90	110·5	10	110	1·738	1·730
Chloride of Strontium	$SrCl+6HO$	133·33	53·3	6	54	66·1	6	66	2·020	2·015
Chloride of Calcium	$CaCl+6HO$	109·92	55·6	6	54	65·5	6	66	1·665	1·680
Chloride of Magnesium	$MgCl+6HO$	102·16	56·0	6	54	65·4	6	66	1·548	1·562

Sulphate of Alumina and Ammonia, Al_2O_3, $3SO_3 + NH_4O$
$SO_3 + 24HO = 454\cdot26$.—20 grains of this salt dissolved in
4100 grains of water with an increase of $10\cdot0$ at $58°$.

Ammoniacal Alum, vol. in solution $227\cdot1$.

The eighth part of an equivalent ($56\cdot78$ grains), immersed
in turpentine, caused a rise in the stem of $34\cdot9$ in one experiment and $35\cdot0$ in another, the temperature being $60°$.

		Sp. gr.
I. Ammonia-alum, vol. of salt . . .	279·2	1·627
II. „ „ „	280·0	1·623
Mean . .	279·6	1·625

Chrome-alum, Cr_2O_3, $3SO_3 + KO$, $SO_3 + 24HO = 504\cdot1$.—
On dissolving 32 grains of this salt in 4100 grains of water,
an increase of $13\cdot7$ was effected at $37°$.

Chrome-alum, vol. in solution $215\cdot8$.

In two experiments 63 grains of this salt thrown into
turpentine caused an increase of $34\cdot5$.

	Sp. gr.
Chrome-alum, vol. of salt . . . 276	1·826

Iron Ammonia-alum, Fe_2O_3, $3SO_3 + NH_4O$, $SO_3 + 24HO =$
$481\cdot03$.—On dissolving $30\cdot06$ grains in 1000 grains of water,
an increase of $14\cdot3$ was obtained at a temperature of $37°$.

Iron-alum, vol. in solution 228.

The eighth part of an equivalent ($60\cdot13$ grains) produced
an increase of $35\cdot0$ measures when thrown into turpentine.

	Sp. gr.
Ammoniacal Iron-alum, vol. of salt . . 280·0	1·718

Pyrophosphate of Soda, $2NaO$, $PO_5 + 10\,HO = 224\cdot15$.—
The eighth part of an equivalent of the crystallized pyrophosphate, 28 grains, dissolved in 1000 grains of water, with
an increase of $11\cdot2$ in one experiment and of $11\cdot3$ in another,
the temperature in both cases being $58°$.

Pyrophosphate of Soda, vol. in solution . . 89·6

      ,,   ,,   ,,   90·4

                                   Mean . . 90·0

On immersing 56·04 grains of the salt in a saturated solution, an increase of 30·5 was obtained in two experiments.

                                         Sp. gr.

Pyrophosphate of Soda, vol. of salt. . 122·0   1·836

By tabulating the results thus obtained we find the following relationship between the class of alums (Table III.).

The peculiarity of the salts described in the table is, that the quotient of the divisor for the potash-alums in the solid state is not the same as in the state of solution, and that the ammoniacal alums possess one volume in solution greater than the corresponding potash-alums, both of which peculiarities will find an explanation as we proceed. Pyrophosphate of soda shares this peculiarity, and is therefore introduced into the table.

We now proceed to describe a class of hydrated salts in which the divisor for the solid volume is certainly not the number 11.

Carbonate of Soda, $NaO, CO_2 + 10 HO = 143·4$.—On dissolving 35·85 grains of this salt in 1000 grains of water, the increase was 22·5 in one experiment at 64°, and 22·9 in a second experiment at 65°.

   I. $NaO, CO_2 + 10 HO$, vol. in solution . . 90·0

  II.   ,,   ,,   ,,   91·6

                                     Mean . . 90·8

On throwing 35·8 grains of the salt into turpentine, the increase was 24·7, 24·5, 24·6, and 24·8 in consecutive experiments with different specimens.

                                       Sp. gr.

   I. Carbonate of Soda, vol. of salt . . 98·8   1·451

  II.   ,,   ,,   ,,   98·0   1·463

 III.   ,,   ,,   ,,   98·4   1·457

 IV.   ,,   ,,   ,,   99·2   1·446

                             Mean . . 98·6   1·454

TABLE III.—Showing the Volumes of certain Alums *.

| Designation | | | | Volume in Solution. | | | Volume in state of Salt. | | | | |
Name.	Formula.	Atomic weight.	Volume in solution by experiment.	9, or volume of water, as unity.	Volume by theory.	Volume of salt by experiment.	11, taken as unity.	Volume of salt by theory.	Specific gravity by theory.	Specific gravity by experiment.
Potash-alum	$AlO_3, 3 SO_3, KO SO_3 + 24 HO$	474·95	217·7	24	216	275·2	25	275	1·727	1·726
Ammonia-alum.......	$AlO_3 3 SO_3 + NH_4O SO_3 + 24 HO.$	454·26	227·1	25	225	279·6	25	275	1·652	1·625
Chrome-alum	$Cr_2O_3 3 SO_3 + KO SO_3 + 24 HO.$	504·1	215·8	24	216	276·0	25	275	1·833	1·826
Iron-alum	$Fe_2O_3 3 SO_3 + NH_4O SO_3 + 24 HO.$	481·03	228·0	25	225	280·0	25	275	1·749	1·718
Pyrophosphate of Soda ..	$2 NaO PO_5 + 10 HO$	224·15	90	10	90	122	11	121	1·852	1·836

* *Vide* Conclusion for explanation of the high volume of the Ammonia-alums.

Rhombic Phosphate of Soda, $2NaO, HO, PO_5 + 24 HO =$ 359·1.—The eighth part of an equivalent of this salt, 44·9 grains, dissolved in 1000 grains of water, with an increase of 27·0 in one experiment and 27·1 in a second; by some mistake the temperature of the solution has not been recorded.

I. Phosphate of Soda, vol. in solution . . 216·0
II.   „   „   „   216·8

Mean . . 216·4

The same quantity of salt thrown into turpentine produced an increase of 29·4 in two experiments, and 29·5 in a third.

					Sp. gr.
I.	Phosphate of Soda,	vol. of salt	. .	235·2	1·527
II.	„	„	„	235·2	1·527
III.	„	„	„	236·0	1·521
			Mean . .	235·5	1·525

Subphosphate of Soda, $3NaO, PO_5 + 24 HO = 381·6$.—The eighth part of an equivalent of this salt, 47·7 grains, dissolved in 1000 grains of water, with an increase of 27·1 in two experiments at 48°.

I., II. Subphosphate of Soda, vol. in solution 216·8.

The same quantity of salt thrown into turpentine produced an increase of 29·4 in two experiments under favourable circumstances, although in another experiment, in which we were not satisfied with the state of hydration of the salt, the increase was only 28·9.

Sp. gr.
Subphosphate of Soda, vol. of salt . . 235·2   1·622

Arseniate of Soda, $2NaO, HO, AsO_5 + 24 HO = 402·9$.— On dissolving 50·36 grains, the eighth part of an equivalent, in 1000 grains of water, an increase of 27·2 was obtained at the temperature of 54°.

Arseniate of Soda, vol. in solution 217·6.

The same quantity of salt thrown into a saturated solution caused an increase of 29·0 in several experiments. This salt loses its water with such facility that it is almost impossible to obtain it in a state well-fitted for experiment. In two specimens of salts, prepared at different times, the volume for the above quantity of salt was 29·7 and 29·8; but as in most cases it was only 29·0, we give the result most generally obtained.

		Sp. gr.
Arseniate of Soda, vol. of salt . . 232		1·736.

Subarseniate of Soda, $3NaO, AsO_5 + 24 HO = 425·2$.— The eighth part of an equivalent, 53·15 grains, of this salt dissolved in 1000 grains of water, with an increase of 27·0 in one experiment and 26·9 in another, at a temperature about 55°.

I. Subarseniate of Soda . .		216·0
II. „ „		215·2
	Mean . .	215·6

The same quantity of salt immersed in turpentine caused an increase of 29·4 and 29·5 in two experiments.

		Sp. gr.
I. Subarseniate of Soda, vol. of salt . .	235·2	1·808
II. „ „ „	236·0	1·801
	Mean . . 235·6	1·804

Cane-Sugar, $C_{12} H_{11} O_{11} = 171·6$.—25·8 grains of sugar dissolved in 3140 grains of water caused an increase of 14·8 at 32°; 42·9 grains, or the fourth of an equivalent, gave an increase of 25·0 at 60° in two experiments.

I. Cane-sugar, vol. in solution . .		98·4
II. „ „ „		100·0
III. „ „ „		100·0
	Mean . .	99·5

300 grains of sugar-candy thrown into alcohol previously saturated with it, caused an increase in the first experiment

of 188·0, in the second of 188·75 ; in a third experiment, 49·65 grains thrown into turpentine caused an increase of 31·0 ; and the same quantity, in a fourth experiment, of 31·1 ; the temperature in all the cases being about 60°.

				Sp. gr.
I.	Cane-sugar, vol. of solid . .	107·5		1·596
II.	,,	,,	107·9	1·590
III.	,,	,,	107·1	1·602
IV.	,,	,,	107·5	1·596
		Mean . .	107·5	1·596

In this section a class of salts presents itself in which the volumes are clearly not represented by any multiple of 11 ; yet they are uniform in their isomorphous relations, and are sensibly multiples of the same number. To discover whether the solid volume has any relation to that occupied by ice, we have determined the specific gravity of the latter with great care. The distilled water, which we converted into ice, was deprived of air by long-continued boiling, and a weighed portion of the ice was quickly immersed in water at 32°, the balance being kept at the same temperature during the weighing of the ice. The rise in the stem of the volumenometer, in which the fragments of ice had previously been placed, indicated the volume due to the quantity of ice immersed. On treating in this manner 54·2 grains of ice, a rise in the stem of 59·0 was produced; and in a second experiment 52·8 grains of ice occasioned an increase of 57·5 ; the temperature in both cases being exactly 32°.

		Sp. vol.	Sp. grav.
I.	Ice volume . .	9·797	0·9186
II.	,,	9·801	0·9183
	Mean . .	9·799	0·9184

As the true specific gravity of ice is a subject of much importance, we place here all the recorded results, as given in Böttger's most useful work on Specific Gravity, and in the first volume of Scoresby's 'Arctic Regions.'

TABLE IV.—Showing the Volumes occupied by certain Phosphates, Arseniates, Carbonate of Soda, and Cane-sugar.

Designation.			Volume in Solution.			Volume of Salt.				
Name.	Formula.	Atomic weight.	Volume in solution by experiment.	9, or volume of water, as unity.	Volume by theory.	Volume of salt by experiment.	9·8, or volume of ice, as unity.	Volume by theory.	Specific gravity by theory.	Specific gravity by experiment.
Carbonate of Soda	NaO, CO_2+10HO	143·4	90·8	10	90	98·6	10	98·0	1·463	1·454
Phosphate of Soda	$2NaO, HO, PO_5+24HO$ ..	359·1	216·4	24	216	235·5	24	235·2	1·527	1·525
Subphosphate of Soda ..	$3NaO, PO_5+24HO$	381·6	216·8	24	216	235·2	24	235·2	1·622	1·622
Arseniate of Soda......	$HO, 2NaO, AsO_5+24HO$	402·9	217·6	24	216	232	24	235·2	1·713	1·736
Subarseniate of Soda ..	$3NaO, AsO_5+24HO$	425·2	215·6	24	216	235·6	24	235·2	1·808	1·804
Cane-sugar	$C_{12}H_{11}O_{11}$	171·6	99·5	11	99	107·5	11	107·8	1·591	1·596

Specific gravity of ice . . 0·888 Dulk.

     „   „   0·937 Irvine.

     „   „   0·945 Williams.

     „   „   0·885 Meineke.

     „   „   0·905 Heinrich Kraft.

     „   „   0·927 Osann.

     „   „   0·950 Roger & Dumas.

     „   „   0·920 Scoresby.

Mean . . . 0·919

The mean of all these experiments, differing only $\frac{1}{1000}$ from our own determination, warrants us in concluding that our result is accurate, and that 9·80 may safely be taken as the specific volume of an atom of ice. Now it must at once strike the observer of the previous experiments, that this number forms the divisor for the volumes of the salts described in the present section.

Connected with the latter group there is a class of salts which come out uniformly with themselves, but the divisor of which is not 11 in the solid state. We subjoin them in the following group.

Sulphate of Magnesia, MgO, $SO_3 + 7HO = 123·86$.—When this salt is dissolved in a large quantity of cold water, the volume observed after solution is always less at ordinary temperatures than that due to the water contained in the salt. That this diminution is due to a contraction caused by an affinity of the salt for water is shown by the fact that anhydrous sulphate of magnesia dissolved in a large quantity of water actually lessens, instead of increasing, the bulk of the water; and to compensate for this contraction, a certain temperature has to be given to the water. In the following experiments with the sulphates of magnesia, zinc, and iron, this circumstance has been attended to, and the temperature is given at which the results come out exact.

31 grains of crystallized sulphate of magnesia were dissolved in 3140 grains of water at 32°, and caused an increase of 15·22; at 85° the increase was 15·75.

MgO, $SO_3 + 7HO$, vol. in solution 63.

Half an equivalent (61·93 grains) being placed in a saturated solution of the salt, caused an increase of 37·5 in one experiment, but in three other experiments the increase was not greater than 37·2; the temperature in all the cases being 54°.

					Sp. gr.
I. $MgO, SO_3 + 7HO$, vol. of salt	. .	75·0	1·651
II.	„	„	„	74·4	1·664
III.	„	„	„	74·4	1·664
IV.	„	„	„	74·4	1·664
		Mean . . .	74·55	1·660

Sulphate of Zinc, $ZnO, SO_3 + 7HO = 143·43$.—This salt possesses the same property as sulphate of magnesia, of causing a contraction when the anhydrous salt is dissolved in a large quantity of cold water.

35·9 grains were dissolved in 3140 grains of water at a temperature of 32°, causing a rise in the stem of 14·03. At 90° the rise was 15·77.

$ZnO, SO_3 + 7HO$, vol. in solution 63.

The half of an equivalent of this salt (71·71 grains), being thrown into a saturated solution, caused an increase of 37 1; and 35·85 grains produced a rise in the stem of 18·6 in two experiments, and of 18·5 in a fourth.

					Sp. gr.
I. $ZnO, SO_3 + 7HO$, vol. of salt	. .	74·2	1·933
II.	„	„	„	74·4	1·928
III.	„	„	„	74·4	1·928
IV.	„	„	„	74·0	1·937
		Mean . . .	74·25	1·931

Sulphate of Iron, $FeO, SO_3 + 7HO = 138·3$.—The fourth part of an equivalent (34·6 grains) dissolved in 3140 grains of water at 32° with an increase of 15·25, which became 15·75 at 80°.

$FeO, SO_3 + 7HO$, vol. in solution 63.

The same quantity of salt thrown into turpentine gave in

one experiment an increase of 18·6, in another of 18·7, and in a third of 18·6.

						Sp. gr.
I.	FeO, $SO_3 + 7HO$, vol. of salt	. .	74·4	1·860
II.	,,	,,	,,	74·8	1·850
III.	,,	,,	,,	74·4	1·860
		Mean . . .	74·5	1·857

Sulphate of Nickel, NiO, $SO_3 + 6HO = 131·74$.—This salt we found to contain only 6 atoms of water instead of 7 atoms, as usually described; but it is known to crystallize with both proportions. On dissolving 35 grains in 1000 of water, the increase obtained was 14·0 at a temperature of 55°.

NiO, $SO_3 + 6HO$, vol. in solution 52·7.

We have not ourselves obtained the specific gravity of this salt, but this has been determined by Kopp, who gives it at 2·037, without, however, describing the character of the hydrate which he examined. It is possible, therefore, that it may not be the same as that which we have examined; but presuming it to be so, the volume of this salt, according to Kopp, would be

Sp. gr.

NiO, $SO_3 + 6HO$, vol. of salt . . 64·6   2·037.

The volumes of the magnesian sulphates with 7 atoms of water are obviously less than those which would result were they multiples of the volume 11. But as we have already seen that the water of hydration does not always enter into combination with the volume 11, but occasionally with that of 9·8, or the volume of ice, the results obtained may be explained on this view. Graham*, in his researches on the phosphates and on the heat of combination, drew attention to the fact that the atoms of water seem to be attached together in *twos*. Millon† more lately has shown that the last two atoms of water in sulphate of magnesia are less

---

* Phil. Trans. part 1, 1837, p. 67.

† Annales de Chimie, 3 série, t. xii. p. 134.

firmly attached than the five remaining atoms; that a magnesian sulphate in fact may be viewed as

$$MgO, SO_3 + 5HO + 2HO.$$

That 5 atoms of water form the natural numbers for the magnesian sulphates we have evidence in the salts of copper and manganese, both of which possess these 5 atoms of water in combination with a volume of 11, at least.

CuO, $SO_3 + 5HO$, vol. of salt 55·4 (B. and J.).
MnO, $SO_3 + 5HO$,   „   57·6 (Kopp).

As, then, the two additional atoms of water are retained by a less feeble affinity than the remaining five, may we not assume that they are present, as in the case of other salts possessing a feeble affinity for water, with the volume of ice, whilst the original 5 atoms possess the higher volume of 11? The following table (see p. 38) will show that this hypothesis gives results by calculation which do not differ widely from those obtained by experiment.

Before leaving this section we would sum up some of the principal facts observed. In the first place, it is of much importance to know that these salts dissolve in water without increasing the bulk more than is due to the water attached to them as crystallized water. The acid and bases entirely disappear in the water which is attached to them; and so closely does this rule prevail, that the atom of basic water in the tribasic arseniates and phosphates has ceased to play the part of water, either in solution or in the solid state. In the condition of solid salts, we find four classes to which we have drawn attention. The first of these is represented by the salts having their water firmly attached, and possess as a divisor for their atomic volume a number equal or approaching to 11; and we have concluded, as the quotient of this divisor is always the same as the number of atoms of water attached to the salt, that 11 is the volume of an atom of water in combination; and hence that the salts have disappeared in this attached water, adding to its weight, but not to its observed bulk.

TABLE V.

Designation			Volume in Solution.			Volume of Salt.				
Name.	Formula.	Atomic weight.	Volume in solution.	θ, or volume of water, as unity.	Volume by theory.	Volume of salt by experiment.	11 as unity for 5 atoms, and the volume of ice for 2 atoms.	Volume by theory.	Specific gravity by theory.	Specific gravity by experiment.
Sulphate of Magnesia ..	$MgO, SO_3+5HO+2HO$	123·86	63	7	63	74·55	5+2	74·6	1·660	1·660
Sulphate of Zine	$ZnO, SO_3+5HO+2HO$	143·43	63	7	63	74·25	5+2	74·6	1·926	1·931
Sulphate of Iron	$FeO, SO_3+5HO+2HO$	138·3	63	7	63	74·5	5+2	74·6	1·854	1·857
Sulphate of Nickel	$NiO, SO_3+5HO+HO$	131·74	52·7	6	54	64·6	5+1	64·8	2·033	2·037

The second class of salts in this section is represented by potash-alum, in which the astonishing result is obtained that the 23 anhydrous atoms of this salt have combined in some way with 24 atoms of water, so as to cease to occupy bulk in solution. The peculiarity of this group is, that an additional 11 becomes attached to the solid salt, so that the quotient of the divisor is 25 instead of 24. This fact, and that connected with the ammoniacal alums in the same group, cannot be discussed with propriety in the present place.

The third group of salts in this section is one of high interest, and is represented by salts having their hydrate water attached by a feeble affinity. In them the volume of the water is exactly the same as that of ice itself. Sugar belongs to this category, not because the $H_{11} O_{11}$ are feebly attached, for it has yet to be shown that they are present *quasi* water. The fact, however, that these 11 atoms of hydrogen and oxygen take up the same space as liquid water in solution, and as ice in the solid state of sugar, and that the 12 atoms of carbon have ceased to occupy space, is a matter of supreme interest. and cannot fail to lead to important results when we come to the consideration of organic compounds.

The fourth class in this section finds its representatives in the sulphates of the magnesian class of metals, and perhaps ought to include the magnesian chlorides also. They possess their constitutional water with the usual volume of 11, while the water feebly attached is present with the volume of ice.

Although, then, we have four distinct groups in the section of salts possessing a large amount of hydrate water, we have only two modifications of volume, the one represented by a number equal or approximating to 11, the other by the volume of ice itself, viz. 9·8.

We now proceed to the consideration of salts which either are destitute of water, or contain it in small proportion only. The volumes affected by them must be volumes peculiar to themselves, and not, as in the present section, to the water with which they are combined.

Section II.

Sulphates with a small proportion of Water of Hydration,
Anhydrous and Double Sulphates.

Sulphate of Potash, KO, $SO_3 = 87\cdot25$.—Half an equivalent
of this salt, dissolved in 3140 grains of water at 37°, increased
$7\cdot2$, and at 80°, $9\cdot0$; the same quantity dissolved in 1000
grains of water at 66°, increased $9\cdot0$.

I., II. KO, SO_3, vol. in solution $18\cdot0$.

A whole equivalent of the salt being placed in a saturated
solution, affected a rise in the stem of $33\cdot0$ at a temperature
of 55°; and a repetition of the experiment gave an increase
of $33\cdot1$.

			Sp. gr.
I. KO, SO_3, vol. of salt	. .	33·0	2·644
II. ,, ,,		33·1	2·636
	Mean . . .	33·05	2·640

Sulphate of Potash and Sulphate of Water, KO, $SO_3 +$
HO, $SO_3 = 136\cdot35$.—The fourth of an equivalent ($34\cdot08$
grains) being dissolved in 1000 grains of water, caused an
increase of $9\cdot0$ at a temperature of 59°; and 33 grains
dissolved in the same quantity of water occasioned a rise of
$8\cdot75$ at 44°.

I. KO, SO_3 + HO, SO_3, vol. in solution	. .	36·0
II. ,, ,, ,,		36·1
	Mean . . .	36·05

Half an equivalent ($68\cdot2$ grains) of the salt, previously
fused, immersed in a saturated solution, produced a rise in
the stem of $27\cdot5$, and a second experiment with the same
quantity, but with salt which had not been fused, of $27\cdot6$,
the temperature on both occasions being 55°.

			Sp. gr.
I. KO, SO_3 + HO, SO_3, vol. of salt	. .	55·0	2·479
II. ,, ,, ,,		55·2	2·470
	Mean . . .	55·1	2·475

Sulphate of Ammonia, $NH_4O,SO_3 + HO = 75·25$.—In three separate experiments, in which 75·25 grains of this salt were dissolved in 1000 grains of water, the increase was exactly 36·0 at 60°.

I., II., III. $NH_4O, SO_3 + HO$, vol. in solution 36·0.

Half an equivalent (37·6 grains) being immersed in a saturated solution at 49°, caused, in two experiments, an increase of 21·5.

Sp. gr.

I., II. $NH_4O, SO_3 + HO$, vol. of salt . . 43·0 1·750.

Sulphate of Ammonia and Sulphate of Water, $NH_4O, SO_3 + HO, SO_3 = 115·35$.—Half an equivalent (57·7 grains) of this salt, dissolved in 1000 grains of water, gave a rise in the stem of 23·0 at 56°, in two separate experiments.

I. II. $NH_4O, SO_3 + HO, SO_3$, vol. in solution 46·0.

The same quantity of salt being placed in a saturated solution, caused an increase of 32·5 in one experiment and of 33·0 in a second, the temperature in both cases being 58°.

				Sp. gr
I. $NH_4O, HO, 2SO_3$, vol. of salt	. . 65·0	1·775
II. ,, ,, ,,	66·0	1·747
	Mean . . . 65·5	1·761

Sulphate of Soda and Sulphate of Water, $NaO, SO_3 + HO, SO_3 = 120·64$.—The fourth of an equivalent (30·16 grains) dissolved in 1000 grains of water, in the first experiment with an increase of 4·6, in the second of 4·7, both at a temperature of 56°.

I. $NaO, SO_3 + HO, SO_3$, vol. in solution 18·4
II.   ,,   ,,   ,,   18·8

Mean . . . 18·6

The same quantity of salt thrown into a saturated solution, caused, in two experiments, an increase of 11·0 at a temperature of 54°.

Sp, gr.

I., II. Bisulphate of Soda, vol. of salt . . . 44·0   2·742.

Ammoniacal Sulphate of Copper, CuO, SO_3, $HO + 2NH_3 =$
123·0.—The fourth of an equivalent (30·8 grains) of this
substance in beautiful large indigo-blue crystals, dissolved in
1000 grains of water, with an increase of 13·3 in one experiment and 13·0 in another, the temperature being 54° and
5 0.

    I. CuO, SO_3, $HO + 2NH_3$, vol. in solution . . 53·2
    II.   „   „   „   52·0
                                    Mean . . 52·6

61·5 grains of this salt, placed in the solution from which
it had been crystallized, caused an increase of 34·3, and on a
repetition of the experiment of 34·4, at a temperature of 60°.

                                                    Sp. gr.
    I. CuO, SO_3, $HO + 2NH_3$, vol. of salt . . 68·6   1·793
    II.   „   „   „   68·8   1·788
                                    Mean . . 68·7   1·790

Sulphate of Copper and Sulphate of Potash, CuO, $SO_3 + KO$,
$SO_3 + 6HO = 221·31$.—The fourth of an equivalent (55·32
grains) dissolved in 3140 grains of water at 32° increased to
16·3, and at 72° to 18·0.

    CuO, $SO_3 + KO$, $SO_3 + 6HO$, vol. in solution 72·0.

The same quantity of salt placed in a saturated solution
caused an increase of 24·7 in one experiment and of 24·6 in
a second, the temperature on both occasions being 55°.

                                                    Sp. gr.
    I. CuO, $SO_3 + KO$, $SO_3 + 6HO$, vol. of salt . 98·8   2·239
    II.   „   „   „   98·4   2·249
                                    Mean . . 98·6   2·244

Sulphate of Copper and Sulphate of Ammonia, CuO, $SO_3 +$
NH_4O, $SO_3 + 6HO = 199·88$.—On dissolving 50 grains of
this salt in 1000 grains of water, an increase was occasioned

in the first experiment of 20·2, in the second of 20·3, both at a temperature of 59°.

I. Sulphate of Copper and Ammonia, vol. in solution 80·8
II.   ,,   ,,   ,,   ,,   81·2

Mean . . 81·0

On immersing the same quantity in a saturated solution, an increase of 26·4 was obtained in the first experiment and of 26·45 in the second, both at a temperature of 59°.

Sp. gr.

I. $CuO, SO_3 + NH_4O, SO_3 + 6HO$, vol. of salt 105·6 1·892
II.   ,,   ,,   ,,   105·8 1·889

Mean . . 105·7 1·891

Sulphate of Zinc and Sulphate of Potash, $ZnO, SO_3 + KO, SO_3 + 6HO = 221·86$.—The fourth of an equivalent of this salt (55·46 grains), on being dissolved in 1000 grains of water, increased to 18 at a temperature of 60° in two experiments.

I., II. $ZnO, SO_3 + KO, SO_3 + 6HO$, vol. in solution 72.

The same quantity immersed in a saturated solution, caused an increase also in two experiments of 24·7, the temperature being 56°.

Sp. gr.

I., II. $ZnO, SO_3 + KO, SO_3 + 6HO$, vol. of salt .. 98·8 2·245

Sulphate of Zinc and Sulphate of Ammonia.—$ZnO, SO_3 + NH_4O, SO_3 + 6HO = 200$.—On dissolving 45 grains of this salt in 1000 of water, an increase of 18·0 was occasioned in three separate experiments at a temperature of 58°.

I., II., III. $ZnO, SO_3 + NH_4O, SO_3 + 6HO$, vol. in solution 80.

On adding the fourth of an equivalent (50 grains) to a saturated solution, an increase of 26·4 was occasioned in the first experiment and of 26·3 in the second, both at a temperature of 55°.

Sp. gr.

I. Sulph. Zinc and Ammonia, vol. of salt . . 105·6 1·894
II. ,,   ,,   ,,   105·2 1·901

Mean . . 105·4 1·898

Sulphate of Magnesia and Sulphate of Potash, MgO, SO_3 + KO, SO_3 + 6HO = 202·29. When a quarter of an equivalent of this salt (50·57 grains) is dissolved in as many as 3140 grains of water, the volume at 32° is only 15·45, but is 18 at 80°. This gives for the salt a very dilute solution—

MgO, SO_3 + KO, SO_3 + 6HO, vol. in solution 63 at 40°.
          „   „   „   72 at 80°.

The same quantity of salt, after immersion in a saturated solution, gave in the first experiment an increase of 24·3 and in the second 24·4, both at a temperature of 57°.

				Sp. gr.
I. Sulph. Magnesia and Potash, vol. of salt	97·2	2·081
II. „ „ „	97·6	2·071
		Mean . . 97·4	2·076

Sulphate of Magnesia and Ammonia, MgO, SO_3 + NH_4O, SO_3 + 6HO = 181·12. The fourth of an equivalent (45·28 grains) being dissolved in 1000 grains of water, caused an increase of 20·0 at 60°; and a repetition of the experiment at the same temperature gave the increase 20·1.

I. Sulph. Magnesia and Ammonia, vol. in solution . .	80·0
II. „ „ „	80·4
	Mean . . 80·2

The same quantity of salt placed in a saturated solution gave on two occasions a rise in the stem of 26·3 at a temperature of 60°.

	Sp. gr.
I., II. Sulph. Magnesia & Ammonia, vol. of salt 105·2	1·721

Sulphate of Iron and Potash, FeO, SO_3 + KO, SO_3 + 6HO = 216·73.—The eighth of an equivalent (27·09 grains), when dissolved in 1000 of water, caused an increase of 9 at a temperature of 65°.

Sulph. Iron and Potash, vol. in solution 72.

The same quantity immersed in a saturated solution occasioned a rise in the stem in two experiments of 12·3 at a temperature of 61°.

Sp. gr.

I., II. Sulph. Iron and Potash, vol. of salt . . 98·4   2·202

Sulphate of Iron and Ammonia, $FeO, SO_3 + NH_4O, SO_3 + 6HO = 195.55$.—On dissolving 33·45 grains of this salt in 1000 of water, the increase in the first experiment was 13·4, in the second 14, both at a temperature of 59°; a third experiment with 66·9 grains gave the increase 28 at the same temperature.

    I. Sulph. Iron and Ammonia, vol. in solution . . 78·3

    II.   ,,   ,,   ,,   81·8

    III.   ,,   ,,   ,,   81·8

                            Mean . . . 80·7

48·89 grains of this salt, being projected into a saturated solution, caused in the first experiment an increase of 26·4, in the second of 26·5.

Sp. gr.

    I. Sulph. Iron and Ammonia, vol. of salt . . 105·6   1·851

    II.   ,,   ,,   ,,   106·0   1·845

                      Mean . . 105·8   1·848

In the last section we gave the volumes occupied by those salts which did not occupy any space of themselves, but merely that due to their combined water. The divisor of the volumes observed in solution was therefore necessarily 9, or the atomic volume of water itself. But in this section we have experimented upon salts which take up space quite independent of their water of crystallization, even when they contain water, and yet the most interesting result follows, that the same divisor, 9, continues for the volumes ascertained by experiment. The volumes in solution of the salts examined, allowing for errors of observation, are therefore always multiples of 9—the atomic volume of water. The volumes of the solids are, like those of the previous section, multiples of one and the same number, that number being also, as in the former case, 11; but the ammonia-salts do not arrange themselves under this divisor for reasons which will be explained presently. The averages of the experiments

TABLE VI.—Showing the Volumes of certain Sulphates with a small proportion of Water of Hydration, Anhydrous and Double Sulphates.

Name.	Designation. Formula.	Atomic weight.	Volume in Solution.			Volume in the state of Salt.				
			Volume in solution by experiment.	9, or volume of water, as unity.	Volume by theory.	Volume of salt by experiment.	11, or supposed vol. of combined water, as unity.	Volume by theory.	Specific gravity by theory.	Specific gravity by experiment.
Sulphate of Potash	KO, SO_3	87·25	18·0	2	18	33·05	3	33	2·644	2·640
Bisulphate of Potash	$KO, SO_3 + HO, SO_3$	136·35	36·05	4	36	55·1	5	55	2·479	2·475
Sulphate of Ammonia	$NH_4O, SO_3 + HO$	75·25	36·0	4	36	43·0	4	44	1·710	1·750
Bisulphate of Ammonia	$NH_4O, SO_3 + HO, SO_3$	115·35	46·0	5	45	65·5	6	66	1·747	1·761
Bisulphate of Soda	$NaO, SO_3 + HO, SO_3$	120·64	18·6	2	18	44·0	4	44	2·742	2·742
Ammoniacal Sulphate of Copper	$CuO, SO_3 HO + 2NH_3$	123·00	52·6	6	54	68·7			.....	1·790
Sulphate of Copper and Potash	$CuO, SO_3 + KO, SO_3 + 6HO$	221·31	72·0	8	72	98·6	9	99	2·235	2·244
Sulphate of Copper and Ammonia	$CuO, SO_3 + NH_4O, SO_3 + 6HO$	199·88	81·0	9	81	105·7	9	99	2·019	1·891
Sulphate of Zinc and Potash	$ZnO, SO_3 + KO, SO_3 + 6HO$	221·86	72·0	8	72	98·8	9	99	2·241	2·245
Sulphate of Zinc and Ammonia	$ZnO, SO_3 + NH_4O, SO_3 + 6HO$	200·00	80·0	9	81	105·4	9	99	2·020	1·897
Sulphate of Magnesia and Potash	$MgO, SO_3 + KO SO_3 + 6HO$	202·29	72·0	8	72	97·4	9	99	2·043	2·076
Sulphate of Magnesia and Ammonia	$MgO, SO_3 + NH_4O, SO_3 + 6HO$	181·12	80·2	9	81	105·2	9	99	1·829	1·721
Sulphate of Iron and Potash	$FeO, SO_3 + KO, SO_3 + 6HO$	216·73	72·0	8	72	98·4	9	99	2·190	2·202
Sulphate of Iron and Ammonia	$FeO, SO_3 + NH_4O, SO_3 + 6HO$	195·55	80·7	9	81	105·8	9	99	1·975	1·848
Sulphate of Nickel and Potash	$NiO, SO_3 + KO, SO_3 + 6HO$	218·99	71·5	8	72	100·0	9	99	2·212	2·190

on all the salts are thrown into the opposite table, into which is also introduced the exact numbers which would have resulted had there been a strict accordance with the law obviously indicated by experiment.

The correspondence between the observed and calculated results in the preceding table, as far as regards the potash-salts, is so striking as to remove any doubt of the basis upon which the calculations are made. It is therefore of interest to consider the results indicated by the table a little more in detail. The first point of remark is, that in every case the ammoniacal salt has one volume greater in solution than the corresponding potash-salt. Sulphate of potash possesses two volumes in solution ; sulphate of ammonia divested of one volume for its atom of water possesses three. These volumes are respectively carried through the whole class of double sulphates. The volumes of these double sulphates are made up of the sum of the volumes of their constituent salts, which appear, therefore, to be united unchanged. We saw in the previous section that the magnesian sulphates dissolve in water without increasing its bulk more than is due to their water of combination. The same takes place in their double sulphates, for subtracting the volumes of the atoms of water which have been carried by the sulphates into their union with sulphate of potash, the remainder shows the volumes belonging to the latter salts, as indicated by direct experiment. This is strikingly exemplified also by bisulphate of soda. Sulphate of soda was shown in the last section to possess no volume in solution, and in this acid salt we find that the sulphate of soda has in solution ceased to occupy space, for the resulting volume of the acid salt is only 18 or 9×2, which is the atomic volume of sulphate of water, as ascertained by the volume occupied by it in bisulphate of potash and bisulphate of ammonia, and as determined also by a calculation, which we have made, of the volume occupied by hydrated sulphuric acid in a *dilute* solution, founded upon recorded specific gravities.

Although the ammoniacal sulphates, on account of their analogy to the potash-salts, have been introduced into the above table, it is obvious that the numbers representing their volumes are too wide from the theoretical numbers to be considered multiples of 11. Hydrated sulphate of ammonia affects four volumes, 11×4, but the anhydrous salt obeys a different law. On immersing in turpentine 33·15 grains of anhydrous NH_4O, SO_3, the increase was 19·6 and 19·5, the mean being 19·55 water grain-measures. This gives 39·1 as the vol. of the equivalent, and $9·8 \times 4 = 39·2$. Anhydrous sulphate of ammonia affects, therefore, 4 vol. of ice; and the double salts consist of the magnesian sulphates with 6 equivalents of water attached to an equivalent of anhydrous sulphate of ammonia, as will be seen from the following table of their solid volumes and specific gravities.

Name.	Solid volume by experiment.	Solid volume by theory.	Specific gravity by experiment.	Specific gravity by theory.
Sulphate of Ammonia	39·1	39·2	1·695	1·691
Sulphate of Copper and Ammonia	105·7	105·2	1·891	1·900
Sulphate of Zinc and Ammonia ..	105·4	105·2	1·887	1·901
Sulphate of Magnesia and Ammonia	105·2	105·2	1·721	1·721
Sulphate of Iron and Ammonia ..	105·8	105·2	1·848	1·858

As one of the members of the group of double salts here described takes up no space of itself, it became of importance to ascertain the volume of the salt when deprived of water, and also the space occupied by the double salt reduced to the same state. In this examination it was quite unnecessary to obtain the volumes in solution, because it was obvious that salts not occupying in solution a greater volume than that due to their water of hydration, would, in their anhydrous condition, take up no space at all. In fact we had ascertained that not only was there no increase in dissolving such salts in water, but that actually there was a contraction if the water were in large proportion to the salt; when this is not the case, the increased expansibility of the solution prevents the contraction being observed.

In the following examination will be found almost all the salts previously described in their hydrated condition, with the exception of the phosphates and arseniates, which we reserve for another paper.

Sulphate of Magnesia, $MgO, SO_3 = 60.86$.—Half an equivalent of this salt (30·43 grains), thrown into turpentine, caused an increase of 11·0, but in a second experiment the increase was 11·5, the temperature in both cases being 65°.

			Sp. gr.
I. MgO, SO_3, vol. of salt	. .	22·0	2·766
II. ,, ,,		23·0	2·646
	Mean . . .	22·5	2·706

Sulphate of Zinc, $ZnO, SO_3 = 80.45$—Half an equivalent of this salt (40·22 grains), projected into turpentine, caused an increase of 11·05, and, in another experiment, of 10·8.

			Sp. gr.
ZnO, SO_3, vol. of salt	. .	22·1	3·639
,, ,,		21·6	3·723
	Mean . . .	21·85	3·681

Sulphate of Copper, $CuO, SO_3 = 79.88$.—Half an equivalent (39·94 grains) of the salt, placed in turpentine, caused in several successive experiments an increase of exactly 11·0.

			Sp. gr.
CuO, SO_3, vol. of salt	. . .	22·0	3·631

Sulphate of Iron, $FeO, SO_3 = 75.3$.—Half an equivalent of this salt (37·65 grains) caused, in two experiments with the same salt, a rise of 12·0, which gives for the equivalent 24·0, and a specific gravity of 3·138.

Sulphate of Cobalt, $CoO, SO_3 = 77.69$.—On immersing 19·42 grains of this salt in turpentine, an increase of 5·5 was obtained in two experiments; this gives for the equivalent 22·0, and for the specific gravity 3·531.

Sulphate of Soda, $NaO, SO_3 = 71.43$.—On throwing a whole equivalent of this salt into turpentine, the increase was only 27·5 in several successive experiments, which gives for the specific gravity 2·597. Karsten found its specific gravity to be 2·631, a result approximating to our own; attention is drawn to this circumstance because both results are anomalous.

Sulphate of Silver, AgO, $SO_3 = 156·48$.—On immersing in turpentine 78·24 grains of this salt, the increase was 14·7, which gives as the volume of the equivalent 29·4, and a specific gravity of 5·322.

Chromate of Silver, AgO, $CrO_3 = 168·49$.—The fourth of an equivalent of this salt (42·12 grains) gave an increase, when thrown into turpentine, of 7·3 in two successive experiments. This gives 29·2 for the volume of the equivalent, and 5·77 for the specific gravity of the salt.

Sulphate of Copper and Potash, CuO, $SO_3 + KO$, $SO_3 = 167·31$.—41·82 grains, the fourth of an equivalent, thrown into turpentine, caused an increase of 14·9 in one experiment and of 15·0 in another, the temperature in both cases being 54°.

				Sp. gr.
CuO, $SO_3 + KO$, SO_3, vol. of salt	. . .	59·6		2·807
„	„	„	. . . 60·0	2·788
		Mean . . .	59·8	2·797

Sulphate of Nickel and Potash, NiO, $SO_3 + KO$, $SO_3 = 164·99$.—41·54 grains caused an increase of 14·2 in one experiment and 14·5 in a second, the temperature in both cases being 54°.

				Sp. gr.
NiO, $SO_3 + KO$, SO_3, vol. of salt	. . .	56·4		2·925
„	„	„	. . . 57·5	2·869
		Mean . . .	56·95	2·897

Sulphate of Zinc and Potash, ZnO, $SO_3 + KO$, $SO_3 = 167·86$.—41·96 grains, the fourth of an equivalent, placed in turpentine, caused an increase of 14·9 in two experiments.

			Sp. gr.
ZnO, $SO_3 + KO$, SO_3, vol. of salt	. .	59·6	2·816

Sulphate of Magnesia and Potash, $MgO, SO_3 + KO, SO_3 =$ 148·29.—37·07 grains, or the fourth of an equivalent, caused in one experiment an increase of 13·9 and in a second of 13·8, the temperature being 55°.

				Sp. gr.
$MgO, SO_3 + KO, SO_3$, vol. of salt	. .	55·6	2·667
,,	,,	,,	55·2	2·686
		Mean . . .	55·4	2·676

Sulphate of Manganese and Potash, $MnO, SO_3 + KO, SO_3$ = 163·07.—40·8 grains of this salt, one fourth of an equivalent, placed in turpentine, caused an increase of 13·5 in one experiment and 13·6 in another, at a temperature of 55°.

				Sp. gr.
$MnO, SO_3 + KO, SO_3$, vol. of salt	. .	54·0	3·020
,,	,,	,,	54·4	2·996
		Mean . . .	54·2	3·008

Sulphate of Copper and Ammonia, $CuO, SO_3 + NH_4O, SO^3$ = 145·88.—36·53 grains of this salt thrown into turpentine caused an increase of 16·7 in one experiment and of 16·6 in another, at a temperature of 60°.

			Sp. gr.
I. Sulphate of Copper and Ammonia	. .	66·6	2·190
II. ,, ,,		66·2	2·204
	Mean . . .	66·4	2·197

Sulphate of Zinc and Ammonia, $ZnO, SO_3 + NH_4O, SO_3 =$ 146·0.—30 grains of this salt, thrown into turpentine, caused an increase of 13·5 at 60°.

		Sp. gr.
Sulphate of Zinc and Ammonia, vol. of salt	. . 65·7	2·222

Sulphate of Magnesia and Ammonia, $MgO, SO_3 + NH_4$ $O, SO_3 =$ 127·12.—The fourth of an equivalent (31·78 grains), placed in turpentine, caused an increase of 16·5 in the first experiment and of 16·4 in the second.

Sp. gr.

I. Sulphate of Magnesia and Ammonia . . 66·0   1·926
II.   ,,   ,,   ,,   65·6   1·938

Mean . . . 65·8   1·932

Sulphate of Alumina, $Al_2O_3, 3SO_3 = 171·95$.—This salt, and the anhydrous alums, offer difficulties to the correct estimation of their specific gravity on account of their great porosity and liability to carry down air. The best mode of obviating this source of error is to introduce a metallic wire previously moistened with turpentine into the volumenometer, and employ this to break the numerous air-bubbles which arise on immersing the salts. The following estimations were taken with great care, but, from this source of error, may possibly be inaccurate.

The eighth part of an equivalent (21·49 grains), immersed in turpentine and treated as described above, gave results varying from 9·8 to 10·0, the mean result being 9·9.

Sp. gr.

$Al_2 O_3, 3SO_3$, vol. of salt . . 79·2   2·171

Sulphate of Alumina and Potash, $Al_2O_3, 3SO_3 + KO, SO_3 = 259·36$.—The eighth part of an equivalent (32·42 grains) of anhydrous alum, immersed in turpentine and treated as described in the case of the sulphate of alumina, gave an increase of 14·5 and 14·6 in two experiments.

Sp. gr.

I. Alum, vol. of salt . . 116·0   2·236
II.   ,,   ,,   116·8   2·220

Mean . . . 116·4   2·228

Ammonia-alum, $Al_2O_3, 3SO_3 + NH_4O, SO_3 = 238·2$.—The eighth part of an equivalent of this salt (29·77 grains), treated as in the previous cases, gave an increase of 14·6 in two experiments.

Sp. gr.

Ammonia-alum, vol. of salt . . 116·8   2·039

Carbonate of Soda, NaO, $CO_2 = 53\cdot47$.—The equivalent of this salt, thrown into turpentine, gave an increase of exactly $22\cdot0$, which makes its specific gravity $2\cdot430$.

Chloride of Magnesium, $MgCl = 48\cdot12$.—The anhydrous chloride of magnesium used in the experiment was made by saturating equal portions of muriatic acid with magnesia and ammonia, mixing together, evaporating to dryness, and heating to redness.

Half an equivalent ($24\cdot06$ grains), thrown into turpentine, caused an increase of $11\cdot0$ in one experiment and of $11\cdot1$ in a second.

				Sp. gr.
I.	MgCl, vol. of salt . . .	22·0	2·187
II.	„	„ . . .	22·2	2·167
		Mean . . .	22·1	2·177

Chloride of Calcium, $CaCl = 55\cdot92$.—This salt was rendered anhydrous by fusing it in a platinum crucible for some time. 28 grains of the fused salt, thrown into turpentine, caused an increase of $11\cdot3$, at a temperature of $63°$.

		Sp. gr.
CaCl, vol. of salt . .	22·5	2·480

Chloride of Cobalt, $CoCl = 65\cdot0$.—On throwing the fourth of an equivalent ($16\cdot25$ grains) of anhydrous chloride of cobalt into turpentine, an increase of $5\cdot5$ was obtained in two experiments, and of $5\cdot6$ in a third trial.

			Sp. gr.
I.	CoCl, vol. of salt . .	22·0	2·954
II.	„	„	22·0	2·954
III.	„	„	22·4	2·902
		Mean . . . 22·13	2·937

TABLE VII.—Showing the Volume occupied by certain Hydrated Salts rendered Anhydrous.

Designation.			Volumes of Anhydrous Salts.				
Name.	Formula.	Atomic weight.	Volume of salt by experiment.	11, taken as unity.	Volume by theory.	Specific gravity by theory.	Specific gravity by experiment.
Sulphate of Magnesia	MgO, SO_3	60·86	22·5	2	22	2·766	2·706
Sulphate of Zinc	ZnO, SO_3	80·43	21·85	2	22	3·656	3·681
Sulphate of Copper	CuO, SO_3	79·88	22·0	2	22	3·631	3·631
Sulphate of Iron	FeO, SO_3	75·3	24·0	2	22	3·423	3·138
Sulphate of Cobalt	CoO, SO_3	77·69	22·0	2	22	3·531	3·531
Sulphate of Soda	NaO, SO_3	71·43	27·5	..	..	....	2·597
Sulphate of Silver	AgO, SO_3	156·48	29·4	..	..	....	5·322
Chromate of Silver	AgO, CrO_3	168·49	29·2	..	..	....	5·770
Sulphate of Alumina	$Al_2O_3, 3SO_3$ ..	171·95	79·2	..	..	....	2·171
Sulphate of Copper and Potash	$CuO, SO_3 +$ KO, SO_3 ..	167·31	59·8	5	55	3·042	2·797
Sulphate of Nickel and Potash	$NiO, SO_3 +$ KO, SO_3 ..	164·99	56·95	5	55	2·998	2·897
Sulphate of Zinc and Potash	$ZnO, SO_3 +$ KO, SO_3 ..	167·86	59·6	5	55	3·034	2·816
Sulphate of Magnesia and Potash	$MgO, SO +$ KO, SO_3 ..	148·29	55·4	5	55	2·694	2·676
Sulphate of Manganese and Potash	$MnO, SO_3 +$ KO, SO_3 ..	163·07	54·2	5	55	2·964	3·008
Sulphate of Copper and Ammonia	$CuO, SO_3 +$ NH_4O, SO_3	145·88	66·4	6	66	2·192	2·197
Sulphate of Zinc and Ammonia..............	$ZnO, SO_3 +$ NH_4O, SO	146·0	65·7	6	66	2·212	2·222
Sulphate of Magnesia and Ammonia	$MgO, SO_3 +$ NH_4O, SO_3	127·0	65·8	6	66	1·924	1·932
Potash-alum	$Al_2O_3 3SO_3 +$ KO, SO_3 ..	259·36	116·4	..	..	....	2·228
Ammonia-alum........	$Al_2O_3, 3SO_3 +$ NH_4O, SO	238·2	116·8	..	..	....	2·039
Carbonate of Soda	NaO, CO_2	53·47	22·0	2	22	2·427	2·427
Chloride of Cobalt	$Co\,Cl$	65·0	22·1	2	22	2·955	2·937
Chloride of Magnesium ..	$Mg\,Cl$	48·12	22·1	2	22	2·187	2·177
Chloride of Calcium......	$Ca\,Cl$	55·92	22·5	2	22	2·542	2·485

The preceding table exhibits various points of great interest as regards isomorphism. Hydrogen has for a long time been recognized by chemists as equivalent to a magnesian metal, and hence the sulphate of a metal of this class should possess the volume of sulphate of water. The volume of bisulphate of potash is 55·0 by experiment, which leaves 22·0 for that of sulphate of water, on deducting the volume of sulphate of potash, which is 33·0; and the same result follows when the volume of sulphuric acid is deducted from bisulphate of soda, if we suppose the sulphate of soda to enter that salt with two volumes. Thus we have :—

Sulphate of water . . . . $= 22 \div 11 = 2$
Sulphate of a magnesian oxide $= 22 \div 11 = 2$.

We now see that bisulphate of potash (sulphate of water and sulphate of potash) is exactly equivalent to the double sulphates of the magnesian class. (*Vide* Section V.)

Bisulphate of Potash $(HO, SO_3 + KO, SO_3) = 55$.

Sulphate of Magnesia and Potash $(MgO, SO_3 + KO, SO_3) = 55$.

It is now comprehensible why bisulphate of soda should have a volume of 44·0 in the solid state, and only of 18·0 in a state of solution, because sulphate of soda, which assumes a volume in the solid state, becomes added to the same volume possessed by sulphate of water, while in the state of solution the proper volume of sulphate of soda disappears altogether.

Bisulphate of ammonia possesses a volume due to a combination of sulphate of water and sulphate of ammonia with a volume of 11×4; and it will be observed that the same result attends the double sulphates of the magnesian metals with sulphate of ammonia.

Bisulphate of Ammonia $(NH_4O, SO_3 + HO, SO_3) = 66$.

Sulphate of Ammonia and Copper $(NH_4O, SO_3 + CuO, SO_3) = 66$.

The cause of this singular result is the mutual convertibility of the primitive volumes 9·8 and 11.

It is very curious to observe the large number of volumes which have disappeared when the salt combines with water.

Thus sulphate of alumina in its anhydrous state possesses a bulk equal to 79·2, which has ceased to occupy space in the hydrated salt; and still more remarkable instances of this are seen in the alums, which add to this the volumes of their alkaline sulphates. A curious result obtained in the examination of the hydrated alums is now explicable. We found that the potash-alums took up in solution only the space due to their water; but that the space occupied by them in the state of salts was one volume in addition to this quantity. In the preceding section we observed that sulphate of potash possessed the singular property of expanding one volume in becoming solid; 9×2 in a state of solution becoming 11×3 in the state of a salt. It is impossible to refrain from accepting this as an explanation of the increase of one in the quotient obtained by dividing the volumes by their proper numbers 9 and 11; 24×9 becoming 25×11.

The difficulties to which we have already alluded prevent us placing much confidence in our results for the anhydrous alums. Sulphate of alumina seems to affect eight volumes of ice, $9·8 \times 8 = 78·4$; in ammonia-alum the latter becomes united to the volume of anhydrous sulphate of ammonia, $9·8 \times 8 + 9·8 \times 4 = 117·6$; while potash-alum should consist of $9·8 \times 8 + 11 \times 3 = 114·4$. It is unnecessary to remark that these theoretical numbers possess only an approximation to our experimental results. (*Vide* remarks on Section V.)

The sulphates of soda and silver and the corresponding chromate are also obviously exceptions to the general rule of the solid volume being a multiple of 11. But in the last section we had similar exceptions in salts which ranged themselves under 9·8, or the volume of ice. The sulphates now under consideration have the same divisor, if sulphate of soda be not considered an exception, as the variation is decidedly too great to be attributed to a mere error of experiment; it ought, however, to be observed that Mohs gives for the specific gravity of this salt 2·462, a number much more in accordance with theory than our own result; but as our experiments have been often repeated, they may perhaps be viewed as an argument in favour of an opinion, deduced

from other considerations, that sulphate of soda has a double
atom $27.5 \times 2 = 55$,. which is 11×5.

Name.	Volume by experiment.	9·8, or volume of ice, as unity.	Volume by theory.	Specific gravity by theory.	Specific gravity by experiment.
Sulphate of Soda	27·5	3	29·4	2·430	2·597
Sulphate of Silver. . . .	29·4	3	29·4	5·322	5·322
Chromate of Silver . .	29·2	3	29·4	5·711	5·770

Section III.

Nitrates, &c.

The nitrates do not, in general, affect a large proportion
of water of hydration, and are therefore well calculated to
show the volume occupied by anhydrous salts. It will be
observed that they present some peculiarities.

Nitrate of Potash, KO, $NO_5 = 101.3$.—The half of an
equivalent of this salt (50·65 grains), being dissolved in 1000
grains of water, gave an increase of 18·05 at 45°.

KO, NO_5, vol. in solution 36·1.

The same quantity of salt, 50·65 grains, thrown into turpentine, caused a rise in the stem of 24·5, 24·4, 24·5 in three
successive experiments.

				Sp. gr.
I.	KO, NO_5, vol. of salt . . .	49·0		2·067
II.	,,	,,	. . . 48·8	2·075
III.	,,	,,	. . . 49·0	2·067
		Mean . . .	48·9	2·070

Nitrate of Ammonia, NH_4O, $NO_5 = 80.3$.—The volume of
nitrate of ammonia in solution was determined by dissolving
40·15 grains of this salt in 1000 grains of water. In one
experiment the increase in the stem was 22·5, the temperature being 57°; in a second the rise was 23·0 at 63°.

I. NH_4O, NO_5, vol. in solution . . . 45·0

II.   „   „   . . . 46·0

Mean . . . 45·5

Half an equivalent of this salt, well dried (40·15 grains), on being immersed in turpentine, produced an increase in three experiments of 24·7, 24·5, 24·5.

		Sp. gr.
I. NH_4O, NO_5, vol. of salt . . . 49·4	1·625
II. „ „ . . . 49·0	1·639
III. „ „ . . . 49·0	1·639
Mean . . . 49·1	1·635

Nitrate of Soda, NaO, $NO_5 = 85·45$.—On dissolving 85·45 grains, or one equivalent, of this salt in 1000 grains of water, an increase of 27·1 was obtained, the temperature being 59°; but on repetition of the experiment at the same temperature the increase was only 26·0.

I. NaO, NO_5, vol. in solution . . . 27·1

II.   „   „   26·0

Mean . . . 26·5

The half of an equivalent of this salt (42·72 grains), well dried, produced an increase, on being thrown into turpentine, of 19·6 in three experiments, and 19·5 in a fourth trial.

		Sp. gr.
I. NaO, NO_5, vol. of salt . . . 39·2	2·180
II. „ „ . . . 39·2	2·180
III. „ „ . . . 39·2	2·180
IV. „ „ . . . 39·0	2·190
Mean . . . 39·1	2·182

Nitrate of Silver, AgO, $NO_5 = 170·0$.—On dissolving 42·5 grains of this salt in 1000 grains of water, an increase of 6·8 was effected at a temperature of 59°.

AgO, NO_5, vol. in solution 27·2.

The same quantity of salt, 4·25 grains, thrown into turpentine, produced an increase of 9·8.

Sp. gr.

AgO, NO₅, vol. of salt . . 39·2   4·336

Nitrate of Lead, PbO, NO₅ = 165·75.—This salt gives very unsatisfactory results on being dissolved in water; at low temperatures the volume for the atom is equal to nearly 18·0, or 9 × 2. But at higher temperatures the volume in solution approaches nearly to 27, or 9 × 3; and although the results do not come out exact, unless corrected for expansion, we are inclined to view the latter as the true result. 83 grains dissolved in water gave an increase of 12·5; in a second experiment of 12·7; both at a temperature of 65°.

PbO, NO₅, vol. in solution . . . 25·0

„   „   . . . 25·4
                              ———
                    Mean . . . 25·2

The fourth part of an equivalent (41·43 grains), immersed in turpentine, gave an increase of 9·7; 82·87 grains gave the increase 19·2; and a third experiment 19·0.

Sp. gr.

I. PbO, NO₅, vol. of salt . . . 38·8   4·272

II.   „   „   . . . 38·4   4·316

III.   „   „   . . . 38·0   4·362
                              ————   ————
              Mean . . . 38·4   4·316

Nitrate of Barytes, BaO, NO₅ = 130·85.—Half an equivalent of this salt (65·42 grains), dissolved in 1000 grains of water, with an increase of 13·5 at a temperature of 60°; and a repetition of the experiment was attended with the same result.

BaO, NO₅, vol. in solution 27·0.

The same quantity of salt, immersed in turpentine, caused an increase of 19·8 in three experiments, and 20·0 and 20·2 in two other experiments, the salts being all different specimens, and decrepitated previously to the experiment.

					Sp. gr.
I.	BaO, NO_5, vol. of salt	. . .	39·6		3·304
II.	,,	,,	. . .	39·6	3·304
III.	,,	,;	. . .	39·6	3·304
IV.	,,	,,	. . .	40·0	3·271
V.	,,	,,	. . .	40·4	3·238

Mean . . . 39·84 　　3·284

Nitrate of Strontia=106·0.—Half an equivalent of this salt (53 grains) was dissolved in 1000 grains of water, with an increase of 13·0, the temperature being 62°; 106 grains dissolved in 1000 grains of water, with an increase of 27·0 at a temperature of 63°.

I. SrO, NO_5, vol. in solution . . . 26·0
II. 　　,,　　　　　　,,　　　. . . 27·0

Mean . . . 26·5

53 grains immersed in turpentine gave an increase of 19·6; and this result was confirmed by a second experiment.

　　　　　　　　　　　　　　　　　Sp. gr.
SrO, NO_5, vol. of salt . . . 3·92　　2·704

Nitrate of Black Oxide of Mercury, $Hg_2O, NO_5 + 2HO = 282·0$.—This salt, in beautiful large transparent crystals, was dissolved in water containing nitric acid, to prevent the formation of a subsalt; 70·5 grains thus treated caused an increase of 13·5.

Protonitrate of Mercury, vol. in solution 54·0.

On immersing the same quantity of salt in turpentine, the increase in three experiments was 14·8, 14·7, and 14·7.

				Sp. gr.
I.	Protonitrate of Mercury, vol. of salt . .	59·2		4·763
II.	,,	,,	58·8	4·796
III.	,,	,,	58·8	4·796

Mean . . . 58·9　　4·785

Nitrate of Copper, CuO, $NO_5 + 3HO = 120·8$.—Half an equivalent (60·4 grains) dissolved in 1000 grains of water with an increase of 22·4 at 60°, and in a second experiment, of 22·6; and in a third experiment, 30·2 grains, dissolved in the same quantity of water, gave an increase of 11·4.

I. CuO, NO_5, vol. in solution . . . 44·8

II. „ „ . . . 45·2

III. „ „ . . . 45·6

Mean . . . 45·2

In two experiments, 60·4 grains thrown into turpentine caused an increase of 29·5, which gives for the equivalent of the salt the volume 59·0, and a specific gravity 2·047.

Nitrate of Magnesia, MgO, $NO_5 + 6HO = 128·8$.—The fourth part of an equivalent of crystallized nitrate of magnesia (32·2 grains) dissolved in 1000 grains of water at 60°, with an increase of 18·1 and 18·3 in two experiments.

I. MgO, $NO_5 + 6HO$, vol. in solution . . 73·2

II. „ „ 72·4

Mean . . . 72·8

The same quantity thrown into turpentine produced an increase of 22·0, which gives for the volume of an equivalent of the salt 88·0, and for the specific gravity 1·464.

Nitrate of Bismuth, BiO, $NO_5 + 3HO = 160·33$. — This salt, being decomposed when thrown into water, is not fitted for determining volume by solution; but when 80·16 grains were thrown into turpentine, the increase was obtained, in two experiments, of 29·2 and of 29·4.

		Sp. gr.
I. BiO, $NO_5 + 3HO$, vol. of salt . . 58·4		2·745
II. „ „ 58·8		2·727
Mean . . . 56·6		2·736

Basic Nitrate of Mercury, 2HgO, $NO_5 + 2HO = 291·0$.— This salt cannot be dissolved in water without the formation of a subsalt, unless the water is used in small proportion; it is therefore unfitted for our experiments, as far as regards the volume in solution. On immersing 68·7 grains in turpentine, an increase of 16·2 was obtained in two successive experiments. This gives 68·6 as the volume of the equivalent, and a specific gravity of 4·242.

Basic Nitrate of Lead, 2PbO, $NO_5 = 277·72$.—This salt is so insoluble that it is difficult to determine its volume in solution with any great degree of accuracy. The sixteenth part of an equivalent dissolved in 1000 grains of water gave an increase of 2·6, which seems to indicate a volume of 9×5.

69·43 grains, being immersed in turpentine, gave an increase of 12·3 in several experiments.

		Sp. gr.
Basic Nitrate of Lead, vol. of salt . . .	49·2	5·645

The same multiple relation of 9 is carried through all the salts of this class dissolved in water. The divisor for the solid volume is, however, different from the salts of the previous sections. Exceptional cases were pointed out in their examination, in which 9·8, or the volume of ice, became the divisor; and in the present group of salts we observe a wonderful uniformity in this respect. (See Table opposite.)

It is almost superfluous to offer any remarks upon this group of salts, especially as we shall have to consider several of them in a future section. It cannot escape attention that the nitrates of soda, silver, lead, strontia, and barytes possess the same atomic volume, as might have been expected from the isomorphism of several of them. Nitrates of soda and potash do not possess the same atomic volume, and therefore their alleged isomorphism, deduced from the observation by Frankenheim * of microscopic crystals of nitrate of potash similar to those of nitrate of soda, is highly questionable. The principal exception to the volumes of the nitrates now described being multiples of ice is that of nitrate of lead, which has a volume of 38·4 instead of 39·2; but this must be due to the nature of the salt, which comes out as unsatisfactorily in a state of solution as in the solid state.

* Poggendorff's Ann., Band xl. S. 447.

TABLE VIII.—Showing the Volumes occupied by certain Nitrates.

Name	Designation. Formula.	Atomic weight.	Volume in Solution.			Volume in the state of Salt.				
			Volume by experiment.	9 taken as unity.	Volume by theory.	Volume of salt by experiment.	9·8, or volume of ice, as unity.	Volume by theory.	Specific gravity by theory.	Specific gravity by experiment.
Nitrate of Potash	KO,NO₅	101·3	36·1	4	36	48·9	5	49·0	2·067	2·070
Nitrate of Ammonia	NH₄O,NO₅	80·3	45·5	5	45	49·1	5	49·0	1·639	1·635
Nitrate of Soda	NaO,NO₅	85·45	26·5	3	27	39·1	4	39·2	2·180	2·182
Nitrate of Silver	AgO,NO₅	170·0	27·2	3	27	39·2	4	39·2	4·336	4·336
Nitrate of Lead	PbO,NO₅	165·75	25·4	3	27	38·4	4	39·2	4·228	4·316
Nitrate of Barytes	BaO,NO₅	130·85	27·0	3	27	39·8	4	39·2	3·338	3·284
Nitrate of Strontia	SrO,NO₅	106·0	26·5	3	27	39·2	4	39·2	2·704	2·704
Nitrate of Black Oxide of Mercury	Hg₂O,NO₅+2HO	282·0	54·0	6	54	58·9	6	58·8	4·796	4·785
Basic Nitrate of Mercury	2HgO,NO₅+2HO	291·0				68·6	7	68·6	4·242	4·242
Basic Nitrate of Lead	2PbO,NO₅	277·72				49·2	5	49·0	5·667	5·645
Nitrate of Bismuth	BiO,NO₅+3HO	160·33				58·6	6	58·8	2·727	2·736
Nitrate of Copper	CaO,NO₅+3HO	120·8	45·2	5	45	59·0	6	58·8	2·054	2·047
Nitrate of Magnesia	MgO,NO₅+6HO	128·8	72·8	8	72	88·0	9	88·2	1·460	1·464

Section IV.

Chlorides, Bromides, and Iodides.

Chloride of Potassium, KCl=74·7.—On dissolving 37·5 grains of this salt in 1000 grains of water, the increase was 13·3 at a temperature of 57°; a second experiment with the same quantity gave the increase 13·5 at 58°; and a third experiment gave 13·7 at 65°.

$$
\begin{array}{lll}
\text{I. KCl, vol. in solution} & \text{. . .} & 26\text{·}5 \\
\text{II. \quad ,, \qquad ,,} & \text{. . .} & 26\text{·}8 \\
\text{III. \quad ,, \qquad ,,} & \text{. . .} & 27\text{·}2 \\
\hline
\text{Mean . . .} & & 26\text{·}8
\end{array}
$$

The whole of an equivalent thrown into turpentine (the salt having been decrepitated) increased 39·6 and 39·3 in two experiments. Half an equivalent (37·35 grains) caused a rise in the stem of 19·6 in two experiments.

			Sp. gr.
I. KCl, vol. of salt	. . .	39·6	1·887
II. ,, ,,	. . .	39·3	1·900
III. ,, ,,	. . .	39·2	1·905
IV. ,, ,,	. . .	39·2	1·905
	Mean . . .	39·3	1·900

Chloride of Ammonium, NH_4Cl=53·66.—Half an equivalent (26·83 grains) dissolved in 1000 grains of water with an increase of 17·5 at a temperature of 60°; in two other experiments, at 63°, the increase was 18·0.

I. NH_4Cl, vol. in solution	. . .	35·0
II. ,, ,,	. . .	36·0
III. ,, ,,	.·. .	36·0
	Mean . . .	35·7

Our experiments on the specific gravity of this salt gave 1·578 as a uniform result, indicating a volume of 34·0.

Bromide of Potassium, KBr=117·6.—The fourth part of

an equivalent, 29·4 grains, on being dissolved in water at 49°, gave in two experiments an increase of 7·2 ; which gives for the volume of the salt in solution 28·8. The same quantity of salt, immersed in turpentine at 63°, caused an increase of 11·0 in two experiments.

Sp. gr.

KBr, vol. of salt . . . 44·0   2·672

Iodide of Potassium, KI = 165·82.—This salt was decrepitated, and on dissolving 41·5 grains gave an increase of 11·0 at 57° ; a second experiment with 83·0 grains gave an increase of 22·0 at 55°.

I. KI, vol. in solution . . . 44

II.   ,,   ,,   ,, . . . 44

Mean . . . 44

On projecting 41·45 grains of this salt, previously decrepitated, into turpentine, an increase of 13·6 and 13·5 was produced in two successive experiments.

Sp. gr.

I. KI, vol. of salt . . . 54·4   3·048

II.   ,,   ,, . . . 54·0   3·070

Mean . . . 54·2   3·059

Chloride of Sodium, NaCl = 58·78.—The whole of an equivalent of this salt, previously decrepitated, dissolved in 1000 grains of water at 60° with a rise of 18·0, and in a second experiment of 18·2 ; in a third experiment 118 grains of salt were dissolved in 1000 grains of water at 62° with an increase of 38·0.

I. NaCl, vol. in solution . . . 18·0

II.   ,,   ,,   ,, . . . 18·2

III.   ,,   ,,   ,, . . . 18·9

Mean . . . 18·3

80 grains of salt were treated as described in the mercurial volumenometer, fig. 3, and the empty part of the tube, after the restoration of the mercury, showed a volume of 40·0. The same quantity thrown into alcohol previously saturated with it gave an increase of 39·5. The whole of an equivalent (58·78 grains), thrown into a saturated solution, caused an increase of 29·3.

				Sp. gr.
I.	NaCl, vol. of salt	. . .	29·4	2·000
II.	„	„	. . . 29·0	2·026
III.	„	„	. . . 29·3	2·006
		Mean . . .	29·23	2·011

Bromide of Sodium, $NaBr + 3HO = 128·7$.—On dissolving 25·7 grains of this salt in water, an increase of 9·2 was occasioned in two experiments at a temperature of 53°.

I., II. $NaBr + 3HO$, vol. in solution 46.

The same quantity of salt put into turpentine caused an increase of 11.

		Sp. gr.
$NaBr + 3HO$, vol. of salt . . . 55		2·340

Chloride of Barium, $BaCl + 2HO = 122·83$.—30·7 grains dissolved in 1000 grains of water increased 7·0 at a temperature of 58°; a second experiment, in which 20 grains of the salt were dissolved, gave an increase of 4·5.

I.	$BaCl + 2HO$, vol. in solution . . .	28·0
II.	„	„	„ . . . 27·6
		Mean . . .	27·8

The fourth of an equivalent, 30·7 grains, being immersed in a saturated solution, gave an increase of 9·7 at a temperature of 60°; and the same quantity in two other experiments gave an increase of 9·8.

				Sp. gr.
I.	BaCl + 2HO, vol. of salt	. . . 38·8		3·166
II.	„	„	. . . 39·2	3·133
III.	„	„	. . . 39·2	3·133
		Mean . . . 39·07		3·144

Perchloride of Mercury, HgCl = 136·9.—The fourth of an equivalent, 34·2 grains, of corrosive sublimate, on being dissolved in 1000 grains of water, gave an increase of 4·6 at a temperature of 62°; a second experiment with the same quantity was attended with the same result.

I., II. HgCl, vol. in solution 18·4.

Half an equivalent (68·45 grains), thrown into a saturated solution of the salt, caused an increase of 11·0 at a temperature of 56°.

		Sp. gr.
I. HgCl, vol. of salt . . . 22		6·223

Chloride of Hydrogen, HCl = 36·47.—It was of interest to ascertain the volume of hydrochloric acid, in order to compare it with other chlorides of the magnesian metals when dissolved in water. It was natural to expect that the volume of muriatic acid in dilute solutions would be different from that possessed by it in its concentrated state; and therefore the following experiments must be viewed in this light. Peligot's salt, the bichromate of the chloride of potassium, on dissolving in water, was decomposed into bichromate of potash and muriatic acid; and the volume of the latter was obtained by deducting that due to the former salt and adding the volume of water. The fourth part of an equivalent of this salt, 44·75 grains, dissolved in 1000 grains of water with an increase of 13·5 at 65°, and of 13·6 in another experiment at 68°. This result gives for the whole volume of the salt when dissolved 54·0 and 54·4, from which must be deducted 45·0 for the volume of bichromate of potash, and 9 must be added on account of the equivalent of water.

I. Muriatic Acid, in dilute solutions . . . 18·0
II.   „   „   „   . . . 18·4
                                    ———
                        Mean . . . 18·2

Chloride of Copper, $CuCl + 2HO = 85·18$.—Half an equivalent, 42·6 grains, was dissolved in 1000 grains of water with an increase of 13·4 at a temperature of 60°; on a second experiment, 47 grains occasioned an increase of 14·0 at a temperature of 58°.

I. $CuCl + 2HO$, vol. in solution . . . 26·8
II.   „   „   „   . . . 25·4
                                    ———
                        Mean . . . 26·1

Half an equivalent, 42·6 grains, being immersed in a saturated solution at 62°, caused an increase of 17·0; a second experiment with the same quantity of salt gave an increase of 16·6.

		Sp. gr.
I. $CuCl + 2HO$, vol. of salt . . . 33·2		2·566
II. „ „ „ . . . 34·0		2·505
Mean . . . 33·6		2·535

Chloride of Copper and Ammonium, $CuCl + NH_4Cl + 2HO = 138·84$.—34·7 grains of this salt, being dissolved in 1000 grains of water, gave an increase of 15·5 in the first experiment and of 15·4 in the second, both at a temperature of 68°.

I. $CuCl + NH_4Cl + 2HO$, vol. in solution . . . 62·0
II.   „   „   „   . . . 61·6
                                    ———
                        Mean . . . 61·8

32·46 grains, thrown into a saturated solution, caused an increase of 16·1 in two experiments at a temperature of 60°, and a repetition of the experiment confirmed this result.

Chloride of Copper and Potassium, $CuCl + KCl + 2HO =$

159·88.—34·7 grains of this salt, being dissolved in 1000 grains of water, caused an increase of 11·5 at 62°.

$CuCl + KCl + 2HO$, vol. in solution 53·0.

The same quantity (34·7 grains), thrown into a saturated solution, caused an increase of 14·3.

Sp. gr.

$CuCl + KCl + 2HO$, vol. of salt. . . 65·9   2·426

Chloride of Tin, $SnCl + 3HO = 121·39$.—One fourth of an equivalent (30·35 grains) was dissolved in 1000 grains of water, acidulated with muriatic acid, with an increase of 9·0 at a temperature of 60°; a second experiment, with the same quantity of salt and at the same temperature, gave an increase of 9·2.

    I. $SnCl + 3HO$, vol. in solution. . . 36·0
    II.   „   „   „   . . . 36·8
                                        Mean . . . 36·4

The same quantity, 30·35 grains, of the salt being immersed in a saturated solution, yielded an increase of 11·0, the temperature being 60°; and exactly the same result attended the repetition of the experiment.

Sp. gr.

$SnCl + 3HO$, vol. of salt . . 44·0   2·759

Chloride of Tin and Ammonium, $SnCl + NH_4Cl + 3HO = 175·05$.—On dissolving 44 grains of this salt in 1000 grains of water, the increase was 18·3 at a temperature of 60°; a second experiment, with the same quantity and at the same temperature, gave an increase of 18·5.

I. Chloride of Tin and Ammonium, vol. in solution 72·7
II.   „   „   „   „   73·5

                                    Mean . . . 73·1

On immersing 43·76 grains of the salt in a saturated solution, an increase of 20·8 was obtained at a temperature of 60°, which gives 83·2 as the volume of the equivalent, and 2·104 as the specific gravity of the salt.

Chloride of Tin and Potassium, $SnCl + KCl + 3HO = 196·09$.—On dissolving 24·3 grains of the salt in 1000 grains of water, an increase of 8·0 was obtained at a temperature of 60° ; and 48·5 grains, dissolved in the same quantity of water, gave an increase of 15·5.

I. Chloride of Tin and Potassium, vol. in solution 64·5
II.   ,,   ,,   ,,   ,,   62·7

Mean . . . 63·6

On throwing the fourth part of an equivalent, 49 grains, into a saturated solution, an increase of 19·5 was obtained at a temperature of 54°.

Sp. gr.
$SnCl + KCl + 3HO$, vol. of salt . . . 78·0   2·514

A. *Chloride of Mercury and Ammonium,* $HgCl + NH_4Cl + HO = 199·8$.—On dissolving 49·95 grains of this salt in 1000 grains of water, an increase was obtained of 16·0, and in a second experiment of 16·2, the temperature being about 60° in both cases.

I. A. Chloride of Mercury and Ammonium, vol. in sol. 64·0
II.   ,,   ,,   ,,   ,,   64·8

Mean . . . 64·4

The same quantity of salt, thrown into a saturated solution at 60°, occasioned an increase of 17·0 in two experiments, which makes the volume of the equivalent 68·0, and the specific gravity 2·938.

B. *Chloride of Mercury and Ammonium,* $NH_4Cl + 2HgCl + HO = 336·4$.—On dissolving 42 grains of this salt in 1000 grains of water, an increase of 10·1 was occasioned in two experiments at 54°, and of 10·2 in a third experiment at 60°.

I., II. $NH_4Cl + 2HgCl + HO$, vol. in solution . . . 80·9
III.   ,,   ,,   ,,   81·6

Mean . . . 81·2

42 grains, or one eighth of an equivalent, thrown into a saturated solution of the salt, caused a rise in the stem of 11 in two experiments.

	Vol. of salt.	Sp. gr.
I., II. B. Chloride of Mercury and Ammonium	88·0	3·822

Chloride of Mercury and Potassium, $KCl + HgCl + 2HO =$ 366·5.—The eighth part of an equivalent, 45·8 grains, being dissolved in 1000 grains of water, caused in two experiments an increase of 10·1 at a temperature of 53°.

	Vol. in solution.
I., II. Chloride of Mercury and Potassium . .	80·8

The same quantity of salt, 45·8 grains, thrown into a saturated solution, caused an increase of 12·0 in one experiment and of 12·4 in two other trials, the temperature in all the cases being 58°.

	Vol. of salt.	Sp. gr.
I. Chloride of Mercury and Potassium . .	96·0	3·818
II. ,, ,, ,,	99·2	3·694
III. ,, ,, ,,	99·2	3·694
Mean . .	98·1	3·735

Chloride of Mercury and Sodium, $NaCl + 2HgCl + 4HO =$ 368·5.—On dissolving 46·06 grains of this salt in 1000 grains of water, the increase was 12·4 at 63°. This gives for the equivalent a volume of 99·2, or 11 equivalents. The same quantity of salt thrown into turpentine produced an increase of 15·3, which gives for the equivalent 122·4, and for the specific gravity 3·011.

A careful consideration of the previous experiments shows that there are two distinct classes of chlorides, &c. The first of these is placed in the next table, Table IX., and possesses 11 as the divisor of the solid.

TABLE IX.

Name.	Designation. Formula.	Atomic weight.	Volume in Solution.			Volume of Solid.				
			Volume by experiment.	9 taken as unity.	Volume by theory.	Volume by experiment.	11 taken as unity.	Volume by theory.	Specific gravity by theory.	Specific gravity by experiment.
Bromide of Potassium	KBr	117·6	28·8	3	27	44·0	4	44	2·672	2·672
Chloride of Ammonium	NH$_4$Cl	53·66	35·7	4	36	34·0	3	33	1·626	1·578
Iodide of Potassium	KI	165·82	44·0	5	45	54·2	5	55	3·015	3·059
Bromide of Sodium	NaBr+3HO	128·70	46·0	5	45	55·0	5	55	2·340	2·340
Chloride of Mercury	HgCl	136·9	18·4	2	18	22·0	2	22	6·223	6·223
Chloride of Hydrogen	HCl	36·47	18·2	2	18	......	...	...	......	......
Chloride of Copper	CuCl+2HO	85·18	26·1	3	27	33·6	3	33	2·581	2·534
Chloride of Tin	SnCl+3HO	121·39	36·4	4	36	44·0	4	44	2·759	2·759
Chloride of Mercury and Sodium	2HgCl+NaCl+4HO	368·5	99·2	11	99	122·4	11	121	3·045	3·011

TABLE IX A.—Showing the Volumes in Solution and in the Solid State of certain Chlorides.

Name.	Formula.	Atomic weight.	Volume in Solution.			Volume of Salt.				
			Volume by experiment.	g, or volume of water, as unity.	Volume by theory.	Volume of salt by experiment.	9·8, or volume of ice, as unity.	Volume by theory.	Specific gravity by theory.	Specific gravity by experiment.
Chloride of Potassium	KCl	74·7	26·8	3	27	39·3	4	39·2	1·905	1·900
Chloride of Sodium	NaCl	58·78	18·3	2	18	29·2	3	29·4	2·000	2·011
Chloride of Barium	BaCl+2HO	122·83	27·8	3	27	39·07	4	39·2	3·133	3·144
Chloride of Copper and Potassium	CuCl+KCl+2HO	159·88	53·0	6	54	65·9	7	68·6	2·331	2·426
Chloride of Copper and Ammonium	CuCl+NH$_4$Cl+2HO	138·84	61·8	7	63	68·8	7	68·6	2·024	2·018
Chloride of Tin and Ammonium	SnCl+NH$_4$Cl+3HO	175·05	73·1	8	72	83·2	8	83·2	2·104	2·104
Chloride of Tin and Potassium	SnCl+KCl+3HO	196·09	63·6	7	63	78·0	8	78·4	2·501	2·514
A. Chloride of Mercury and Ammonium	HgCl+NH$_4$Cl+HO	199·8	64·4	7	63	68	7	68·6	2·912	2·938
B. Chloride of Mercury and Ammonium	2HgCl+NH$_4$Cl+HO	336·4	81·2	9	81	88	9	88·2	3·814	3·824
Chloride of Mercury and Potassium	2HgCl+KCl+2HO	366·5	80·8	9	81	98·1	10	98·0	3·739	3·736

In the second class (Table IX. A) the primitive volume is 9·8, or, as in the case of the double chlorides of tin, the metallic salt enters into combination with the volume 11; while NH_4Cl remains a multiple of 9·8. It is interesting to observe that NH_4Cl affects in combination as a solid the same number of volumes which it has as a liquid.

The results of the experiments detailed in this section afford strong proofs of the law of multiple proportions, and exhibit at the same time that remarkable alteration of the divisor of the solid volumes which we have already noticed so frequently. Thus, while many of the chlorides and bromides are multiples of 11, we have decided exceptions in chlorides of potassium and sodium, which possess for their divisor the volume of ice, viz. 9·8; and this reappears in the double salts.

It is impossible, however, not to see that these results are somewhat singular, for in the double salts the chloride of potassium forces the double salt with which it is associated to assume the multiples of 9·8, and then exhibits its natural isomorphous relation to chloride of ammonium, which *per se* it did not possess. Chloride of ammonium, anomalous in being a multiple of 11 in the solid state, assumes four volumes, multiples of 9·8, in the double chlorides, and then presents the same number for its solid volume as chloride of potassium. The isomorphism of potassium and sodium is so entirely hypothetical, that it will not excite surprise to find the volumes of the chlorides so different. We were less prepared to detect the difference between iodide and chloride of potassium, but have confirmed it by an examination of iodide of ammonium, 50 grains of which dissolved in 1000 of water with an increase of 18·7, which gives $\frac{54\cdot1}{9} = \text{six}$ volumes for the equivalent—a result confirmatory of our determination of five volumes for iodide of potassium, the increase of one volume being in conformity with the usual behaviour of ammoniacal salts. We shall return to the consideration of the chlorides in a future section.

Section V.

Chromates.

The chromates present a class of salts which offer some peculiarities with regard to their volumes, in elucidating which we had occasion to repeat our experiments very often, and therefore give the mean of the results, instead of taking up unnecessary space in the Transactions of the Society by describing each experiment individually.

Chromic Acid, $CrO_3 = 52\cdot19$.—The chromic acid used in our experiments was obtained by adding sulphuric acid to bichromate of potash. It was in beautiful distinct crystals of nearly a quarter of an inch in length, being the finest and purest specimen which we have obtained in many preparations of this kind.

The half of an equivalent, 26·09 grains, dissolved in 1000 grains of water with an increase of 9·0 at 72°; this gives 18·0 as the volume of chromic acid in solution.

The same quantity of acid, thrown into the solution from which it had been crystallized, gave an increase of 9·7 and 9·8 in two experiments.

		Sp. gr.
I. Chromic Acid, volume . .	19·4	2·690
II. „ „ „	19·6	2·663
Mean . .	19·5	2·676

Yellow Chromate of Potash, $KO, CrO_3 = 99\cdot50$.—On dissolving 50 grains of this salt in 1000 grains of water, the increase was 9·0 at a temperature of 58°; this gives 17·9 as the volume of the equivalent in solution.

The mean result of *ten* experiments, on immersing 49·75 grains in turpentine, was an increase of 18·55, which gives 37·1 for the volume of the equivalent, and 2·682 as the specific gravity of the salt.

Sesquichromate of Potash, $2KO, 3CrO_3 = 251\cdot09$.—This salt, which will be described in a future communication by one of us, is obtained by boiling a solution of bichromate of potash with an excess of finely pounded litharge. The oxide

of lead removes only one fourth of the chromic acid of the
bichromate, and the solution on cooling deposits the sesquichromate in flattened prisms of a paler but more resplendent
colour than the bichromate of potash. On dissolving the
fourth part of an equivalent, 62·77 grains, in 1000 grains of
water, the increase in four experiments at 58° was exactly
18·0; this gives 72·0 as the volume of the equivalent in
solution.

The mean of six experiments, placing the fourth of an
equivalent, 62·77 grains, in turpentine, was an increase of
23·7, which gives 94·8 as the volume of the equivalent, and
2·648 as the specific gravity of the salt.

Bichromate of Potash, KO, $2CrO_3 = 151·70$.—On dissolving
76 grains of this salt in 1000 grains of water, an increase of
22·5 and 23·0 were obtained in two experiments at 60° and
65°.

I. KO, $2CrO_3$, vol. in solution . . 44·9
II.   ,,   ,,   ,,   45·8
————
Mean . . . 45·3

Half an equivalent of this salt, 75·84 grains, immersed in
turpentine gave an increase, the mean of *ten* experiments,
of 28·9, which gives 57·8 as the volume of an equivalent,
and 2·624 as the specific gravity of the salt.

Terchromate of Potash, KO, $3CrO_3 = 203·92$.—This salt
was obtained by mixing a solution of bichromate of potash
with nitric acid and crystallizing. On dissolving 51 grains
of the salt in 1000 grains of water, an increase was occasioned
of 18·0 at 60°; this gives 71·9 as the volume of the equivalent in solution.

On immersing 50·98 grains in turpentine, the increase was
19·3 in two experiments, and 19·0 in a third trial.

		Sp. gr.
I. KO, $3CrO_3$, vol. of salt . . 77·2	2·641
II. ,, ,, ,, 77·2	2·641
III. ,, ,, ,, 76·0	2·683
Mean . . 76·8	2·655

Bichromate of Chloride of Potassium, $KCl + 2CrO_3 =$ 179·08.—The fourth part of an equivalent, 44·77 grains, being dissolved in 1000 grains measure of a dilute solution of muriatic acid, gave an increase of 15·7 in two experiments at 57°; this result makes the volume of an equivalent in solution 62·8.

The mean of various experiments on this salt gave an increase of 18·15 on immersing the above quantity of salt in turpentine, which yields 72·6 as the volume of the equivalent and 2·466 as the specific gravity of the salt.

The results now described show that the chromates form a group different from the classes of salts hitherto given.

TABLE X.—Showing the Volumes occupied by certain Chromates.

Designation.			Volume of Salt in Solution.			Volume of Salt.	
Name.	Formula.	Atomic weight.	Volume in solution by experiment.	9 taken as unity.	Volume by theory.	Volume of salt.	Specific gravity.
Chromic Acid	CrO_3	52·19	18·0	2	18·0	19·5	2·676
Chromate of Potash......	KO, CrO_3 ..	99·50	17·9	2	18·0	37·1	2·682
Sesquichromate of Potash .	$2KO, 3CrO_3$	251·07	72·0	8	72·0	94·8	2·648
Bichromate of Potash	$KO, 2CrO_3$..	151·70	45·3	5	45·0	57·8	2·624
Terchromate of Potash....	$KO, 3CrO_3$..	203·92	71·9	8	72·0	76·8	2·655
Bichromate of Chloride of Potassium	$KCl + 2CrO_3$	179·08	62 8	7	63	72·6	2·466

An inspection of the previous table will show clearly that the chromates differ from the salts described in the former sections. In the volumes in solution there is no difference; they are multiples of 9, and follow the usual law of the sum of the volumes, being made up of the volumes of the con-

stituents of the salt.   Chromate of potash possesses two
volumes in solution, exactly as is the case with its analogue
sulphate of potash.   The latter salt affects three volumes in
the solid state, and so naturally should chromate of potash.
In bichromate of potash we see these three volumes appearing
in solution, united to two volumes possessed by the chromic
acid attached to the chromate of potash; in sesquichromate
of potash they again reappear, and so also in terchromate of
potash.   The fact that the number of volumes possessed in
the solid state by the lowest member of a series of salts
passes over into the higher members when in solution, finds
examples in the carbonates and oxalates, and is not peculiar
to the chromates.

The solid volumes of the chromates possess decided peculiarities, being neither multiples of 11 nor of 9·8.   Chromic
acid itself is obviously twice the volume of ice, $9·8 \times 2 = 19·6$,
the experimental number being 19·5.   But all the other
salts in this group refuse to arrange themselves under either
of the heads which we have found to explain most of the
salts in the previous sections.   In an exception of this kind
we are entitled to make an assumption, which will in all
probability be near the truth, if by means of it we can bring
into one uniform system a whole group of anomalous salts.
Sesquichromate of potash is of great importance in the
history of the chromates, from its frequent occurrence,
although hitherto it has been altogether neglected by chemists.
Chromic acid is actually able to displace sulphuric acid from
sulphate of potash, in order to gratify its love for the potash
in the peculiar condition of the sesquichromate.   In numerous
instances of decomposition, as will be pointed out by one of
us in another paper, this sesquichromate appears.   The
sesquichromate is not formed readily, if indeed it is ever
formed, by crystallizing chromate of potash with chromic
acid in the proportion of sesquichromate, the result being
bichromate of potash and chromate of potash, which crystallize separately.   Here, then, is a remarkable point in the
constitution of the chromates, which can only be explained
by supposing that sesquichromate of potash contains a double

atom of chromate of potash united to one of chromic acid. The decomposition of bichromate of potash by oxide of lead necessarily implies that its atom should be doubled; $2KO, 3CrO_3 + CrO_3$ boiled with litharge, gives $2KO, 3CrO_3 + PbO, CrO_3$.

We have found the volume of KO, CrO_3 to be 37·1, not 33·0 as in the case of sulphate of potash. Karsten obtained the specific gravity 2·640, which gives the volume 37·6; and Thomson states the specific gravity to be 2·612, which gives the volume 38·1; the mean of all these experiments is 37·6, which, multiplied by 2, for the reasons already stated, gives as the volume of $2 (KO, CrO_3)$, 75·2. The natural volume of chromate of potash, deduced from its analogy to sulphate of potash, would be 11×3, or on the double atom $11 \times 6 = 66$. Now the assumption we make to explain this class of salts is, that the double atom of chromate of potash enjoys its anomalous character by adding to its natural volume the volume of ice, thus $66·0 + 9·8 = 75·8$, which is not very far from the volume ascertained by experiment. This assumption of a volume of ice, in addition to other volumes of 11, has been shown to exist in the magnesian sulphates, and therefore its hypothetical existence in the chromates is by no means extravagant. Sesquichromate of potash must then be the double chromate of potash united to an equivalent of chromic acid, $75·8 + 19·6 = 95·4$, which is not very far from 94·8, the volume determined by experiment. Bichromate of potash would consist of a double atom of chromate of potash and 2 of chromic acid, or $75·8 + 39·2 = 115·0$, which agrees pretty closely with the experimental determination of 115·6; and terchromate of potash, in like manner, is 1 atom of double chromate of potash with 4 of chromic acid, or $75·8 + 78·4 = 154·2$, which is almost exactly the same as 154·4, found in the two consecutive experiments, and not far distant from 153·6, the mean of the three experiments.

This view received confirmation from the volume of Peligot's salt, which certainly consists of the volume of KCl, when in combination, added to that of 2 atoms of chromic acid, $33·0 + 39·2 = 72·2$, a number very close to the experimental result

72·6. It is quite true that we have made a gratuitous assumption at the outset of our explanation; but it is not surprising to find an unusual law prevailing in a class of salts so anomalous as the chromates. When the experimental numbers, and those calculated on the assumption, are so near as we have shown them to be, there is, we think, a good argument for the truth of the hypothesis.

TABLE X. A.

Name.	Formula.	Volume by experiment.	Volume by theory.	Specific gravity by theory.	Specific gravity by experiment.
Chromic Acid	CrO_3	19·5	19·6	2·663	2·676
Chromate of Potash, doubled	$(2KO, 2CrO_3)$	75·2	75·8	2·627	2·646
Sesquichromate of Potash .	$(2KO, 2CrO_3)+CrO_3$	94·8	95·4	2·658	2·648
Bichromate of Potash	$(2KO, 2CrO_3)+2CrO_3$	115·6	115·0	2·638	2·624
Terchromate of Potash ..	$(2KO, 2CrO_3)+4CrO_3$	154·4	154·2	2·644	2·641
Bichromate of Chloride of Potassium........	$KCl+2CrO_3$	72·6	72·2	2·480	2·466

A singular result obtained in the examination of the anhydrous double sulphates seems to be explained by the behaviour of the chromates. We found sulphate of copper and potash, and sulphate of magnesia and potash, to affect a volume of 59·8 instead of 55·0; and we ascertained, by many experiments, that this high number was not due to an error of observation. Now if we suppose the KO, SO_3 in these salts to behave like KO, CrO_3 in assuming one volume of ice on the double atom, then $2KO, SO_3 = 75·8 + 2MO, SO_3 = 44·0 = \frac{119·8}{2} = 59·9$, a number almost identical with the experimental result. On this view, then, anhydrous double sulphates are constituted on the type of the red chromate of potash, the *two volumes* of CrO_3 being replaced by the *two volumes* of MO, SO_3. Anhydrous alum was found to have a volume of 116·4 instead of 111·4, but would be reconciled with theory if we supposed it to contain the peculiar KO, SO_3, analogous to KO, CrO_3; in this case the theoretical volume would be 116·3.

SECTION VI.

Carbonates.

Carbonate of Potash, $KO, CO_2 = 69·4$.—On dissolving 34·7 grains of carbonate of potash in 1000 grains of water, the increase was 4·6 at 62°; the atomic volume in solution is therefore 9·2. The same quantity of salt thrown into turpentine caused, in various experiments, an increase of 16·5; this makes the volume of the equivalent 33·0, and the specific gravity of the salt 2·103.

Bicarbonate of Potash, $KO, HO, 2CO_2 = 100·6$.—The fourth part of an equivalent (25·1 grains) dissolved in 1000 grains of water at 61° with an increase of 8·9, and, in another experiment, of 9·0. The mean of these results, 8·95, gives the volume of the equivalent in solution, 35·8. The same quantity of salt, thrown into turpentine, gave an increase of 12·0 in various experiments, which gives for the specific volume of the salt 48·0, and for its specific gravity 2·092. As this salt was one of the very few substances used in this inquiry not prepared by ourselves, we take the mean of our own result and the only other recorded specific gravity of which we are aware, viz. that by Gmelin, 2·012, and adopt 49·0 as the correct volume and 2·052 as the specific gravity.

Bicarbonate of Ammonia, $HO, NH_4O, 2CO_2 = 79·3$.—This salt was made by exposing the carbonate of the shops to the air until it ceased to emit smell, and then crystallizing the remainder. On dissolving 19·82 grains, the fourth of an equivalent, in 1000 grains of water, the increase was 9·0 at 55°, and 9·4 in another experiment at 62°. The mean result gives 36·8 as the volume of the salt in solution.

On immersing 19·82 grains of the salt in turpentine, an increase of 12·5 was effected, which gives as the volume of the salt 50·0, and for its specific gravity 1·586.

Bicarbonate of Soda, NaO, HO, $2CO_2 = 84\cdot64$.—On dissolving $42\cdot32$ grains of this salt in 1000 grains of water at $67°$, an increase of $9\cdot0$ was obtained; this gives for the volume of an equivalent in solution $18\cdot0$. On immersing the same quantity of salt in turpentine, the increase was $19\cdot4$ and $19\cdot2$ in two experiments.

					Sp. gr.
I.	NaO, HO, $2CO_2$, vol. of salt	. .	$38\cdot8$	$2\cdot181$
II.	„	„	„	$38\cdot4$	$2\cdot204$
			Mean . . .	$38\cdot6$	$2\cdot192$

Although we have examined other carbonates, we purposely avoid bringing them into the present paper, because they involve considerations upon which we are at present engaged in minute study, and do not wish to hazard without sufficient proof. We subjoin a few carbonates here examined in a tabular form.

TABLE XI.—Showing the Volumes occupied by the Alkaline Carbonates.

Designation.			Vol. in Solution.			Vol. of Salt.	
Name.	Formula.	Atomic weight.	Volume in solution.	9 taken as unity.	Volume by theory.	Volume of salt by experiment.	Specific gravity by experiment.
Carbonate of Potash	KO, CO_2	$69\cdot4$	$9\cdot2$	1	9	$33\cdot0$	$2\cdot103$
Carbonate of Soda	NaO, CO_2	$53\cdot47$	..	..	..	$22\cdot0$	$2\cdot427$
Bicarbonate of Potash....	$HO, KO, 2CO_2$ ..	$100\cdot6$	$35\cdot8$	4	36	$49\cdot0$	$2\cdot052$
Bicarbonate of Soda......	$HO, NaO, 2CO_2$..	$84\cdot64$	$18\cdot0$	2	18	$38\cdot6$	$2\cdot192$
Bicarbonate of Ammonia..	$HO, NH_4O, 2CO_2$	$79\cdot3$	$36\cdot8$	4	36	$50\cdot0$	$1\cdot586$

The results shown in this Table will appear perplexing, unless the facts already observed in the previous sections be borne in mind. We find in carbonate of potash an astonishing difference between the liquid and solid volume; and this is still more marked in the case of carbonate of soda, which ceases to occupy volume in solution. Both of these salts have 11 as the divisor of their solid volume, KO, CO_2 affecting three, and NaO, CO_2 two volumes. In the last section we saw that the three volumes possessed by chromate of potash in its solid state passed over into bichromate of potash; and in bicarbonates of potash and ammonia we observe the same circumstance, except that the volumes change from multiples of 11 to multiples of 9·8, and in solution are one less than in the state of a salt. It is probably owing to this circumstance that we do not in this case observe the usual increase of one volume in the ammoniacal over the corresponding salt of potash. The bicarbonates of potash, soda, and ammonia are probably multiples of 9·8, or the volume of ice.

	Vol. by experiment.	Vol. by theory.
Bicarbonate of Potash	49·0...2·052	$9·8 \times 5 = 49·0...2·052$
,, Ammonia	50·0...1·586	$9·8 \times 5 = 49·0...1·618$
,, Soda . .	38·6...2·192	$9·8 \times 4 = 39·2...2·159$

Section VII.

Oxalates.

The oxalates offered an interesting group of salts for examination, especially on account of the accurate determination of their composition and hydration by Graham.

Oxalate of Water, HO, $C_2O_3 + 2HO = 63·26$.—32 grains of oxalic acid, dissolved in 1000 grains of water, caused an increase of 18·5 at a temperature of 55°; the same quantity, being subjected to a second experiment, caused an increase of 19·0; and a third experiment, in which 21 grains were dissolved in 9½ ounces of water, occasioned an increase of 12·0 at 40°.

I. HO, $C_2O_3 + 2HO$, vol. in solution . . . 36·5

II. ,, ,, ,, . . 37·5

III. ,, ,, ,, . . 36·0

Mean . . . 36·6

A whole equivalent thrown into turpentine caused in various experiments an increase of 39·0, which gives for its specific gravity 1·622. Richter states the specific gravity to be 1·507; but it is impossible that he can have operated upon a pure specimen, as we have repeated the experiments upon this acid very frequently.

Oxalate of Potash, $KO, C_2O_3 + HO = 92·39$.—A quantity of salt, 42·5 grains, being dissolved in 1000 grains of water, gave an increase of 13·0; and the same result attended a repetition of the experiment, the temperature in both cases being at 60°.

$$KO, C_2O_3 + HO, \text{vol. in solution } 28·2.$$

46·2 grains of the same salt, being put into a saturated solution, caused a rise in the stem of 22·0; a repetition of the experiment with the same quantity gave an increase 21·9, the temperature in both cases being 61°.

					Sp. gr.
I.	$KO, C_2O_3 + HO$, vol. of salt	. . .	44·0		2·100
II.	„ „ „	. . .	43·8		2·109
			Mean . . .	43·9	2·104

Oxalate of Ammonia, $NH_4O, C_2O_3 + HO = 71·43$.—Half an equivalent of this salt (35·71 grains) was dissolved in 1000 grains of water, with an increase of 18·0 at a temperature of 55°; and a repetition of the experiment, with the same quantities and at the same temperature, gave exactly the same result.

$$I., II. \ NH_4O, C_2O_3 + HO, \text{vol. in solution } 36.$$

35·71 grains, being immersed in a saturated solution, gave in the first experiment an increase of 24·5, in the second of 24·4; the first experiment being at 48°, the second at 50°.

					Sp. gr.
I.	$NH_4O, C_2O_3 + HO$, vol. of salt	. . .	49·0		1·458
II.	„ „ „	. . .	48·8		1·464
			Mean . . .	48·9	1·461

Binoxalate of Potash, KO, $C_2O_3 + HO$, $C_2O_3 + 2HO =$ 146·63.—To determine the volume of this salt, 18·33 grains were dissolved in 1000 grains of water with a rise of 6·8 at a temperature of 57°; and the same result attended a repetition of the experiment; in a third experiment, 25 grains at the same temperature caused an increase of 9·0.

I., II. Binoxalate of Potash, vol. in solution . . . 54·4
III.   „   „   „   . . . 52·8
                                                    ———
                                          Mean . . . 53·6

Half an equivalent of the salt (73·31 grains), being immersed in a saturated solution, caused an increase of 37·4 in the first experiment and of 37·2 in the second, the temperature in both cases being 55°.

		Sp. gr.
I. Binoxalate of Potash, vol. of salt . . . 74·8		1·960
II. „ „ „ . . . 74·4		1·971
Mean . . . 74·6		1·965

Oxalate of Copper and Potash, KO, $C_2O_3 + CuO, C_2O_3 + 2HO = 177·25$.—On account of the sparing solubility of this salt, 11·08 grains, or the sixteenth part of an equivalent, were dissolved in water, and caused an increase of 3·4 in two experiments at a temperature of 59°.

I., II. KO, $C_2O_3 + CuO$, $C_2O_3 + 2HO$, vol. in solution 54·4.

The fourth of an equivalent (44·3 grains), placed in a saturated solution, caused an increase in one experiment of 19·5; in another of 18·9; and in a third of 19·7; all at a temperature varying from 54° to 57°.

			Sp. gr.
I. Oxalate of Copper and Potash, vol. of salt 78·0			2·272
II. „ „ „ 75·6			2·344
III. „ „ „ 78·8			2·249
Mean . . 77·5			2·288

Binoxalate of Ammonia, NH_4O, $C_2O_3 + HO$, $C_2O_3 + 2HO$ = 125·69.—31·42 grains of this salt, dissolved in 1000 grains of water, caused in the first experiment an increase of 18·0 at 60°; in the second, of 18·2 at 61°; in a third experiment, of 17·8 at 54°; in a fourth, 42 grains dissolved in 4000 grains of water increased 24·0 at 53°.

I. Binoxalate of Ammonia, vol. in solution . . . 72·0
II.   ,,   ,,   ,,   . . . 72·8
III.   ,,   ,,   ,,   . . . 71·2
IV.   ,,   ,,   ,,   . . . 71·8

Mean . . . 71·9

The half of an equivalent (62·84 grains), being immersed in a saturated solution, caused an increase of 40·3 in two experiments, and of 40·0 in a third.

					Sp. gr.
I. Binoxalate of Ammonia, vol. of salt . .				80·6	1·559
II. ,,	,,	,,	. .	80·6	1·559
III. ,,	,,	,,	. .	80·0	1·571
		Mean . . .		80·4	1·563

Oxalate of Copper and Ammonia, NH_4O, $C_2O_3 + CuO$, $C_2O_3 + 2HO = 156·38$.—The solution of 18·3 grains gave an increase of 8·6 at 65°; this gives 73·3 as the volume of this salt when in solution.

On immersing 20 grains in turpentine, an increase of 10·4 was obtained, which gives for the volume of the equivalent 81·3, and for the specific gravity of the salt 1·923.

Quadroxalate of Potash, KO, $C_2O_3 + 3HO$, $C_2O_3 + 4HO = 255·11$.—32·0 grains dissolved in water gave an increase of 15·0 at 60°; and a second experiment, in which 16·0 grains were dissolved in 1000 grains of water, gave the increase of 7·2 at a temperature of 44°.

I. Quadroxalate of Potash, vol. in solution . . 119·4
II.   ,,   ,,   ,,   . . 114·8

Mean . . . 117·1

63·8 grains, the fourth part of an equivalent, thrown into a saturated solution, caused a rise of 35·1 in two experiments

Sp. gr.

I., II. Quadroxalate of Potash, vol. of salt 140·4   1·817

Quadroxalate of Ammonia, NH_4O, $C_2O_3 + 3HO$, $C_2O_3 + 4HO = 234·15.$—On dissolving 20 grains of this salt in 2500 grains of water at 50°, the increase is 11·5, which gives 134·5 as the volume of the equivalent in solution.

58·5 grains of the salt, thrown into a saturated solution, caused in the first experiment an increase of 36·8, in the second of 36·9, both at a temperature of 62°.

Sp. gr.

I. Quadroxalate of Ammonia, vol. of salt	147·2	1·591
II. „ „ „	147·6	1·586
	Mean . . 147·4	1·589

The volumes of the oxalates can only be explained by an attentive consideration of the previous results. We have already seen numerous instances in which the primitive volumes 9·8 and 11·0 become mutually convertible : this is strikingly the case with the salts of the present section. Hydrated oxalic acid has a volume $9·8 \times 4$; oxalate of potash possesses the volume 11×3, and passes with this volume into the binoxalate and quadroxalate of potash, the oxalic acid in the binoxalate being associated as two volumes of ice, although the water of crystallization possesses the volume 11. Quadroxalate of potash is to be viewed as anhydrous binoxalate *plus* 2 equivalents hydrated oxalic acid, the latter having become 11×4 instead of $9·8 \times 4$. The same explanation applies to the binoxalate and quadroxalate of ammonia, the only difference being that anhydrous oxalate of ammonia, $9·8 \times 4$, takes the place of oxalate of potash. On these views the following table is constructed (p. 88).

The examination of the volumes occupied by the oxalates presents several points of great interest. The volume of the oxalic acid itself is a multiple of the volume of ice, or $9·8 \times 4$.

Oxalate of potash in its solid state possesses four volumes, 11×4, but loses one volume on passing into solution, as usually is the case with neutral salts of potash. As one of

TABLE XII.—Showing the Volumes occupied by certain Oxalates.

Name.	Formula	Atomic weight.	Vol. in Solution.			Volume in state of Salt.				
			Volume in solution by experiment.	g, or volume of water, as unity.	Volume by theory.	Volume by experiment.	11 and 9·8 taken as unity, as above described.	Volume by theory.	Specific gravity by theory.	Specific gravity by experiment.
Oxalic Acid	HO, C_2O_3+2HO	63·26	36·6	4	36	39·0	4	39·2	1·616	1·622
Oxalate of Potash	KO, C_2O_3+HO	92·39	28·2	3	27	43·9	4	44·0	2·100	2·104
Oxalate of Ammonia	NH_4O, C_2O_3+HO	71·43	36·0	4	36	48·9	5	49·0	1·458	1·461
Binoxalate of Potash	$KO, 2C_2O_3+3HO$	146·63	53·6	6	54	74·6	7	74·6	1·965	1·965
Binoxalate of Ammonia	$NH_4O, 2C_2O_3+3HO$	125·67	71·9	8	72	80·4	8	80·8	1·555	1·563
Oxalate of Copper and Potash	$KO, C_2O_3+CuO, C_2O_3+2HO$	177·25	54·4	6	54	77·5	7	77·0	2·301	2·288
Oxalate of Copper and Ammonia	$NH_4O, C_2O_3+CuO, C_2O_3+2HO$	156·98	73·3	8	72	81·3	8	80·8	1·935	1·923
Quadroxalate of Potash	$KO, 4C_2O_3+7HO$	255·11	117·1	13	117	140·4	13	140·6	1·814	1·817
Quadroxalate of Ammonia	$NH_4O, 4C_2O_3+7HO$	234·15	134·5	15	135	147·4	14	146·8	1·595	1·589

these volumes is due to its combined water, the proper number of volumes in anhydrous oxalate of potash is three, and these it carries into binoxalate of potash, which is therefore a simple combination of oxalate of potash and hydrated oxalic acid, the crystalline water of the latter having assumed the volume 11.

	In solution.	As a salt.
KO, C_2O_3	18	33
$HO, C_2O_3 + 2HO$	36	41·6
Binoxalate of Potash . . .	54	74·6

The only difference between the volumes of this salt and those of its constituents, when uncombined, is that the crystalline water of the hydrated oxalic acid has assumed the volume 11. Quadroxalate of potash consists of anhydrous binoxalate of potash united to hydrated oxalic acid, as Graham has already announced in his researches on the oxalates. The three volumes affected by oxalate of potash in its solid state pass into solution with it in quadroxalate of potash, just as we saw in the case of chromate and bichromate of potash; and the attached oxalic acid affects 11×4 instead of $9·8 \times 4$.

	In solution.	As a salt.
I. Anhydrous Binoxalate of Potash . .	45	52·6
II. Hydrated Oxalic Acid.	72	88·0
Quadroxalate of Potash	117	140·6

The assumption of two volumes in solution above those of binoxalate of potash was already characteristic of binoxalate of ammonia; and the same increase is seen in the quadroxalate, showing clearly that that salt must contain its ammonia *quasi* binoxalate and not as oxalate of ammonia. It is very possible that the volumes in solution of quadroxalate of ammonia should be 14, instead of 15; but the temperature 31°, at which it comes out 14 volumes, is so low, that it is more natural to keep the volumes we have given in the table.

It is interesting to observe how closely oxalate of copper relates itself to oxalate of water.

	Volumes in solution.	Volumes as salt.
Oxalate of Copper and Potash . . .	6	7
„ Water and Potash. . . .	6	7
„ Copper and Ammonia . .	8	8
„ Water and Ammonia . .	8	8

Thus even in the apparently anomalous behaviour of bin-oxalate of ammonia, in assuming two volumes more than the corresponding salt of potash, we find oxalate of copper and ammonia imitating its example. The reason of their increase will be explained in the next section.

Section VIII.

Subsalts and Ammoniacal Salts.

The salts which we have hitherto examined have been those soluble in water, and having a constitution to a certain degree well defined. We have now to consider the insoluble subsalts, and, in some cases, their neutral insoluble types, and also to ascertain how far the results thus obtained serve to throw light on the constitution of ammoniacal salts.

Subsulphate of Copper, CuO, SO_3, $4HO + 3CuO = 234 \cdot 9$.—This well-known salt was made by adding ammonia to a solution of sulphate of copper. The fourth part of an equivalent (58·7 grains), thrown into water, caused an increase of 19·1 and 19·0 in two successive experiments.

			Sp. gr.
I. Subsulphate of Copper, vol. of salt . .	76·4		3·074
II. „ „ „ . .	76·0		3·090
Mean . . .	76·2		3·082

Subsulphate of Zinc, ZnO, SO_3, $3ZnO$, $4HO = 237 \cdot 3$.—This salt is apt to combine with more water than four atoms, but may be obtained with four by drying at 212°. On placing 29·66 grains, the eighth part of an equivalent, in turpentine, an increase of 9·5 was obtained; and on treating 22·8 in a similar manner, the rise in the stem was 7·3. Both of these experiments exactly agree in making

		Sp. gr.
Subsulphate of Zinc, vol. of salt . . .	76·0	3·122

Sulphate of Protoxide of Mercury, Hg_2O, SO_3 $=251\cdot0$.— This salt was prepared in the usual way by digesting one part of mercury in $1\frac{1}{2}$ part of sulphuric acid. The fourth part of an equivalent, $62\cdot75$ grains, thrown into turpentine, increased $8\cdot3$.

		Sp. gr.
Hg_2O, SO_3, vol. of salt . . .	$33\cdot2$	$7\cdot560$

Sulphate of Peroxide of Mercury, HgO, $SO_3 = 149\cdot6$.— The salt used in this experiment was prepared by heating five parts of sulphuric acid, mixed with a little nitric acid, with four parts of mercury until the whole became a dry saline mass. On immersing $37\cdot5$ grains of the salt thus prepared in turpentine, an increase of $5\cdot8$ was obtained, which gives $23\cdot1$ for the volume of the equivalent, and $6\cdot466$ for the specific gravity of the salt.

Subsulphate of Mercury, HgO, $SO_3 + 2HgO = 368\cdot46$.— The last salt thrown into water and washed with warm water is converted into the beautiful yellow powder known as turpeth mineral. On throwing $57\cdot4$ grains of this salt thus prepared into water, an increase of $6\cdot9$ was obtained, which gives $44\cdot3$ as the volume of the equivalent, and $8\cdot319$ as the specific gravity of the salt.

Chromate of Lead, PbO, $CrO_3 = 163\cdot97$.—On throwing $81\cdot98$ grains of the chromate of lead, previously well dried, into turpentine, an increase of $14\cdot5$ was effected; this gives $29\cdot0$ as the volume of the equivalent, and $5\cdot653$ as the specific gravity of the salt.

Subchromate of Lead, PbO, $CrO_3 + PbO = 275\cdot7$.—This salt was prepared by projecting chromate of lead into melted nitre, and afterwards washing out all soluble matter. On immersing $68\cdot92$ grains, the fourth part of an equivalent, an increase of $11\cdot0$ was obtained in two experiments. This gives $44\cdot0$ for the volume of the equivalent, and $6\cdot266$ as the specific gravity of the salt.

Sesquibasic Chromate of Lead, $2(PbO, CrO_3) + PbO$ $= 439\cdot67$.—The mineral melanchroit is of the composition expressed by the above formula, and has a specific gravity of $5\cdot75$ according to Hermann; this gives the number $76\cdot5$ as the atomic volume of the compound.

Subnitrate of Copper, CuO, NO_5, $HO + 2CuO = 182\cdot17$.—
The fourth part of an equivalent (45·54 grains) caused an
increase of 16·5 in two experiments, and of 16·4 in a third.

					Sp. gr.
I.	CuO, NO_5, $HO + 2CuO$, vol. of salt			. . 66·0	2·760
II.	,,	,,	,,	. . 66·0	2·760
III.	,,	,,	,,	. . 65·6	2·777
			Mean . . .	65·87	2·765

A. *Subnitrate of Bismuth*, BiO, NO_5, $HO + 2BiO = 300\cdot4$.—
This salt was prepared in the same manner as the subnitrate
of copper, viz. by heating the nitrate to 400° or 500°. The
fourth part of an equivalent (75·1 grains), thrown into water,
caused, in various experiments, an increase of 16·5, which
gives 66·0 as the atomic volume, and 4·551 as the specific
gravity of the salt.

B. *Subnitrate of Bismuth*, BiO, $NO_5 + 2BiO = 291\cdot4$.—
This salt was prepared by adding nitrate of bismuth to a
large quantity of water; the white powder which falls by
this treatment is composed, according to Phillips, of three
equivalents of oxide of bismuth united to one of nitric acid.
It is therefore the same salt as the one last described,
deprived of its constitutional water. On immersing 72·85
grains of the salt in water, a rise in the stem of 13·9 was
effected, and 36·42 grains treated in the same way gave an
increase of 6·9.

			Sp. gr.
I.	3BiO, NO_5, vol. of salt	. . . 55·6	5·241
II.	,,	,, . . . 55·2	5·279
	Mean . . .	55·4	5·260

Subnitrate of Mercury, HgO, NO_5, $HO + 2HgO = 391\cdot49$.—
This salt was obtained in a yellow powder by adding the
crystallized subnitrate of mercury to water, and washing it,
according to the directions of Kane, with hot, but not boiling
water. The fourth part of an equivalent (97·87 grains),
thrown into turpentine, caused an increase of 16·4, which
gives 65·6 as the atomic volume of this salt, and 5·967 as the
specific gravity.

Ammoniacal Sulphate of Copper, CuO, SO_3, $HO + 2NH_3$ = 123·0.—This salt has already been described in a previous section; it had a volume of 54·0 or 9 × 6 in a state of solution, and of 68·6 or 9·8 × 7 in the solid state. The salt examined in that case was in fine large indigo-blue crystals, and was prepared by ourselves. Another portion, made by Mr. Morson in small crystals, we found to possess a volume of 68·0 and a specific gravity of 1·809. When this salt is heated, it loses one equivalent of water and one of ammonia, being converted into a green powder, the formula of which is CuO, $SO_3 + NH_3$; 24·27 grains of this, thrown into turpentine, caused an increase of 9·8, which gives 39·2 as the volume of the equivalent, and 2·476 as the specific gravity of the salt. The latter salt, on being moistened with water, absorbs three equivalents, and therefore assumes the atomic weight 124·07; the fourth part of which, 31·0 grains, thrown into turpentine, caused an increase of 15·9, making the atomic volume of CuO, $SO_3 + NH_3 + 3HO$, 63·6, and its specific gravity 1·950.

Ammonia-Sulphate of Zinc.—Kane describes several ammonia-sulphates of zinc, obtained by passing a stream of ammonia through a hot solution of sulphate of zinc, until the precipitate at first formed is redissolved. The solution thus obtained deposited transparent crystals in a few hours; but these effloresced so quickly after being dried, that we did not determine their specific gravity. The effloresced crystals have, according to Kane, the formula

$$ZnO, SO_3 + 2NH_3 + 2HO = 132·8.$$

We fear, however, that we have not been successful in procuring this salt in its proper state, as the determination of its volume varied between 57·5 and 64·0—results so discordant that it would not be safe to take their mean as a correct result. On heating this salt, it loses water and ammonia, being converted into ZnO, $SO_3 + NH_3$; 26·7 grains of which (the fourth of an equivalent), thrown into turpentine, caused an increase of 10·8, which gives 39·5 as the volume of the salt, and 2·479 for its specific gravity.

Ammonia-Sulphate of Mercury, HgO, $SO_3 + HgAd + 2HgO$

=486·0.—This salt, which Kane calls Ammonia Turpeth, was prepared by heating turpeth mineral with ammonia until it became changed to a heavy white powder. The eighth part of an equivalent (60·75 grains), immersed in water, caused an increase of 8·3 in two experiments; this makes the volume of the compound 66·4, and its specific weight 7·319.

Ammonia-Sulphate of Silver, $AgO, SO_3 + 2NH_3 = 190·86$.— This salt was obtained in the usual way, by dissolving sulphate of silver in ammonia and crystallizing. The first specimen tried was in small, indistinct crystals; in the second instance the crystals were large and well defined. 25·62 grains gave an increase of 8·6; and 37·7 grains of the better specimen of salt gave the increase 13·2.

					Sp. gr.
I. $AgO, SO_3 + 2NH_3$, vol. of salt	. . .	64·0	2·979
II. ,, ,, ,,	. . .	66·8	2·857
	Mean . . .	65·4	2·918

Ammonia-Chromate of Silver, $AgO, CrO_3 + 2NH_3 = 202·8$. —This salt was obtained in fine large crystals, in the same manner as the last salt. On immersing 25·35 grains in turpentine, the increase was 8·3; and on treating 50·7 grains in the same way, the increase was 16·5.

					Sp. gr.
I. $AgO, CrO_3 + 2NH_3$, vol. of salt	. . .	66·4	3·054
II. ,, ,, ,,	. . .	66·0	3·073
	Mean . . .	66·2	3·063

Ammonia-Nitrate of Copper, $CuO, NO_5 + 2NH_3 = 128·4$.— On dissolving 64·2 grains, half an equivalent, in 1000 grains of water, the increase was 32·0 in two experiments at a temperature of 60°; this makes the atomic volume in solution 64·0. On putting the same quantity into turpentine, there was a rise in the stem, in three experiments, of 34·0, 34·0, and 34·8.

					Sp. gr.
I. $CuO, NO_5 + 2NH_3$, vol. of salt	. . .	68·0	1·888
II. ,, ,, ,,	. . .	68·0	1·888
III. ,, ,, ,,	. . .	69·6	1·845
	Mean . . .	68·5	1·874

Ammonia-Subnitrate of Mercury, $HgO, NO_5 + 2HgO + NH_3$ $= 399\cdot7$.—This salt was prepared by adding a dilute solution of ammonia to nitrate of mercury, and was of a pure milk-white colour, as described by Kane. On throwing 40 grains of this compound into water, an increase of $6\cdot7$ was obtained; this gives a volume of $67\cdot0$ on the equivalent, and $5\cdot970$ as the specific gravity of the salt.

Chloride of Copper, $CuCl = 67\cdot18$.—The volume of hydrated chloride of copper was shown to be 33 or 3×11; but we have not yet examined the bulk occupied by the anhydrous chloride. The chloride was deprived of its water by a heat considerably below that of redness, in order to prevent the formation of any subchloride. On throwing $33\cdot59$ grains, or half an equivalent, into turpentine, the increase in two experiments was exactly $11\cdot0$, which gives $22\cdot0$ as the volume of the salt, and $3\cdot054$ as its specific gravity.

Ammonia-Chloride of Copper, $CuCl + 2NH_3 + HO = 110\cdot3$. —This salt was made by passing a stream of ammonia through a solution of chloride of copper until the precipitate formed had completely redissolved. The crystals, which deposited as the solution cooled, were dried in a receiver containing slaked lime, so as to prevent the carbonic acid of the atmosphere acting upon the ammonia; but in spite of this precaution, the crystals had slightly effloresced on the surface. The effloresced matter was removed, and the pure crystals employed. $27\cdot6$ grains of them, when thrown into turpentine, produced in two experiments an increase of $16\cdot5$, making the volume of the salt $66\cdot0$, and its specific gravity $1\cdot672$. On dissolving the same quantity of salt ($27\cdot6$ grains) in 1000 grains of water, the rise was $15\cdot9$ at $62°$, making the volume of the salt when in solution $63\cdot6$.

On exposing this salt to heat, water and ammonia are expelled, and a green powder remains, having the formula $CuCl + NH_3$. $21\cdot07$ grains of this salt thrown into turpentine produced an increase of $9\cdot6$, making the volume of the equivalent $38\cdot4$, and the specific gravity of the salt $2\cdot194$.

Subchloride of Copper, $Cu_2Cl = 98\cdot89$.—The subchloride used in the experiment was made by adding protochloride of

Table XIII.

Showing the Volumes occupied by certain Subsalts and Salts of Ammonia.

Name.	Designation. Formula.	Atomic weight.	Volume in Solution.			Volume of Salt.				
			Volume of salt in solution.	9 taken as unity.	Volume by theory.	Volume of salt by experiment.	11 taken as unity.	Volume by theory.	Specific gravity by theory.	Specific gravity by experiment.
Subsulphate of Copper	$CuO, SO_3, 3CuO+4HO$	234·9	...	...	...	76·2	7	77	3·051	3·082
Subsulphate of Zinc	$ZnO, SO_3 \; 3ZnO+4HO$	237·3	...	...	...	76·0	7	77	3·082	3·122
Protosulphate of Mercury	Hg_2O, SO_3	251·0	...	...	...	33·2	3	33	7·606	7·560
Persulphate of Mercury	HgO, SO_3	149·6	...	...	...	28·1	2	22	6·800	6·466
Subsulphate of Mercury	HgO, SO_3+2HgO	308·46	...	...	...	44·3	4	44	8·374	8·319
Chromate of Lead	PbO, CrO_3	163·97	...	...	...	29·0	:	:	...	5·653
Subchromate of Lead	PbO, CrO_3+PbO	275·7	...	...	...	44·0	4	44	6·266	6·266
Melanchroit	$2PbO, 2CrO_3+PbO$	439·67	...	...	...	76·5	7	77	5·710	5·750

Name	Formula									
Subnitrate of Copper	$CuO, NO_5, HO + 2CuO$	182·17	:	:		65·9	6	66	2·760	2·765
A. Subnitrate of Bismuth	$BiO, NO_5, HO + 2BiO$	300·4	:	:		66·0	6	66	4·551	4·551
B. Subnitrate of Bismuth	$BiO, NO_5 + 2BiO$	291·4	:	:		55·4	5	55	5·298	5·260
Subpernitrate of Mercury	$HgO, NO_5, HO + 2HgO$	391·49	:	:	54	65·6	6	66	5·932	5·967
A. Ammonia-sulphate of Copper	$CuO, SO_3, HO + 2NH_3$	123·0	54	6	54	68·7	:	:		1·790
B. Ammonia-sulphate of Copper	$CuO, SO_3 + NH_3$	97·08	:	:		39·2	:	:		2·467
Hydrate of ammonia-sulphate of Copper	$CuO, SO_3 + NH_3 + 3HO$	124·07	:	:		63·6	:	:		1·950
Ammonia-sulphate of Zinc	$ZnO, SO_3 + NH_3$	97·8	:	:		39·5	:	:		2·479
Ammonia-turpeth	$HgO, SO_3 + HgAd + 2HgO$	486·0	:	:		66·4	6	66	7·383	7·319
Ammonia-sulphate of Silver	$AgO, SO_3 + 2NH_3$	190·86	:	:		65·4	6	66	2·880	2·918
Ammonia-chromate of Silver	$AgO, CrO_3 + 2NH_3$	202·8	:	:	63	66·2	6	66	3·073	3·063
Ammonia-nitrate of Copper	$CuO, NO_5 + 2NH_3$	128·4	64	7	63	68·5	:	:		1·874
Ammonia-nitrate of Mercury	$HgO, NO_5 + 2HgO + NH_3$	399·7	:	:		67·0	:	:		5·970
Chloride of Copper	$CuCl$	67·18	:	:		22·0	2	22	3·054	3·054
A. Ammonia-chloride of Copper	$CuCl + 2NH_3 + HO$	110·3	63·6	7	63	66·0	6	66	1·671	1·671
B. Ammonia-chloride of Copper	$CuCl + NH_3$	84·11	:	:		38·4	:	:		2·194
Subchloride of Copper	Cu_2Cl	98·89	:	:		29·2	:	:		3·376
Subchloride of Mercury	Hg_2Cl	238·33	:	:	33	33·2	3	33	7·221	7·178
Subchloride and Amide of Mercury	$Hg_2Cl + Hg_2Ad$	458·1	:	:		66·8	6	66	6·941	6·858
Chloride and Amide of Mercury	$HgCl + HgAd$	254·5	:	:		44·6	4	44	5·784	5·700
Basic Chloride and Amide of Mercury	$HgCl + HgAd + 2HgO$	473·3	:	:		65·9	6	66	7·171	7·176

tin to a solution of chloride of copper. During the desicca-
tion of the salt it became slightly green, showing that a
little chloride had been formed by the absorption of oxygen;
but the change was so slight as probably not to interfere
materially with the result; 42·2 grains, thrown into turpentine, caused an increase of 12·5, which gives 29·2 as the
volume, and 3·376 as the specific gravity of the salt.

Subchloride of Mercury, $Hg_2Cl = 238·33$.—The fourth part
of an equivalent (59·58 grains), thrown into turpentine,
caused an increase of 8·3.

	Sp. gr.
Calomel, vol. of salt . . . 33·2	7·178

Hassenfratz states the specific gravity to be 7·176, a result
very near our own determination.

Subchloride and Amide of Mercury, $Hg_2Cl + Hg_2Ad$
$= 458·1$.—The eighth part of an equivalent (57·26 grains),
thrown into water, caused an increase of 8·3 and 8·4 in two
experiments.

		Sp. gr.
I. Black compound of Calomel . .	67·2	6·816
II. „ „ „ . . .	66·4	6·899
Mean . . .	66·8	6·858

The salt used in the experiments was prepared in the usual
way, by acting upon calomel with ammonia.

Chloride and Amide of Mercury, $HgCl + HgAd = 254·5$.—
The excellent researches of Kane, so often alluded to, have
shown that the above formula represents the composition of
white precipitate. It must be dried by a pretty strong heat,
to get rid of all its hygrometric water. On projecting 63·8
grains, the fourth of an equivalent, into water, an increase of
11·2 was obtained in two experiments; this gives 44·6 as
the specific volume of the compound, and 5·700 as its specific
gravity.

Basic Chloride and Amide of Mercury, $HgCl + HgAd$
$+ 2HgO = 473·3$.—This yellow compound was made in the
usual way, by boiling white precipitate with water. On
throwing 59·2 grains into water, the rise was 8·2 in one
experiment and 8·3 in another.

Sp. gr.

I. The above salt, volume . . . 65·5   7·220
II.   ,,   ,,   . . . 66·3   7·132

Mean . . . 65·9   7·176

The important researches of Graham have shown that water plays a most important part in the constitution of salts; and that salts with an excess of base may be viewed as hydrates in which oxide of hydrogen becomes replaced by a metallic oxide. The previous experiments will be found to give this view the fullest confirmation. Sulphate of zinc crystallizes with seven atoms of water and affects a volume of 74·6; and sulphate of copper assumes the same state of hydration, when crystallized with the latter salt, although *per se* it assumes only five atoms. Placing together the sub-sulphates and hydrated sulphates of these metals, we perceive not only a close similarity in their formulæ, but also in their volumes, as ascertained by experiment.

Differences.

$ZnO, SO_3, 3HO, 4HO,$ vol. 74·6   } 1·4
$ZnO, SO_3, 3ZnO, 4HO,$,, 76·0
$CuO, SO_3, 3HO, 4HO,$,, 74·6   } 1·6
$CuO, SO_3, 3CuO, 4HO,$,, 76·2

The difference between the two states of the sulphates is probably greater as stated than it actually is. We have already shown that the magnesian sulphates with seven atoms of water do not possess a volume of 77·0, because two of the atoms possess a volume of 9·8 instead of 11·0; and perhaps a similar circumstance tends to reduce the volume of the subsulphates. Similar instances of replacement of water by a metallic oxide are seen in other parts of the table (pp. 96, 97). We have already shown that nitrates of copper and bismuth possess a volume of 58·8 or 9·8 × 6. We have also seen instances in which 9·8, the volume of ice, in feeble compounds, became changed into the volume 11, when the salt entered into combination. In this point of view, the subsalts MO, $NO_5 + HO + 2MO$ become assimilated to the hydrated nitrates MO, $NO_5 + HO + 2HO$, the number of volumes in both cases

H 2

being the same, the only difference being, that in the former case the salts are multiples of 11, and in the latter of 9·8, or the volume of ice. The hydrated type affects six volumes, and so do the subnitrates, as will be seen by the following table :—

$$CuO, NO_5, HO + 2CuO, \quad \text{volume } 65·9 \text{ or } 11 \times 6.$$
$$BiO, NO_5, HO + 2BiO, \qquad ,, \qquad 66·0 \text{ or } 11 \times 6.$$
$$HgO, NO_5, HO + 2HgO, \qquad ,, \qquad 65·6 \text{ or } 11 \times 6.$$

We have further evidence of the equivalency of water to the metallic oxide in anhydrous nitrate of bismuth, which has a volume of 55·0 or 66−11; the formula for the salt being $BiO, NO_5 + 2BiO$. The conversion of the volume 9·8 into 11 is by no means uncommon, and is again seen in the subchromate of lead. Chromate of lead has a volume sensibly the multiple of 9·8.

	By experiment.	By calculation.
Chromate of Lead .	29·0 . . 5·653	$9·8 \times 3 = 29·4 . . 5·577.$

Boullay gives the specific gravity of oxide of lead as 9·5, which indicates the volume 11·7, a number not far from 11, which we must take as the unit volume. Subchromate of lead consists of one equivalent of the neutral chromate united to one of oxide of lead; but the three volumes of ice in the former have changed in the subsalt to 11×3, and the same is the case in the mineral melanchroit, which contains two equivalents of chromate of lead united to one of oxide of lead.

Subchromate of Lead . $PbO, CrO_3 + PbO \quad = 44 \quad$ or $11 \times 4.$
Melanchroit . . . . $2(PbO, CrO_3) + PbO = 76·5$ or $11 \times 7.$

In these salts we clearly see that oxide of lead takes up the volume and plays the part of an atom of water, although we are ignorant of their hydrated types. The same function of an oxide is seen in turpeth mineral, in which the $2HgO$, attached to HgO, SO_3, assumes the volume of two atoms of water, $22 + 22 = 44$. There can be little doubt, from the previous examples, of the equivalency of CuO, ZnO, BiO, HgO, and PbO, not only to each other, but also to water; and this will be still more strongly seen by placing the volumes of these and other anhydrous magnesian sulphates along with the volume of sulphate of water itself, as deduced from bi-sulphate of potash.

Sulphate of Water, vol. by experiment 22·0.
      „   Zinc,   „   21·8.
      „   Copper,   „   22·0.
      „   Iron,   „   24·0.
      „   Cobalt,   „   22·0.
      „   Mercury,   „   23·1.

The only cases in which there is an appreciable difference from sulphate of water are those of sulphates of iron and mercury, neither of which salts can be obtained without difficulty perfectly pure in an anhydrous state.

But if the equivalency of the magnesian metals to each other and to hydrogen be left in any doubt by the preceding table, this doubt would be entirely removed by a considera-tion of the magnesian chlorides. The strongest muriatic acid obtained has, according to Thomson, a specific gravity of 1·203, and contains 40·66 per cent. of dry muriatic acid, which is equal to 5·91, obviously six atoms of water to one of muriatic acid, as pointed out by Kane. The atomic weight of this compound divided by its specific gravity is $\frac{90·47}{1·203} = 75·2$, which is not far from 72·0 or 9×8, considering that the result remains uncorrected for expansion; this gives a volume of 18·0 or 9×2 for muriatic acid. The acid, which possesses a constant boiling-point and distils over unchanged, has a specific gravity of 1·094, and contains 19·19 per cent. of absolute acid, according to Davy, and 22·44 per cent. according to Thomson. The mean of their results indicates the acid to contain 16·4, or nearly 16 atoms of water. Now $\frac{180·47}{1·094} = 165$, which is not far from 162, the volume of 9×18, making for the volume of muriatic acid in strong solutions 18·0 or 9×2, a result the same as that obtained by the last calculation. These results, and that given in a previous section, along with the fact that hydrochloric-acid gas has twice the volume of steam, leave no doubt that muriatic acid affects two volumes; and converting the liquid into solid volume, we have a volume of 22·0 or 11 × 2 as the atomic volume of solid muriatic acid. By contrasting this volume with the experimental results on the magnesian chlorides, we find a very great similarity.

Chloride of Hydrogen, volume 22·0 or 11 × 2.

,,	Cobalt,	,,	22·2	,,
,,	Magnesium,	,,	22·1	,,
,,	Calcium,	,,	22·4	,,
,,	Copper,	,,	22·0	,,
,,	Mercury,	,,	22·0	,,

In dilute solutions muriatic acid affects only one volume; and this has been shown to be also the case with chlorides of copper and cobalt. Whether nitrate of water and nitrate of a magnesian oxide possess the same volume, it is difficult to decide. Nitrate of water in the acid of specific gravity 1·42 seems to affect four volumes; and this acid, $HO, NO_5 + 3HO$, is constituted on the same type as $CuO, NO_5 + 3HO$; yet $\frac{90·2}{1·42} = \frac{63}{9} = 7$, which gives four volumes for HO, NO_5, while nitrate of copper certainly does not possess more than three volumes. Nitrate of water calculated on weak acids has three volumes; but there being no good fixed point upon which to make the calculation, we must leave at present this point undetermined.

An important question now arises as to the truth of the supposition that two atoms of a magnesian metal are equal to one of the family of which potassium stands as the type. In calomel and chloride of ammonium we have a direct case in point, and the similarity of the volumes is very striking.

Difference.

Chloride of Ammonium, NH_4Cl. . 34·0 ⎫
Calomel, Hg_2Cl . . . . 33·2 ⎬ 0·8
⎭

In this case we have taken chloride of ammonium, because KCl assumes the volume of four atoms of ice.

Subchloride of copper, like NH_4Cl, possesses three volumes according to Karsten's experiments and our own, but these three volumes are multiples of 9·8, and not of 11·0.

By experiment.   By calculation.

Subchloride of Copper, vol. 29·2 . 3·376 9·8 × 3 = 29·4 . 3·363.

Another illustration is furnished in sulphate of protoxide of mercury and sulphate of potash.

Difference.

Protosulphate of Mercury, vol. 33·2 ⎱
Sulphate of Potash,   ,, 33·05 ⎰ 0·15

These are instances in which two atoms of a magnesian metal are at once shown to be equivalent to one of a metal of the potash family; but it does not thereby preclude the possibility of two atoms of a magnesian *oxide* being equivalent to one atom of potash. For example, a magnesian sulphate, MgO, SO_3, affects a volume 22, or 11 × 2; while the same salt united to an atom of constitutional water has the volume 33, or MgO, SO_3, HO becomes equal to KO, SO_3, which also possesses a volume of 33. The most striking case, however, is seen when crystallized subnitrate of lead is compared with nitrate of potash.

Nitrate of Potash, KO, NO_5 . . . vol. 49·0.
Subnitrate of Lead, $PbO, NO_5 + PbO$, vol. 49·0.

The fact that two atoms of a magnesian oxide are equivalent to one of potash appears to find its explanation in the circumstance that we uniformly find the salt of potash assuming one volume greater than the corresponding salt of magnesia. Hence, as the volume of the oxides corresponding to the latter body is equal to unity, the equivalency of two of their atoms to one of potash becomes a matter of necessity.

To sum up these remarks, we conceive (1) that Graham has taken the correct view in supposing subsalts to represent hydrated salts, in which water has been replaced by a metallic oxide; and (2) that the volume of two atoms of a metal of the magnesian family, in which we include hydrogen, is equal in volume to one of the potassium group, or two atoms of the former oxide, when combined, to one of the latter. We are now in a condition to consider the salts of ammonia.

It is quite unnecessary to remind chemists that there are two rival theories regarding the constitution of ammoniacal salts. One of them, proposed by the profound Berzelius, is that the salts of ammonia contain a hypothetical radical termed ammonium, consisting of one equivalent of nitrogen and four equivalents of hydrogen. Sulphate of ammonia is to be viewed as sulphate of oxide of ammonium, the latter hypothetical body being equivalent to potash; and hence the isomorphism between the salts of potash and ammonia. The other view of the constitution of ammonia is that proposed by Kane, and so elaborately supported by him in his paper on subsalts and ammoniacal compounds *. Dr. Kane supposes that an ammoniacal salt is formed on the type of a magnesian salt carrying along with it constitutional water.

$$\text{Sulphate of Copper . . . } CuO, HO, SO_3.$$
$$\text{Sulphate of Ammonia. . } HO, NH_2H, SO_3.$$

On this view, amide of hydrogen is equivalent to, and plays the part of, an atom of water. If this be the case, amidogene must be analogous to oxygen, and ammonia and a magnesian oxide must possess the same atomic volume. At present all this is purely hypothetical, and must be subjected to the test of experiment before we can admit it as a safe foundation on which to rear a theory. The means of deciding this question seemed to present itself in an examination of the amides of mercury, and of the crystallized salts of copper and zinc, in which the ammonia is present *quasi* ammonia; and such compounds have been described in the beautiful researches of Kane on this subject. Wöhler's white precipitate, $HgCl + NH_3$, seems to be constituted in the most simple manner, and possesses a volume of 33·0, which, deducting the volume 22·0 for HgCl, leaves 11·0, or unity, as the volume of NH_3. But again, white precipitate, HgCl

* Transactions of the Royal Irish Academy, vol. xix. part 1.

$+ \text{Hg NH}_2$, affects a volume of 44·6 by experiment, which, deducting 22·0 for HgCl, leaves HgNH_2 also equal to 22·0, and yet the latter compound should correspond in volume to NH_2H. The heavy yellow powder obtained by boiling white precipitate with water has a volume of 66·0, and is constituted according to the formula $(\text{HgCl} + \text{HgAd}) + 2\,\text{HgO}$; so that deducting 44·0, the ascertained volume of the double amide and chloride, 22 or 11×2 remains for *two* atoms of HgO, giving the same result as in the former subsalts, viz. the equivalency of HgO to HO, but not to HgNH_2; and another proof of this is afforded in the reduction of the volume of ammonia turpeth. From this circumstance, the view would appear probable that amide and chloride of mercury are equivalent, and hence would follow the equivalency of chlorine to amidogene. This receives further support from the volume of the double subamide and subchloride of mercury, $\text{Hg}_2\text{Cl} + \text{Hg}_2\text{Ad}$, which has a volume of 66·8, according to experiment. Calomel itself possesses the volume 33·2, which, deducted from that of the salt just described, gives 33·6 as the volume of Hg_2Ad, showing the complete equivalency of the latter to the subchloride.

It has been shown that chloride of mercury and chloride of hydrogen are equivalent, and it now remains to be shown by direct proof that amide of hydrogen (ammonia) is embraced in the same category. Ammonia-chloride of copper, $\text{CuCl} + \text{NH}_3$, was found with a volume of 39·2 or $9·8 \times 4$; chloride of copper itself affects two volumes, which leaves for AdH, as deduced from this salt, also two volumes. But the ammonia in $\text{CuCl} + 2\,\text{NH}_3 + \text{HO}$, if we were to suppose the salts constituted in a manner so simple as expressed by their empirical formula, would only have a volume of 33·0 for two atoms, or $1\frac{1}{2}$ volume for each.

In proceeding further it will be seen that we involve ourselves in inextricable difficulties, if we insist upon the equivalency of NH_2H to HO; or suppose the ammoniacal salts, such as those described, to be constituted on the type of the hydrated salts. Thus, ammonia-sulphate of copper, $\text{CuO, SO}_3 + 2\,\text{NH}_3 + \text{HO}$, has a volume of 68·6 or $9·8 \times 7$ in its solid state, and of 54 or 9 × 6 when in solution. Deducting 19·6 for CuO, SO₃, and 9·8 for HO, there is again left 39·2, or 9·8 × 4 for *two* atoms of ammonia. The simple salt, CuO, SO₃ + NH₃, has a volume of 39·2, which leaves 19·6, or 9·8 × 2, for one atom of ammonia ; but the same salt, when combined with three atoms of water, yields the volume 63·6, which would lead us to suppose that one atom of water is equal to one atom of ammonia. We also find ammonia with the volume 11, or unity, when calculated from the observed volume of hydrated sulphate of ammonia. But in the ammonia-chromate of silver, AgO, CrO₃ + 2 NH₃, and in its corresponding sulphate, we find, on deducting 33·0, or 11 × 3, for the salts themselves, the residual 33·0 for *two* atoms of ammonia. Again, however, we become perplexed by finding that the ammonia in ammonia-pernitrate of mercury possesses the volume of an atom of water. Thus, then, by considering the volumes of the ammoniacal salts as containing their ammonia *quasi* ammonia, and as constituted on the type of the hydrated salts, we obtain the contradictory and absurd result that ammonia, though often taking a volume equal to unity, sometimes possesses a volume of 1½, and occasionally two volumes. It is pretty certain, from these contradictory results, that the salts are not constituted on the hydrated type.

Graham has thrown out the ingenious idea *, that the salts now referred to may actually contain an ammonium in which the fourth equivalent of hydrogen is replaced by an equivalent of a magnesian metal. Thus CuO, SO₃ + NH₃ is constituted, according to Graham, NH₃Cu, O,SO₃, on the type of sulphate of ammonia, NH₃H, O, SO₃. There is nothing whatever opposed to this view in Kane's researches, as he himself admits, the only difference being that he considers the said salts to contain oxide of copper and water united to amide of hydrogen, instead of to *cuprammonium* and oxide of ammonium, according to the views of Berzelius and Graham. While, therefore, Kane admits that amide of hydrogen is very

* Graham's ' Elements of Chemistry,' p. 416.

closely allied to chloride of hydrogen, he claims for the former
body an equally close alliance to water, by asserting that it
is equivalent to a magnesian oxide, although it is difficult
to conceive why chloride of hydrogen has not a right to a
similar claim. Amide and chloride of mercury have undoubtedly the same volume, viz. 22·0, and chloride of hydrogen also enjoys the same number; but water does not in any
case do so. On this point alone, then, are we at issue with
Kane, for there are many proofs that there is extreme probability in the view propounded by him of the presence
of NH_2H and HO in ammoniacal salts. On the former
view alone do we contest the accuracy of the opinion, leaving
for future consideration and research, to which we are now
devoting ourselves, a more defined notion of the reason why
NH_2H and HO are equivalent in many instances, *not* in *all*,
to potash. We have already stated the incongruous results
which would flow from the conception that ammonia was
simply attached to the salts examined. It is true that Kane
gives to some of them a constitution more intimate; and
when he does so his theory accords with our results. But his
conception of the equivalency of NH_2H to HO has led him
in other instances to attach the ammonia to the salt in place
of water; and it is from these cases that we dissent. If he
merely means that NH_2H can replace HO in a compound,
as KO,SO_3 does in a magnesian sulphate, then we cease to
differ, because the resulting compounds do not remain in
strict parallelism; the only point we argue against being
that HO and NH_2H are equivalent. Thus we have, supposing
all of them to affect the primitive volume 9·8 :—

$$HgO, NO_5 + HgO + 2HO = 6 \text{ vol.}$$
$$HgO, NO_5 + HO + 2HgO = 6 \text{ vol.}$$
$$HgO, NO_5 + NH_3 + 2HgO = 7 \text{ vol.}$$

The first two members of the above series have the same
number of volumes, because HgO and HO are equivalent,
and the last salt should affect the same, if $NH_2H = HO$. But
if we consider the last salt as equal to nitrate of ammonia
in which HgO replaces HO, then it becomes intelligible.

NH_2H, HO, NO_5 affects $5+2$ of $HO=7$.

NH_2H, HgO, NO_5 affects $5+2$ of $HgO=7$.

On the same principle we would arrange the other ammoniacal compounds. Thus CuO, SO_3+NH_3 obviously ought to be arranged NH_2H, CuO, SO_3 corresponding to NH_2H, HO, SO_3, anhydrous sulphate of ammonia, and both affect, as they should do on this formula, four volumes. We observed a very decided peculiarity in sulphate of ammonia; for while in its hydrated condition the NH_4O, SO_3 could only be equal to three volumes, in its anhydrous state, or when in combination with salts, it assumed four volumes. The latter peculiarity attends the *alpha* ammonia-sulphate of copper, and is shared also by ammonia-sulphate of zinc, while the hydrate assimilates itself to NH_4O, SO_3+HO.

$NH_2H, HO, SO_3 \quad =39\cdot2$.

$NH_2H, CuO, SO_3 \quad =39\cdot2$.

$NH_2H, ZnO, SO_3 \quad =39\cdot2$

$(NH_2H,CuO,SO_3=39\cdot2)+(HO=9\cdot8)+(NH_3=19\cdot6)=68\cdot6$.

$NH_2H, HO, SO_3 \quad +HO=44$.

$(NH_2H, CuO, SO_3+HO=44)+(2HO=19\cdot6)=63\cdot6$.

In ammonia-nitrate of copper we have an instance in which the ammonia may be present either as nitrate of ammonia or as ammonia; for if we suppose the volume $68\cdot5$, which obviously indicates $9\cdot8\times7=68\cdot6$, to be made up of CuO, NO_5+2NH_3, we must assume that two atoms of ammonia are equal to four atoms of ice, for we already have seen that CuO,NO_5 affects three volumes. On the supposition that the compound contains a substance equivalent to nitrate of ammonia the volumes are equally intelligible.

$NH_2H, HO, NO_5 \quad =49\cdot0$.

$NH_2H, CuO, NO_5=49\cdot0+NH_2H=19\cdot6=68\cdot6$.

Perhaps, however, the clearest instances are seen in the ammoniacal chromate and sulphate of silver. AgO, CrO_3 and AgO, SO_3 affect a volume of $9\cdot8\times3$; and supposing a transformation into multiples of 11, of which we have seen frequent instances, $2NH_3=33\cdot0$, or $NH_3=16\cdot5$, or $1\frac{1}{2}$ times the number which we assume as unity. But on the supposition that AgO takes the place of HO, the difficulty ceases.

$$NH_2H, HO, SO_3 = 44.$$
$$NH_2H, AgO, SO_3 = 44 + NH_2H = 66.$$
$$NH_2H, AgO, CrO_3 = 44 + NH_2H = 66.$$

Perhaps the most anomalous salt in the whole series examined is the chloride of ammonium, which actually decreases one volume in becoming solid, 9×4 in solution being 11×3 in the state of salt. Chloride of potassium refuses to share this anomaly, and we accordingly find it $9 \cdot 8 \times 4$, and NH_4Cl associates itself to KCl in the double salts. Four volumes for NH_4Cl is undoubtedly what we should expect from its composition, and from that number being affected in solution and in its double salts. We also see the three volumes entering into *alpha* ammonia-chloride of copper, although the *beta*, according to our results, seems, singularly enough, to affect the proper four volumes.

$$NH_2H, HCl = 33.$$
$$NH_2Cu, HCl = 33 + NH_2H = 22 + HO = 11 = 66.$$
$$NH_2Cu, HCl = 38 \cdot 4 \text{ or } 9 \cdot 8 \times 4 = 39 \cdot 2.$$

The double amides and chlorides, as we have already shown, affect the same number of volumes as NH_4Cl when in solution, and might be placed on the same type as NH_2Hg, HCl. Without denying that NH_2H and HO may be so intimately associated in the ammoniacal salts as to form the hypothetical body oxide of ammonium, we would call attention to the facts, which show that the resulting volumes of the ammoniacal salts are made up of the volumes of the hydrated acid and amide of hydrogen. It by no means militates against that view, that in hydrated sulphate of ammonia we have one volume in solution less, and also in the state of a solid, than should result from the combination of these two. CuCl has undoubtedly *per se* two volumes, just as HCl has in a concentrated state, or as NH_2H has in combination. But the CuCl in $CuCl + 2HO$ possesses only one volume, the other having disappeared in the water; and HCl itself has only one volume in dilute solutions. The disappearance of one volume in combination with water is by no means so surprising as the disappearance of the volumes of 23 atoms of the constituents of alum in the water in which it is dissolved, especially when we find the salt under consideration, sulphate of ammonia, vindicating its proper volume when in combination. The oxalate of ammonia has its *proper* volume, just as has anhydrous sulphate of ammonia; the only exception is the decidedly anomalous salt chloride of ammonium, although this also ceases to be anomalous in the double chlorides. By placing together the volumes of the hydrated acids and those of the ammoniacal salts, it will be seen that the latter are made up of the volumes of the hydrated acid united to amide of hydrogen affecting two volumes, like HCl :—

Sulphate of Ammonia, $HO, SO_3 = 2 + NH_2H = 2 = 4$.
Nitrate of Ammonia, $HO, NO_5 = 3 + NH_2H = 2 = 5$.
Oxalate of Ammonia, $HO, C_2O_3 = 2 + NH_2H = 2 = 4$.

All the ammoniacal salts which we have described in this section may be arranged in a similar way with a like result.

We do not profess to have resolved the cause of the equivalency $HO + NH_3$ to KO; nor do we insist that they do not enter into more intimate union to form NH_4O. It must not be left out of consideration, however, that in almost every instance the ammoniacal salt affects one volume in solution more than the corresponding salt of potash, and that the number of volumes of the latter becomes augmented by one in passing from the liquid to the solid state, while the number of volumes of the ammoniacal salt remains unchanged. It requires a more minute knowledge of the constitution of salts than we now possess to decide the question at issue.

Conclusion.

Although we have examined many other salts than those described in the previous pages, with results quite confirmatory of our views, we do not feel warranted in extending our memoir, already much too long. We therefore conclude by summing up, in the form of propositions, the laws which we consider regulate the volumes of salts. At the same time we do so with strict reference to the salts which we have described, deprecating their hasty generalization, being ourselves quite satisfied that there are peculiarities in other cases, which must be subjected to close examination. This being only the first of several memoirs on the same subject which we intend to lay before the Society, we do not present this investigation as being in itself complete.

Prop. 1.—*Compounds dissolved in water increase its volume for every equivalent either by 9 or by multiples of 9.*

This, in other words, signifies that the volumes of salts in solution are either equal to each other or are multiples of each other; for 9, being the volume of nine grains, or an equivalent of water, is merely assumed as the standard of comparison.

a. Certain salts, such as the magnesian sulphates, the alums, &c., dissolve in water without increasing its bulk more than is due to the liquefaction of the water which they themselves contain; the anhydrous salt taking up no space in solution.

b. Anhydrous salts, or salts containing a small proportion of water, affect a certain number of volumes in solution, which pass along with them unchanged into their union with other salts.

c. The volume occupied by double salts when dissolved is the sum of volumes occupied by their constituents when separate, with the exception of certain cases described in the previous sections.

Prop. II.—*The volume occupied by a salt in the solid state has a certain relation to the volume of the same salt when in solution; and has also a fixed relation to the volume occupied by any other salt.*

a. The volume of an equivalent of any salt is either 11, or a multiple of 11, or of a number very nearly approaching the number 11.

b. Or the volume of a salt is 9·8, or a multiple of 9·8, or, in other words, of the volume occupied by an equivalent of solid water (ice).

c. Or the volume of a salt is made up of a certain multiple of the number 11, added to a certain multiple of the number 9·8.

On each of these heads we would offer a few remarks.

With two assumptions we have been enabled to connect with each other the volumes occupied by all the salts examined by us in the previous sections. These assumptions are, that the divisor of the volumes of the salts is either 11 or a number very nearly approaching to it, or that the divisor is 9·8, the volume of ice itself.

We have been guarded in stating positively that the first divisor is absolutely 11, because we do not in the present memoir enter into the connection between this number and the volume of ice, 9·8. To show, however, that our experiments agree with those of recent accurate experimenters, and that the number 11, which we have at present to announce empirically, cannot be wide from the truth, we append the theoretical and experimental results upon the alums, which we stated to possess twenty-five volumes, in which therefore any considerable error in the number 11 would be multiplied by 25, and plainly show itself in the results. Notwithstanding this severe test, it will be seen that the theoretical and experimental numbers are actually within the errors of the balance.

	Theoretical sp. gr.	By Kopp's* experiments.	By our experiments.	Mean of experiments.
Potash-alum . .	1·727	1·724	1·726	1·725
Chrome-alum . .	1·833	1·848	1·826	1·837

The number 11 must then be very near the truth, if it be not absolutely the truth. We now append an equally severe test for our view that the volumes of many salts are multiples of 9·8, the number representing the volume of ice. If there be an error in this number, it must become very notable in the phosphates and arseniates, when multiplied by 24, or in carbonate of soda when multiplied by 10. Perhaps sugar itself will form as severe a test as could be desired, for we proceed on the extraordinary fact that the 12 atoms of carbon in sugar have ceased to occupy space, and that the bulk of

* Ann. der Pharm. Bd. xxxvi. S. 10.

an atom of sugar is just the bulk of $H_{11} O_{11}$, or its 11 atoms of hydrogen and oxygen, *quasi* water frozen, into ice.

	Theoretical sp. gr.	Sp. gr. according to our experiments.	Other authorities.
Carbonate of Soda ...	1·463	1·454	1·423, Haidinger.
Phosphate of Soda ...	1·527	1·525	1·514, Tünnerman.
Subphosphate of Soda	1·622	1·622	None.
Arseniate of Soda ...	1·713	1·736	1·759, Thomson.
Subarseniate of Soda	1·808	1·804	None.
Cane-sugar	1·591	1·596	1·600, Schübler & Renz.

Thus even in salts so difficult to obtain in a proper degree of hydration, free from mechanical water, as those given in the above table, the difference between the theoretical and experimental numbers is not greater than might have been expected.

We give one other class of salts to illustrate position *c* in Prop. II., there being in these salts a certain number of volumes represented by 11, and a certain number by 9·8, CnO, SO_3 representing the number of volumes with the divisor 11.

	Theoretical sp. gr.	Sp. gr. by our experiments.	Sp. gr. by other authorities.
Sulphate of Copper	2·270	2·254	2·274, Kopp.
„ Zinc ...	1·926	1·931	1·912, Hassenfratz.
„ Iron ...	1·854	1·857	1·840, idem.
„ Magnesia	1·660	1·660	1·660, idem.
„ Nickel ...	2·033	...	2·037, Kopp.

We have selected these three classes of salts as being the most severe tests which we could apply to our theory, and any chemist who has had experience in this subject will at once admit that the theoretical and experimental numbers are as near each other as the estimation of the specific gravities by any two different experimenters. We do not rest the claims of our theories on our own experiments, but are willing to admit the accuracy of other experimenters, especially of Karsten, Hassenfratz, Kopp, and others, who have preceded us on this subject *; while at the same time we believe that our methods of taking specific gravities have enabled us to introduce more uniformity into the results. The simplicity of the methods themselves is due to Bishop Watson, who was the first to take specific gravities by the increase in the stem of an instrument; and to Holker the suggestion is due of using a saturated solution instead of water employed by Watson.

We conceive that the primitive volume 9·8 is transformable into the primitive volume 11, and *vice versâ*; and for this reason we sometimes see sulphate of ammonia 9·8 × 4; at other times, in combination as in bisulphate of ammonia or the anhydrous double sulphates, it is 11 × 4; and numerous other instances of transformation are presented in the previous sections.

The liquid volume being to the solid volume either as 9 : 11 or as 9 : 9·8, these numbers, used as the divisor for the liquid and solid volumes respectively, usually yield the same quotient. Thus the liquid volume of sulphate of copper is 45, its solid volume is 55. $\frac{45}{9}=5$, and $\frac{55}{11}=5$; so that we may say the salt affects the same number of volumes in the liquid and in the solid state. In the same manner subphosphate of soda has a volume of 216 in solution and of 235 in the state of a salt. Now $\frac{216}{9}=24$ and $\frac{235}{9\cdot8}=24$, so that the number of volumes affected in solution and in the solid state are exactly the same. This is a general rule, and a powerful argument

* The only decided difference which we found from other experimenters is in the case of the hydrated salts. Thus our determination of the volumes of the double magnesian sulphates and sulphate of potash (Table VI.) differs from Kopp's experiments as 99 : 103. These salts contain from 3 to 4 per cent. of mechanical water, as Graham long ago pointed out (Trans. R. S. E. vol. xiii. p. 12), and the neglect of this in Kopp's experiments has probably caused the difference. We take this opportunity of stating that when more than one specific gravity is given by us, the salts have been prepared at different times; in many instances this is not the case, but in much the largest proportion it is so.

for the accuracy of our position. The rule has exceptions in salts of potash, in which the volumes are increased by one volume on becoming solid; thus KO, SO_3 equals $\frac{18}{9}=2$ in solution, and $\frac{33}{11}=3$ in the solid state. This is not an accidental variation, but an actual augmentation of one volume, as is proved by the potash-alums, in which KO, SO_3 has ceased to occupy space in solution, but on the crystallization of the alum the volume becomes increased by one, obviously owing to this peculiarity of KO, SO_3; thus alum in solution $\frac{216}{9}=24$, becomes $\frac{275}{11}=25$ in the state of a salt.

This peculiarity is very striking, especially in the case KO, CO_2, which, with a volume of $\frac{33}{11}=3$ as a solid, becomes $\frac{9}{9}=1$ as a liquid. Let us endeavour to conceive the extraordinary amount of power exerted in this case; the water of the volumenometer, on dissolving an equivalent of KO, CO_2, descends from 33 to 9, so that a bulk of solid matter $=24$ grains of water disappears within it. If we would compare the force to that which would be required to compress the water into this diminished bulk, we must deal in numbers of a magnitude truly immense. We have always been accustomed to view as an exception the expansion of water on becoming solid, but now we see, with Longchamp, that the rule is universal; the salt (muriate of ammonia excepted?) takes up more space as a solid than it does in its liquid state in solution.

We have stated that we desire not to be held responsible for any rash generalization of these laws, which we do not extend at present beyond the salts examined by us. Let us consider the volumes of ammonia-alums, as an example of the danger of applying either of the laws without a proper comprehension of them. These volumes are certainly above 275, the volumes of the potash-alums, and are between 279 and 280, according to our experiments and those of Kopp. Now let us suppose that the four volumes of NH_4O, SO_3 are represented in the alums, and that only $Al_2 O_3$, $3SO_3$ has ceased to occupy space, as it in fact does when hydrated, then an ammonia-alum $Al_2O_3,3SO_3 + NH_4O, SO_3 + 24HO$ may be viewed as $9\cdot8 \times (24 + 4) = 279\cdot4$, and the specific gravities would countenance this idea.

	Sp. gr. by theory.	Sp. gr. by our experiments.	Sp. gr. by Kopp's experiments.
Ammonia-alum	1·626	1·625	1.626
Ammonia-iron-alum...	1·721	1·718	1·712

These results certainly approach the theoretical number very closely; and the theory may represent the truth. But at the same time it is difficult to believe that the ammonia-alum is constituted on a different type from the potash-alum. We might suppose that the only variation between them is the difference between the volumes of KO, SO_3 and NH_4O, SO_3, or the difference between 11×3 and $9\cdot8 \times 4$. This difference, 6·2, added to the volume of potash-alum 275 = 281·2, which is not very wide from the experimental results, and would give the specific gravity by theory for ammonia-alum 1·616, and for ammonia-iron-alum 1·711. These are points which require further inquiry.

We do not refer here to the minor views embraced in the preceding investigation, being anxious principally for inquiry and confirmation into the three main theories propounded. With one assumption for the volume in solution, and with two assumptions for the volumes of solids, we have been enabled to explain, as we trust, the specific gravities detailed in the previous sections. We might perhaps with propriety indulge in speculation, and apply these laws in explanation of isomorphism and dimorphism, but we prefer the safer course of trusting to experimental investigation, part of which we shall in a short time lay before the Society in an inquiry upon the expansion of solutions, and on some other points connected with this important subject.

Researches on Atomic Volume and Specific Gravity.
By LYON PLAYFAIR, *Esq., Ph.D., and* J. P. JOULE, *Esq.*

[' Memoirs of the Chemical Society,' vol. iii. p. 57.]

SERIES II.—*On the Relation in Volumes between Simple Bodies, their Oxides and Sulphurets, and on the Differences exhibited by Polymorphous and Allotropic Substances.*

IN our former memoir we gave a sketch of the progress of discovery in this branch of scientific research, but we unintentionally omitted the name of Dumas. Pierre's * interesting memoir has clearly shown that, twenty years since, the French chemist pointed out the fact that isomorphous groups possess the same atomic volume, and various chemists, amongst whom we have already cited Thomson, Kopp, Schröder, and Persoz, afterwards cultivated the same field. The purpose of our former memoir was to point out that the volumes of salts are related to each other by simple laws; and in the communication which we have now the honour to present to the Society, it is our object to extend, confirm, and simplify our views as to other solid substances. The correct expression of a law is not generally attained at its first discovery, although the truth of the law may be sufficiently indicated. In our present memoir we do not treat of substances in solution, our object being to confirm the multiple relation which we have already pointed out in the case of solid salts, and to exhibit the connexion between the two units 9·8 and 11·0, which we then assumed as the submultiples.

We commence with the volumes of the metallic elements, which are to a certain extent well suited for a correct estimation of specific gravity. But at the same time it must be borne in mind, that the force of cohesion exercises upon them an influence so strong, as to make their exact density depend on the circumstances under which they are examined. Thus iridium, possessing, after fusion, the sp. gr. 20·0, affects

* Millon and Reiset, ' Annuaire de Chimie' for 1846.

only 16·0 when obtained by the reduction of its oxide ; and osmium under similar circumstances possesses respectively the specific gravities 19·5 and 10·0.   The effect of hammering certain metals is also well known to be a means of increasing their density.   These points must be kept in view in considering the results obtained by experiment ; for they obviously indicate that the force of cohesion is able to diminish the natural volumes which the metals would otherwise enjoy.

In most cases it was unnecessary for us again to determine the density of the metals, as that has frequently formed a subject of special examination.   But since the more recent discoveries in electricity furnish the means of procuring metals in a state of greater purity than formerly, we have occasionally availed ourselves of this power, in order to obtain results of the most unexceptionable character.   In such cases the metal was precipitated upon a platinum wire of known weight, and after removing it from the liquid in which precipitation was effected, and washing it with distilled water, the wire with the adhering metal was plunged into strong alcohol, then wrapped in bibulous paper and allowed to remain in a warm place until the smell of alcohol had disappeared.   By this means the surface was preserved perfectly clear and untarnished, which is not the case in the ordinary modes of drying.   The specific gravity was then determined by the usual hydrostatic method, care being taken to use recently boiled water for the purpose of removing the air adhering to the metal.   The temperature of observation is always understood to be 40°.

The specific gravity of the oxides and sulphurets was taken in an instrument similar to that described in our first memoir (p. 405), but susceptible of much greater accuracy.   In our former instruments, as we stated, we only measured to the tenths of a grain ; in those used in our present researches we could read off with the greatest facility the hundredth part of a grain.   This improvement was effected by using tubes of a smaller bore, and graduating them with the instrument invented by Professor Bunsen.   The capacity of the instrument was determined at 40°, the temperature of the maximum density of water, and all our estimations in the present memoir refer to this temperature. Turpentine was used as the liquid in the volumenometer, and it was restored to precisely the same temperature as at the commencement of the experiment by immersing the instrument in water.

Section I

Specific Gravities and Volumes of Metallic Elements.

Manganese, Mn = 27·7.—Bergmann states the sp. gr. of this metal as 6·861 and 7.1. But the later estimations of John and Bachmann, being uniform, deserve a preference : the former gives its sp. gr. as 8·013, the latter as 8·030. The volume of the equivalent is therefore as follows :—

$$
\left.\begin{array}{l}
\text{I. Mn, volume} = 3\text{·}46 \\
\text{II. } \quad \text{,,} \quad \text{,,} \quad = 3\text{·}45
\end{array}\right\} \overset{\text{Mean.}}{3\text{·}455}
$$

Iron, Fe = 28·0.—According to Brisson the specific gravity of this metal is 7·788, according to Karsten 7·790. The volume, according to both results, is 3·59.

Cobalt, Co = 29·5.—The specific gravity of this metal has engaged the attention of various chemists, and the following results have been obtained :—

8·513 Berzelius.
8·558 T. H. Henry*.
8·485 Brunner.
8·500 Mitscherlich.
8·538 Haüy.

The mean of these results, 8·519, gives 3·46 as the atomic volume.

Nickel, Ni = 29·5.—There are various determinations of the specific gravity of this metal in a fused state.

7·807 Brisson
8·279 Richter.
8·402 Tourte

---

* Upon a well-fused button weighing 171 grains and determined at our request by Mr. Henry.

8·380   Tupputi.

8·637   Brunner.

8·477   Baumgartner.

The mean result, 8·33, gives the volume of the equivalent 3·54.

Zinc, Zn = 32·3.—The following specific gravities are recorded by Berzelius and Brisson, to which we add the result of one of our own experiments on an electrotype specimen, of which a detailed account will be found as we proceed.

6·862   Berzelius.

6·861   Brisson.

6·869   P. and J.

The mean result, 6·864, gives 4·71 as the volume of zinc.

Cadmium, Cd = 55·8.—The following specific gravities are the most uniform of those recorded with regard to this metal:—

8·659   Herapath.

8·635   Karsten.

8·604   Stromeyer.

8·670   Children.

The mean result, 8·642, gives the volume 6·46.

Copper, Cu = 31·6.—Our estimations of the specific gravity of this metal agree closely with those of other experimenters.

8·830   Berzelius.

8·900   Herapath.

8·721   Karsten.

8·895   Hatchett.

8·884   P. & J.

8·941   P. & J.

The mean of these results, 8·862, gives a volume for the equivalent of 3·56.

Chromium, Cr = 28·0.—Two estimations exist of the specific gravity of this metal. Richter states it to be 5·90; Thomson makes it 5·1. The mean, 5·5, gives as the volume 5·09.

Aluminium, Al = 13·72.—Wöhler has lately given the

specific gravity of this metal, determined on small quantities, as 2·5 and 2·67. Adopting the latter number as most likely to be nearest the truth, we have the atomic volume 5·13.

Bismuth, Bi=71·07.—The following estimations are selected as being the most uniform:—

9·670   Musschenbroek.
9·654   Karsten.
9·822   Brisson.
9·831   Herapath.
9·882   Thenard.
9·800   Leonhard.

The mean result, 9·776, yields the volume 7·27.

Tin, Sn=58·90.—The specific gravities recorded by various observers are uniform and agree with the density of a beautiful specimen which we prepared by precipitation.

7·291   Kupffer.
7·290   Karsten.
7·291   Brisson.
7·295   Musschenbroek.
7·285   Herapath.
7·248   P. & J.

The mean of these results, 7·283, indicates the volume 8·09 for the equivalent of tin.

Arsenic, As=75·4.—The specific gravity of this metal seems to vary under certain circumstances, as the results obtained by different observers deviate considerably from each other.

5·763   Brisson. Stromeyer.
5·884   Turner.
5·959   Guibourt.
5·700   Guibourt.
5·672   Herapath.
5·628   Karsten.
5·766   Mohs.

The mean, 5·767, yields the volume 13·07 ; but, if we select the two estimations by Karsten and Herapath, the mean specific gravity will be 5·65 and atomic volume 13·35.

Antimony, Sb = 129·2.—The following observers have determined the specific gravity of this metal :—

6·712	Hatchett.
6·733	Boeckmann.
6·702	Brisson.
6·852	Musschenbroek.
6·860	Bergmann.
6·610	Breithaupt.
6·700	Karsten.
6·646	Mohs.

The mean result, 6·727, gives the volume of the equivalent 19·21.

Molybdenum, Mo = 47·96.—According to Bucholz, the specific gravity of this metal is 8·615 and 8·636. Hielm states it at 7·500, a number obviously too low. The mean result of the former chemist gives 8·625 as the specific gravity and 5·56 as the volume of the equivalent.

Tungsten, W = 94·8.—D'Elhuyart describes the specific gravity as 17·60; Bucholz as 17·40; Allan and Aiken as 17·22. The mean of these results gives 17·40 as the specific gravity of the metal, and 5·45 as the volume of the equivalent.

Titanium, Ti = 24·3.—Wollaston describes the specific gravity as 5·3, Karsten as 5·28. The mean result, 5·29, gives the atomic volume as 4·59.

Tellurium, Te = 64·2.—The following are the recorded specific gravities of this metal :—

6·115	Klaproth.
6·138	Magnus.
6·343	Reichenstein.
6·258	Berzelius.
6·130	Berzelius.

The mean result, 6·196, gives 10·36 as the atomic volume.

Lead, Pb = 103·6.—The specific gravities recorded are as follows; we also add two estimations by ourselves :—

11·207	Boeckmann.
11·352	Herapath.
11·352	Brisson.
11·330	Kupffer.
11·388	Morveau.
11·388	Karsten.
11·445	Musschenbroek.
11·275	P. & J.
11·298	P. & J.

The average specific gravity, 11·337, gives 9·14 as the atomic volume.

Mercury, $Hg = 101·43$.—The specific gravity of solid (frozen) mercury is stated by Kupffer and Cavallo to be about 14·0, which makes the atomic volume 7·24.

Silver, $Ag = 108$.—According to Brisson the specific gravity of this metal is 10·474, according to one of our own experiments 10·522; the mean result, 10·498, gives 10·29 as the atomic volume.

Gold, $Au = 199·2$.—Brisson describes the specific gravity as 19·258, which gives the atomic volume 10·34.

Platinum, $Pt = 98·8$.—Brisson states the specific gravity of this metal to be 20·98, a result which is nearly the mean of all the specific gravities recorded. The atomic volume is therefore 4·71.

Palladium, $Pd = 53·3$.—Breithaupt describes the specific gravity of a chemically pure specimen as 10·923, a result closely agreeing with the observations of Cloud and Wollaston. The atomic volume must therefore be 4·88.

Rhodium, $R = 52·2$—Cloud took the specific gravity of a specimen fused by the oxyhydrogen blowpipe and found it to be 11·2. Wollaston states it at 11·0. The mean, 11·1, gives the atomic volume 4·7.

Osmium, $Os = 99·7$.—The specific gravity of the metal, as obtained by the reduction of its oxide with hydrogen, is 10·0; this gives the atomic volume 9·97.

Iridium, $Ir = 98·84$.—The following are recorded specific gravities of this metal, neglecting the determination of

Berzelius, which will afterwards come under special consideration.

$$19\cdot50 \quad \text{Mohs.}$$
$$18\cdot68 \quad \text{Children.}$$
$$23\cdot55 \quad \text{Breithaupt.}$$
$$22\cdot11 \quad \text{Breithaupt.}$$

The mean result, $20\cdot96$, gives $4\cdot71$ as the atomic volume.

Potassium, $K = 39\cdot15$.—According to Gay Lussac and Thenard the specific gravity of this metal is $0\cdot865$; its atomic volume is therefore $45\cdot26$.

Sodium, $Na = 23\cdot3$.—Gay Lussac and Thenard describe the specific gravity of sodium as $0\cdot972$; Davy states it at $0\cdot935$. The mean, $0\cdot953$, yields the atomic volume $24\cdot45$.

Barium, $Ba = 68\cdot7$.—The specific gravity of Barium is about $4\cdot0$, according to Davy; hence its atomic volume would be $17\cdot17$.

Strontium, $Sr = 43\cdot8$.—Gehler states the specific gravity to be between $4\cdot0$ and $5\cdot0$. The mean, $4\cdot5$, yields the atomic volume $9\cdot73$.

In the previous part of this paper all the metals are included, the specific gravities of which have been determined with any degree of certainty. An inspection of the numbers representing their atomic volumes will show the existence of a decided multiple relation. In our former paper we represented $9\cdot8$, the volume of ice, as a primitive volume for salts *. This number, however, must be made up of the numbers representing the volumes of oxygen and hydrogen respectively. The number $9\cdot8$ cannot itself represent the primitive volume, but a certain submultiple of it may do so. A little consideration of the results now detailed shows that this submultiple is the number $1\cdot225$, or the eighth part of the volume of ice—$\frac{9\cdot8}{8} = 1\cdot225$. The divisor 8 is the cube of 2, so that the atom of ice (considered as a globe or cube) will have exactly twice the linear dimensions of the atom possessing the volume $1\cdot225$. It will, however, be observed that in

---

* That number being deduced from the sp. gr. $0\cdot9184$ determined by us, a number almost identical with that recently found by Brunner, $0\cdot918$.

several cases the above volumes correspond more exactly to the multiple of a volume of $\frac{1\cdot225}{2} = \cdot6125$. But as $\frac{1\cdot225}{2}$ is rather an exceptional than a general bulk, the convenience in illustration of the law of multiple proportions will be considerable if in the mean time we assume 1·225 as the standard number for comparison. We do not state it as the absolute unit, but as a general and convenient standard bulk for illustrating the law of multiple proportions.

TABLE I.—Showing the Atomic Volumes and Specific Gravities of the Metallic Elements.

Designation.		Volume and Specific Gravity.				
Name.	Atomic weight.	Vol. by experiment.	1·225, or ½ vol. of ice, as standard for comparison.	Volume by theory.	Specific gravity by experiment.	Specific gravity by theory.
Manganese......	27·7	3·455	3	3·67	8·021	7·547
Iron	28·0	3·59	3	3·67	7·789	7·629
Cobalt	29·5	3·46	3	3·67	8·519	8·038
Nickel	29·5	3·54	3	3·67	8·33	8·038
Copper	31·6	3·56	3	3·67	8·862	8·610
Aluminium	13·72	5·13	4	4·9	2·67	2·80
Zinc	32·3	4·71	4	4·9	6·864	6·590
Cadmium	55·8	6·46	5	6·12	8·642	9·117
Chromium	28·0	5·09	4	4·9	5·500	5·714
Bismuth........	71·07	7·27	6	7·35	9·776	9·669
Tin	58·90	8·09	6½	7·96	7·283	7·400
Arsenic	75·40	13·07	11	13·47	5·767	5·597
Antimony	129·2	19·21	16	19·6	6·727	6·591
Molybdenum	47·96	5·56	4½	5·51	8·625	8·704
Tungsten	94·8	5·45	4½	5·51	17·4	17·205
Titanium	24·30	4·59	4	4·9	5·29	4·959
Tellurium	64·2	10·36	8½	10·41	6·196	6·167
Lead	103·6	9·14	7½	9·18	11·337	11·285
Mercury........	101·43	7·24	6	7·35	14·0	13·8
Silver..........	108·0	10·29	8½	10·41	10·498	10·375
Gold	199·2	10·34	8½	10·41	19·258	19·135
Platinum	98·8	4·71	4	4·9	20·98	20·163
Palladium	53·3	4·88	4	4·9	10·923	10·877
Rhodium	52·2	4·7	4	4·9	11·1	10·657
Osmium........	99·7	9·97	8	9·8	10·0	10·173
Iridium	98·84	4·71	4	4·9	20·96	20·171
Potassium	39·15	45·26	37	45·32	0·865	0·863
Sodium	23·3	24·45	20	24·5	0·953	0·951
Barium	68·7	17·17	14	17·15	4·0	4·005
Strontium	43·8	9·73	8	9·8	4·5	4·469

The table exhibits clearly a simple multiple relation, without, however, giving precise accordance between the theoretical and experimental results in many of the metals *. This want of accordance might either be due to the force of cohesion, which prevents the metals assuming their natural volumes, or it might be owing to the unlike conditions and impurity in the specimens operated upon. To determine these points we instituted the following experiments on simple bodies.

In order to render our conclusions still more worthy of attention, we have repeated the estimation of many specific gravities, carefully guarding against any impurity in the substances under examination as well as any irregularity in their crystalline condition. In the case of the metals, no method appeared to us more likely to give unexceptionable results than that of crystallizing them from their solutions by electricity. By this means we have obtained the specific gravity of silver, copper, zinc, lead, and tin as follows :—

Specific Gravity of Silver.—A solution of the cyanide of silver was furnished with two electrodes, viz. a plate of silver and a piece of fine silver wire, the former being in connection with the positive, the latter with the negative end of a single constant cell of Daniell's arrangement. After five or six days, we found that the fine silver wire was covered to a considerable thickness with a beautiful crystalline deposit of silver of

---

* In the case of alloys, where the cohesion is diminished, the results generally are very exact, and the coincidence in many cases strikingly identical with the theoretical expression. Thus Sn_2Pb has, according to Regnault, a sp. gr. 8·777, according to Kupffer 8·637, mean 8·707. $\frac{221\cdot4}{8\cdot707}=25\cdot42$. V being the unit volume 1·225,

$$SnV \times 2 + PbV = 25\cdot10.$$

SnPb has, according to Regnault, sp. gr. 9·387, according to Kupffer 9·436, mean 9·411. $\frac{162\cdot5}{9\cdot411}=17\cdot26$.

$$SnV + PbV = 17\cdot15.$$

Again $SnPb_2$ has, according to Kupffer, sp. gr. 10·078. $\frac{266\cdot1}{10\cdot078}=26\cdot4$.

$$SnV + PbV \times 2 = 26\cdot33.$$

the most perfect whiteness. Having carefully washed the silver, we suspended it by an excessively fine platinum wire to the hydrostatic scale of a very excellent balance, and weighed it in distilled water of the temperature of 40°. We then carefully dried it and weighed it in air.

Weight in water at 40° . . . . . 84·41 grains
Weight in air. . . . . 93·26   „
Sp. gr. reduced to a vacuum . . . 10·537

Another similar experiment gave the following result:—

Weight in water at 49°. . . . . 122·93 grains
Weight in air . . . . 135·84   „
Sp. gr. reduced to water at 40°. 10·522

I. Sp. gr. of silver . . . 10·537
II.   „   „   . . . 10·522
                          ─────────
                 Mean . . . 10·530

Specific Gravity of Copper.—A piece of sheet copper was employed as the negative element of a Daniell's cell for three or four days, due care being taken to supply the cupreous solution with fresh crystals of sulphate of copper from time to time. The electrotype copper having been detached and washed, was then weighed in distilled water at 49°; and then again, after drying, in air.

Weight in water . . . . 149·07 grains
Weight in air . . . . 167·97   „
Sp. gr. reduced . . . .   8·834

A second experiment conducted in the same way, gave

Weight in water . . . . 261·1 grains
Weight in air. . . . . 293·97   „
Sp. gr. reduced . . . . 8·941

In a third experiment, we found—

Weight in water at 44° . . . . . . 264·508 grains
Weight in air . . . . 297·838   „
Sp. gr. reduced . . . .   8·934

I. Sp. gr. of copper . . . . 8·884
II.   „   „   . . . . 8·941
III.   „   „   . . . . 8·934
                                        ———
                            Mean . . . 8·920

Specific Gravity of Zinc.—A saturated solution of pure sulphate of zinc was furnished with electrodes of zinc, and placed in connection with a battery consisting of three Daniell's cells. In about twenty-four hours a good plate of tough zinc was deposited on the negative electrode. The plate was weighed in water which had been previously boiled in order to prevent its acting on the zinc.

Weight in water . . . . 181·14 grains
Weight in air . . . . 212·0   „
Reduced sp. gr. . . . . 6·869

In a second experiment we obtained—

Weight in water . . . . 161·14 grains
Weight in air . . . . 188·02   „
Reduced sp. gr. . . . . 6·992

A third experiment gave—

Weight in water . . . . 176·53 grains
Weight in air . . . . 206·16   „
Reduced sp. gr. . . . . 6·956

I. Sp. gr. of zinc . . . . 6·869
II.   „   „   . . . . 6·992
III.   „   „   . . . . 6·956
                                      ———
                            Mean . . . 6·939

Specific Gravity of Lead.—A very fine platinum wire was made negative (by means of a single constant cell) in a saturated solution of acetate of lead; a plate of lead formed the positive electrode. In about twenty hours a most beautiful lead-tree was formed on the platinum wire. The wire with its leaden foliage was then carefully removed from the solution, well washed in hot water, and weighed in distilled water at 54° which had been recently boiled to remove carbonic-acid gas. The fine platinum wire upon which the lead was deposited served as the means of suspension on the hydrostatic scale. Lastly, the lead was carefully removed from the water, immersed in alcohol, wrapped up in bibulous paper, and put on a warm stove to dry. By thus operating, the lead was dried without losing much of its brilliancy. A very small correction had to be applied for the platinum wire, which weighed 0·3 of a grain.

Exp. 1.—Weight in water at 54° . . . 74·74 grains
Weight in air . . . . 82·0 „
Sp. gr. reduced to water at 40°, and to a vacuum . } 11·275

Exp. 2.—Weight in water at 60° . . . 62·80 grains
Weight in air . . . . 69·03 „
Sp. gr. reduced . . . . . . 11·070

Exp. 3.—Weight in water at 54° . . . 64·92 grains
Weight in air . . . . 71·22 „
Sp. gr. reduced . . . . . . 11·298

Exp. 4.—Weight in water at 62° . . . 62·82 grains
Weight in air . . . . 68·92 „
Sp. gr. reduced . . . . . . 11·28

I. Sp. gr. of lead . . . . 11·275
II. „ „ . . . . 11·070
III. „ „ . . . . 11·298
IV. „ „ . . . . 11·280

Mean . . . 11·231

Specific Gravity of Tin.—The experiments with tin were made in a manner exactly similar to those with lead, a solution of chloride of tin being employed. The branching crystals of this metal were of extreme beauty.

Exp. 1.—Weight in water . . . . 35·74 grains
Weight in air . . . . . 41·46 „
Reduced sp. gr. . . . . 7·245

Exp. 2.—Weight in water . . . . 36·35 grains
Weight in air . . . . . 42·06 „
Reduced sp. gr. . . . . 7·363

Exp. 3.—Weight in water . . . . 40·58 grains
     Weight in air . . . . . 46·99   ,,
     Reduced sp. gr. . . . . 7·330
Exp. 4.—Weight in water . . . . 34·063 grains
     Weight in air . . . . . 39·477   ,,
     Reduced sp. gr. . . . . 7·288

   I. Sp. gr. of tin . . . . 7·245
   II.   ,,   ,, ... 7·363
   III.   ,,   ,, ... 7·330
   IV.   ,,   ,, ... 7·288

      Mean . . . 7·306

There are several other metals which we have not yet
tried, and others, as for example iron, which we have not
been able to deposit in a satisfactory state; but the instances
we have given are, we think, sufficient to show that the
specific gravities of the electrotype metals are not widely
different from those obtained in the usual way. And therefore while we claim a superior degree of accuracy for our
results with the electrotype metals, we think that if a law
be established for the volumes of the metals, it will be indicated with tolerable accuracy by the results obtained in the
ordinary way.

Specific Gravity of Sulphur.—A quantity of flowers of
sulphur was melted in a glass vessel ; and after a crust had
formed on the surface on cooling, the remainder was poured
off. The crystals of sulphur, weighing 79·07 grains, were
then thrown into a specific-gravity bottle partly filled
with water. The bottle with its contents was slightly
warmed, and then boiled under the exhausted receiver of an
air-pump, in order to remove the air which is so liable to
adhere to sulphur. The bottle was then filled up with
water, reduced to a known temperature, and weighed. The
weight of the bottle when filled with pure water at 50½°
being 644·22 grains, and its weight when filled at the same
temperature with water and 79·07 grains of sulphur being
683·97, it follows that 39·32 grains of water were displaced
by the sulphur.

Weight of water displaced by 79·07 grains of sulphur =
39·32 grains.

Sp. gr. of the crystals, 2·010.

Specific Gravity of Platinum Sponge.—The experiments
with this substance were conducted in exactly the same
manner as that with sulphur.

Exp. 1.—Weight of water displaced by
   109·02 grains of platinum sponge   5·15 grains
   Sp. gr. of the metal . . . . 21·169

Exp. 2.—Weight of water displaced by
   107·32 grains of platinum sponge   5·052 grains
   Sp. gr. of the metal . . . . 21·243

   I. Sp. gr. of platinum sponge . . . 21·169
   II.   ,,   ,,   ,, . . . 21·243

                               Mean . . . 21·206

Specific Gravity of Phosphorus.—Weight of water displaced
by 44·76 grains of phosphorus = 24·87 grains.

Sp. gr. of phosphorus, 1·800.

TABLE II.—Showing the Volumes of Metals and some
other Simple Bodies.

Designation.		Volume.					
Name.	Atomic weight.	Volume by experiment.	Volume increased by $\frac{1}{20}$.	1·225, or $\frac{9\cdot8}{8}$, as unity.	Volume by theory.	Sp. gr. by experiment.	Observer.
ilver	108·3	10·285	10·799	9	11·025	10·530	Our own exper'ts.
opper	31·65	3·548	3·725	3	3·675	8·920	,,
inc	32·31	4·656	4·889	4	4·900	6·939	,,
ead	103·73	9·236	9·698	8	9·800	11·231	,,
in	58·92	8·064	8·467	7	8·575	7·306	,,
ulphur	16·03	7·975	8·373	7	8·575	2·010	,,
latinum	98·84	4·661	4·894	4	4·900	21·206	,,
hosphorus ..	31·44	17·47	18·343	15	18·375	1·800	,,
on	28·0	3·59	3·769	3	3·675	7·800	Various authorities.
ickel	29·62	3·42	3·591	3	3·675	8·660	,,
ntimony	64·62	9·42	9·891	8	9·800	6·860	,,
olid Mercury .	101·43	7·245	7·607	6	7·350	14·000	,,
old	99·6	5·107	5·362	4½	5·512	19·500	,,
alladium	53·36	2·64	4·872	4	4·900	11·500	,,
iamond	12·0	3·478	3·651	3	3·675	3·450	,,

The volumes given in the above table are evidently in a simple multiple ratio with themselves, and by adding one twentieth they all come nearly into the category of the number 1·225. Whether the contracted state of the metals in the solid state be owing to the attraction of cohesion we have not examined. We think it highly probable, however, that if the volumes were reduced to the zero of temperature and compared with ice, also at the zero, they would be found in a strictly correct multiple ratio. The coefficients of dilatation which we have examined seem to favour this view, as also does the high rate of expansion of ice found by Brunner; as, however, we have not yet made many experiments on the dilatation of solid bodies by heat, we prefer deferring the consideration of this point to future communications.

Specific Gravity of Metals in a Finely Divided state.

Under the impression that the contracted state of the metals in their usual solid condition was in a great measure owing to the attraction of cohesion, we now entered with great interest upon the examination of metals obtained by reducing their oxides by hydrogen gas. In these experiments we employed a specific-gravity bottle, on account of the facilities which it presented in the case of those metals which ignite on exposure to air.

Specific Gravity of Copper obtained by reducing the Oxide.— A stream of dried hydrogen gas was transmitted through a tube of German glass heated to redness and containing oxide of copper. After the reduction seemed to be quite completed, the operation was carried on a quarter of an hour longer, in order to prevent the slightest trace of oxide being left. The tube with its contents having cooled, the copper was removed, pounded into small fragments, weighed, and thrown into a specific-gravity bottle partly filled with water. This done, the bottle with its contents was slightly warmed, and then placed under the exhausted receiver of an air-pump, in order to boil away any air that might possibly lodge among the lumps of metal. The bottle was then filled up

with water, and placed in a dish of water in order to reduce
it to a given temperature. The perforated stopper was then
inserted, and the bottle dried and weighed. The weight of
the bottle, when filled with water only, being known, it was
easy to find the quantity of water displaced by the copper
in the manner described in the experiment with sulphur
already cited.

    Exp. 1.—Weight of water displaced by
            212·6 grains of copper=25·22 grains
            Sp. gr. reduced, 8·428
    Exp. 2.—Weight of water displaced by
            156·8 grains of copper=18·48 grains
            Sp. gr. reduced, 8·483.
    Exp. 3.—Weight of water displaced by
            199·55 grains of copper=23·865 grains
            Sp. gr. reduced, 8·360

        I. Sp. gr. of copper . . 8·428
        II.   ,,   ,,   . 8·483
        III.   ,,   ,,   . . 8·360
                            ———
             Mean . . . 8·424

Specific Gravity of Cobalt obtained by reducing the Oxide.—
The experiments with this metal were conducted in a similar
manner to those with copper, except that turpentine was
employed in the specific-gravity bottle instead of water, and
that the metal was weighed in the tube in which it was
obtained, and shaken into the bottle with as little exposure
to the air as possible. On account of the great expansion
of turpentine, it was necessary to take particular care in
reducing the bottle to a given temperature. The specific
gravity of the turpentine was found to be 0·8764 at 40°.

    Exp. 1.—Weight of turpentine displaced by
          52·5 grains of cobalt . . . . . . 5·57 grains
          Sp. gr., referred to water at 40° . 8·26
    Exp. 2.—Weight of turpentine displaced by
          65·26 grains of cobalt . . . . . 7·41 grains
          Sp. gr. referred to water at 40   . 7·718

       I. Sp. gr. of cobalt . . 8·260

       II.   „   „   . . 7·718

                       Mean . . . 7·989

Specific Gravity of Nickel obtained by reducing the Oxide.—
The experiments with this metal were conducted precisely as
those with cobalt.

     Exp. 1.—Weight of turpentine displaced by

             91 58 grains of nickel . 10·1 grains

         Sp. gr. . . . . 7·861

     Exp. 2.—Weight of turpentine displaced by

             70·28 grains of nickel . 7·81 grains

         Sp. gr. . . . . 7·803

         I. Sp. gr. of nickel . . 7·861

         II.   „   „   . . 7·803

                  Mean . . . 7·832

Specific Gravity of Iron obtained by reducing the Peroxide.

       Weight of turpentine displaced by

           46·7 grains of iron . . 7·52 grains

       Sp. gr. . . . . 7·130

*Specific Gravity of Arsenic obtained from Arsenietted
Hydrogen at a low temperature.*

       Weight of turpentine displaced by

          1·031 grains of arsenic . . 0·171 grain

       Sp. gr. . . . . 5·230

Specific Gravity of Flowers of Sulphur.—The sulphur
having been thrown into a specific-gravity bottle, a small
portion of water was added to it, and the bottle well shaken
so as to cause the sulphur to be thoroughly wetted. The
bottle was then placed under the exhausted receiver of an
air-pump, so as to remove the air lodging in the sulphur.
Then more water was added, and the same operation repeated.
When at last we were satisfied that the sulphur was entirely
deprived of air, we filled the remainder of the bottle with
water. reduced it to a known temperature, and weighed it.

Exp. 1.—Weight of water displaced by
        158·2 grains of sulphur . . . . 82·82 grains
        Sp. gr. reduced to water at 40° . 1·910
Exp. 2.—Weight of water displaced by
        167·12 grains of sulphur . . . 87·22 grains
        Sp. gr. reduced . . . . 1·916

  I. Sp. gr. of flowers of sulphur . . 1·910
  II.   ,,   ,,   . . 1·916

                Mean . . . 1·913

Specific Gravity of Platinum obtained by Exposing the Oxide to a bright red heat.

    Weight of water displaced by
       111·92 grains of platinum 6·3 grains
    Sp. gr. . . . . 17·766

This does not differ much from the specific gravity 17·890
of the same body, as obtained by Scholz.

Specific Gravity of Uranium obtained by Wöhler's Method.

    Weight of turpentine displaced by
       65·475 grains of uranium 6·74 grains
    Sp. gr. . . . . 8·425

Specific Gravity of Magnesium.—The specimen examined
was prepared and presented to one of us by Professor Liebig.

Exp. 1.—Weight of turpentine displaced by
       5·235 grains of magnesium . 2·033 grains
       Sp. gr. . . . . 2·233.
Exp. 2.—Weight of turpentine displaced by
       4·144 grains of magnesium . 1·60 grains
       Sp. gr. . . . . 2·246

    I. Sp. gr. of magnesium . 2·233
    II.   ,,   ,,   . 2·246

         Mean . . . 2·240

TABLE III.—Showing the Volumes of several Metals in a finely divided state, and that of Flowers of Sulphur *.

Designation.		Volume.				
Name.	Atomic weight.	Volume by experiment.	1·225 as unity.	Volume by theory.	Sp. gr. by theory.	Sp. gr. by experiment.
Copper	31·65	3·757	3	3·675	8·612	8·424
Cobalt	29·57	3·701	3	3·675	8·047	7·989
Nickel	29·62	3·782	3	3·675	8·060	7·832
Iron	27·18	3·812	3	3·675	7·396	7·130
Arsenic	37·67	7·203	6	7·350	5·125	5·230
Sulphur	16·03	8·380	7	8·575	1·870	1·913
Platinum ..	98·84	5·563	$4\frac{1}{2}$	5·512	17·931	17·766
Uranium ..	217·26	25·790	21	25·725	8·445	8·425
Magnesium	12·69	5·665	$4\frac{1}{2}$	5·512	2·302	2·240

The results above tabulated are, we think, sufficient to render it evident that the specific volumes of the metals in a perfectly divided state are strict multiples of the volume 1·225.

Specific Gravity of Melted Metals.

We now sought an additional confirmation of our views in the volumes of metals in the melted state. And as no doubt could be entertained that this condition would completely obviate the influence of cohesion or that of any particular arrangement of particles, we felt that a complete confirmation or refutation of all the views we have advanced would depend upon the results arrived at with the fused metals. The instrument we employed when the metals could be fused at a low temperature, consisted simply of a small glass matrass, fig. 4, formed by blowing a bulb at one end of a tube about one quarter of an inch diameter. The content

* The volumes of metals in this table are not always the same as those in Table I.; the reasons of the difference will be given in a future part of this Memoir.

of the bulb was ascertained by finding the weight
of mercury which could fill it up to a given point
of the stem, say x. A quantity of mercury, equivalent in volume to five grains of water, was then
added, and another mark was made at y. This
done, the matrass was filled with small pieces of
the metal under examination, and carefully exposed to heat. As the metal melted, additional
portions were introduced until the liquid metal
stood somewhere between the marks x and y, the
exact position being determined to within one tenth of the
space. The whole having then been allowed to cool, the
matrass was broken and the metal weighed.

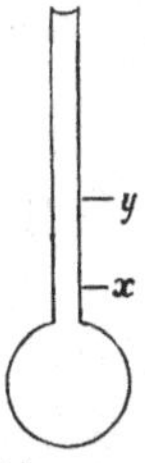

Fig. 4.

About ⅓ actual size.

Specific Gravity of Melted Lead.

Exp. 1.—Capacity of matrass in grains of
water at 40° . . . . 124·17
Capacity corrected for expansion
of glass . . . . 125·18
Weight of the lead . . . . . . 1316 grains
Sp. gr. of the melted metal . . . 10·513

Exp. 2.—Capacity of the matrass . . . . . 64·37
Capacity corrected for expansion 64·90
Weight of the lead . . . . 678·2 grains
Sp. gr. of melted lead . . . . . 10·45

Exp. 3.—Capacity of matrass . . . . . . 43·83
Capacity corrected for expansion 44·20
Weight of lead . . . . 466·9 grains
Sp. gr. . . . . 10·563

I. Sp. gr. of melted lead . . 10·513
II.   „   „   . . 10·450
III.   „   „   . . 10·563

Mean . . . 10·509

Specific Gravity of Melted Tin.

Exp. 1.—Capacity of matrass . . . 116·9
Corrected for expansion . 117·57
Weight of tin . . . . . . 817·0 grains
Sp. gr. . . . . 6·949

Exp. 2.—Capacity of matrass . . .   151·52
           Corrected for expansion .   152·38
           Weight of tin . . . . . .   1053·5 grains
           Sp. gr. . . . .   6·913
Exp. 3.—Capacity of matrass . . .   133·38
           Corrected for expansion .   134·15
           Weight of tin . . . . . .   931·0 grains
           Sp. gr. . . . .   6·94

    I. Sp. gr. of melted tin . .   6·949
    II.   ,,   ,,   . . 6·913
    III.   ,,   ,,   . . 6·940

                Mean . . . 6·934

Specific Gravity of Melted Bismuth.

Exp. 1.—Capacity of matrass . . .   98·65
           Corrected for expansion .   99·28
           Weight of bismuth. . . .   974·0 grains
           Sp. gr. . . . .   9·811
Exp. 2.—Capacity of matrass . . .   129·2
           Corrected for expansion .   130·0
           Weight of bismuth. . . .   1268·3 grains
           Sp. gr. . . . .   9·756
Exp. 3.—Capacity of matrass . . .   141·88
           Corrected for expansion .   142·78
           Weight of bismuth . . .   1414·3 grains
           Sp. gr. . . . .   9·905
Exp. 4.—Capacity of matrass . . .   103·52
           Corrected for expansion .   104·2
           Weight of bismuth. . . .   1013 grains
           Sp. gr. . . . .   9·721

    I. Sp. gr. of melted bismuth . . 9·811
    II.   ,,   ,,   . . 9·756
    III.   ,,   ,,   . . 9·905
    IV.   ,,   ,,   . . 9·721

                Mean . . . 9·798

Specific Gravity of Melted Zinc.

Exp. 1.—Capacity of matrass . . . 45·8
   Corrected for expansion . 46·26
   Weight of zinc . . . . . . 301·7 grains
   Sp. gr. . . . . 6·522

Exp. 2.—Capacity of matrass . . . 55·37
   Corrected for expansion . 55·92
   Weight of zinc . . . . . . 364·1 grains
   Sp. gr. . . . . 6·511

Exp. 3.—Capacity of matrass . . . 61·84
   Corrected for expansion . 62·45
   Weight of zinc . . . . . . 406·2 grains
   Sp. gr. . . . . 6·504

   I. Sp. gr. of melted zinc . . . 6·522
   II. „   „   . . . 6·511
   III. „   „   . . . 6·504
           —
       Mean . . . 6·512

Specific Gravity of Melted Potassium.

   Capacity of matrass . . . 95·68
   Corrected . . . . 95·76
   Weight of potassium . . 80·7 grains
   Sp. gr. . . . . 0·8427

Specific Gravity of Melted Phosphorus.

Corrected capacity of matrass . . . 96·0
Weight of phosphorus . . . . . . 167·3 grains
Sp. gr. . . . . 1·744

Specific Gravity of Melted Sulphur.

Exp. 1.—Capacity of matrass . . . 151·03
   Corrected for expansion . 151·40
   Weight of sulphur . . . . 273·5 grains
   Sp. gr. . . . . 1·806

Exp. 2.—Capacity of matrass . . . 127·84
   Corrected for expansion . 128·16
   Weight of sulphur . . . . 232·7
   Sp. gr. . . . . 1·815

Exp. 3.—Capacity of matrass . . . . 131·74
    Corrected for expansion . . 132·07
    Weight of sulphur . . . . . 237·9 grains
    Sp. gr. . . . . 1·801
Exp. 4.—Capacity of matrass . . . . 117·53
    Corrected for expansion . . 117·8
    Weight of sulphur . . . . . 213·2 grains
    Sp. gr. . . . . 1·809
Exp. 5.—Capacity of matrass . . . . 156·16
    Corrected for expansion . . 156·55
    Weight of sulphur . . . . . 282·4 grains
    Sp. gr. . . . . 1·804

  I. Sp. gr. of melted sulphur . . 1·806
  II.   ,,   ,, . . 1·815
  III.   ,,   ,, . . 1·801
  IV.   ,,   ,, . . 1·809
  V.   ,,   ,, . . 1·804

        Mean . . . 1·807

Specific Gravity of Sulphur in the Viscid Melted state.
    Capacity of matrass . . . . 156·03
    Corrected for expansion . . 156·50
    Weight of sulphur . . . . . 273·5 grains
    Sp. gr. . . . . 1·748

Specific Gravity of Melted Antimony.—We found that the temperature at which this metal enters into fusion was too high for even German glass to bear without changing figure ; consequently we had now to make a complete change in our manner of operating. We took two capsules of white earthenware, one of which was deep and narrow, the other wide and shallow. When capsule *b* was filled with melted metal, capsule *a* was placed upon its top, so that its convex surface pressed out part of the metal from the capsule *b*. When all had cooled, the weight of the metal was ascertained and

Fig. 5.

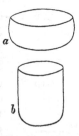

compared with the weight of mercury capable of being held
between the capsules.   We have estimated the expansion of
the capsules at $\frac{1}{600}$ of their bulk for 180°, which is the mean
of the expansions of brown earthenware and stoneware as
given by Dalton*.

Exp. 1.—Capacity between capsules . .   45·88
        Corrected for their expansion .   46·33
        Weight of antimony . . . . . . 308·0 grains
        Sp. gr. . . . .   6·646
Exp. 2.—Capacity between capsules . .   46·01
        Corrected for their expansion .   46·47
        Weight of antimony . . . . . . 303·4 grains
        Sp. gr. . . . .   6·529

    I. Sp. gr. of melted antimony . . 6·646
    II.   ,,   ,, . . 6·529

                Mean . . . 6·587

Specific Gravity of Melted Copper.
Capacity between capsules . .   49·97
Corrected for their expansion .   50·88
Weight of copper . . . . . . 370·0 grains
Sp. gr. . . . .   7·272

Specific Gravity of Melted Silver.
Exp. 1.—Capacity between capsules . .   55·61
        Corrected for their expansion .   56·55
        Weight of silver . . . . 516·5 grains
        Sp. gr. . . . .   9·131
Exp. 2.—Capacity between capsules . .   54·62
        Corrected for their expansion .   55·54
        Weight of silver . . . . 515·5 grains
        Sp. gr. . . . .   9·281

* 'New System of Chemistry,' Part 1, p. 44.

I. Sp. gr. of melted silver . . 9·131
II.   „   „   . . 9·281

Mean . . . 9·206

Specific Gravity of Fluid Mercury.—The density of this
metal at 39°·2 is 13·588 according to Kupffer. Hence its
specific gravity near the point of congelation will be 13·694.

TABLE IV.—Showing the Volumes of several Metals and
other Simple Bodies after entering into Fusion.

Designation.		Volume.				
Name.	Atomic weight.	Volume by experiment.	1·225 as unity.	Volume by theory.	Sp. gr. by theory.	Sp. gr. by experiment.
Mercury	101·43	7·407	6	7·350	13·800	13·694
Tin	58·92	8·497	7	8·575	6·871	6·934
Bismuth	71·07	7·254	6	7·350	9·670	9·798
Lead..........	103·73	9·871	8	9·800	10·584	10·509
Zinc	32·31	4·961	4	4·900	6·594	6·512
Antimony	64·62	9·810	8	9·800	6·594	6·587
Silver	108·30	11·764	9½	11·637	9·306	9·206
Copper........	31·65	4·352	3½	4·287	7·383	7·272
Potassium	39·30	46·620	38	46·550	0·844	0·843
Phosphorus	31·44	18·028	15	18·375	1·711	1·744
Sulphur	16·03	8·871	7¼	8·881	1·805	1·807
Viscid Sulphur.	16·03	9·170	7½	9·187	1·745	1·748

By inspecting the above table it will be seen that the
atomic volumes of the metals are in general in a simple
multiple ratio one to another, and are at the same time in a
simple ratio with the volume of ice, and with the volume 11
given in our former paper. Perhaps the most curious and
least expected results are those with copper and silver. These
metals appear to assume in the melted state half a volume
more than they possess in the solid state. This is highly
interesting in the case of copper, as it explains the reason
why CuO has a volume 3½ + 2 instead of 3 + 2. We will
not, however, insist strongly upon this point until we shall
have repeated the experiments with copper and silver more

frequently, and with a more exact apparatus than we have hitherto been able to employ. The specific gravity of sulphur is also very interesting on account of the states into which it enters; of these we have examined five, viz. :—

	Sp. gr.	Volume.
Viscid melted state	1·748	9·170
Clear amber melted	1·807	8·871
Flowers	1·913	8·380
Crystals	2·010	7·975
Waxy by pouring the viscid sulphur into water	1·921	8·340

To the relations between the different states of this element we shall have to return.

Having now examined three of the conditions in which the metals present themselves, we shall conclude the section by recapitulating the principal points at which we think we have arrived.

1st. The metals, when obtained in a finely divided state, so as to be deprived of cohesion, exhibit volumes which are multiples of the unit 1·225.

2nd. The volumes of metals rendered fluid by heat are in like manner multiples of the volume 1·225.

3rd. The volumes of metals in the solid crystalline condition are, in general, multiples of a number about $\frac{1}{20}$ less than 1·225. We believe that this may *partly* arise from the peculiar molecular arrangement, but that the contraction is *principally* occasioned by the operation of the force of cohesion. Our first table, which gives volumes, as we have already observed, only approximating to multiples of 1·225, though serving to point out the existence of a law, would show a much greater coincidence between the theoretical and actual results than it does at present, if we were to apply a small correction to the observed volumes as we have done in Table II. In some instances of Table I. the volume is $\frac{1·225}{2}$ less than it would have been after applying the empirical correction of $\frac{1}{20}$. May we not therefore suppose that the contraction occasioned by cohesion is itself in some measure guided by the number 1·225 ? We think this is not improbable, but until we shall have given the subject a more minute investigation, it would be premature to insist strongly upon such a view.

We now proceed to test the law with a class of compounds which do not so readily yield to the solicitation of cohesion.

SECTION II.

Oxides of the Metals.

Protoxide of Manganese, $MnO = 35·7$.—The oxide examined was made by passing hydrogen over the hydrate of the protoxide while it was heated to redness. 18·83 grains of the oxide thus prepared gave an increase in the stem of the volumenometer of 3·5, making the sp. gr. 5·38 and the atomic volume 6·63.

Sesquioxide of Manganese, $Mn_2O_3 = 79·4$.—The oxide used in our experiments was prepared by calcining the nitrate. 24·53 grains gave an increase of 5·37 in one experiment and 5·31 in another.

I. Sp. gr. . . . 4·568
II.   „   . . . 4·619

The mean result, 4·593, yields the volume 17·29. Our result does not differ widely from that obtained by Leonhard as the sp. gr. of Braunite, the native sesquioxide, viz. 4·82.

Manganoso-Manganic Oxide, $Mn_2O_3 + MnO = 115·1$.— This common oxide of manganese was obtained by calcining the pure carbonate in open air. 23·73 grains gave an increase in two experiments of 5·0 and 5·1 respectively.

I. Sp. gr. . . . 4·746
II.   „   . . . 4·653

The mean, 4·700, gives the volume 24·49, a result almost identical with that obtained by Leonhard for Hausmanite, viz. 4·72.

Peroxide of Manganese, $MnO_2 = 43.7$—Turner has determined the sp. gr. of pyrolusite, native peroxide, and states it to be 4·81, which makes the atomic volume 9·09.

Varvicite, $2(MnO_2) + Mn_2O_3 + HO = 175.8$.—According to Turner this mineral possesses sp. gr. 4·531, which gives 38·8 for the atomic volume.

Peroxide of Iron, $Fe_2O_3 = 80.0$.—The specific gravity of this oxide varies according to the temperature to which it has been exposed. Two specimens, heated to redness, gave the following results:—24·70 gave an increase of 5·37, and the same quantity in the second experiment gave an increase of 5·19.

I. Sp. gr. . . . 4·60
II.   „   . . . 4·759

The mean, 4·679, gives the atomic volume 17·09.

But the same oxide, raised by means of a wind-lamp to a heat approaching whiteness, gave a different result. 24·70 of the heated oxide gave an increase of only 4·81, making sp. gr. 5·135 and atomic vol. 15·57. Boullay has obviously examined the oxide in this last state, for he describes the sp. gr. as 5·225.

Magnetic Oxide of Iron, $Fe_2O_3 + FeO = 116$.—A specimen artificially prepared by Mercer's method of precipitating boiling solutions in atomic proportions of the sulphates of the protoxide and peroxide of iron, gave us a sp. gr. of 5·453, The native ore gives similar results. The recorded specific gravities of the magnetic oxide are as follows :—

5·400   Boullay.
5·453   P. and J.
4·900   Leonhard.
5·200   Leonhard.
4·960   Gerolt.
5·094   Mohs.
_____
Mean . . . 5·168

The mean, 5·168, gives the atomic volume 22·44.

Protoxide of Cobalt, $CoO = 37·5$.—$19·87$ grains gave an increase of $3·55$, making the sp. gr. $5·597$ and the atomic volume $6·70$; the same oxide heated to whiteness possessed the sp. gr. $5·75$ on operating on $30·13$ grains.

Peroxide of Cobalt, $Co_2O_3 = 83$.—This oxide was prepared by heating the nitrate. $24·94$ grains gave an increase of $5·18$, making the sp. gr. $4·814$ and the volume $17·24$. Herapath describes an oxide of cobalt examined by him as having the sp. gr. $5·322$, which would give a volume $15·6$, and correspond to the strongly ignited peroxide of iron.

Protoxide of Nickel, $NiO = 37·5$.—$39·74$ grains increased $7·10$, making sp. gr. $5·597$, and atomic volume $6·70$. Genth describes the abnormal oxide of nickel discovered by him as having a sp. gr. of $5·745$.

Peroxide of Nickel, $Ni_2O_3 = 83·0$.—The oxide used in our experiments was prepared by calcining the nitrate. $27·33$ grains gave an increase of $5·68$, making the sp. gr. $4·814$, and atomic vol. $17·24$; a result closely agreeing with that obtained by Herapath, who found the sp. gr. $4·846$, and the consequent volume $17·12$. The mean gives $4·830$ as the sp. gr. and $17·18$ as the atomic volume.

Oxide of Zinc, $ZnO = 40·3$.—Boullay describes the sp. gr. as $5·60$; Karsten as $5·734$; and Mohs as $5·432$. The mean sp. gr., $5·588$, gives the atomic volume $7·21$.

Suboxide of Copper, $Cu_2O = 71·2$.—This oxide was prepared according to Bottger's plan, by boiling a solution of oxide of copper in sugar and potash, and was obtained of a beautiful red colour. $20·8$ grains gave the increase $3·62$, making sp. gr. $5·746$. The result obtained by Karsten with the native suboxide, $5·751$, and that of Le Royer and Dumas, $5·75$, agree closely with our determination. The mean sp. gr., $5·749$, gives $12·38$ as the atomic volume.

Protoxide of Copper, $CuO = 39·6$.—This oxide was prepared by igniting the nitrate, avoiding too high a temperature. $22·12$ grains gave an increase of $3·75$, making sp. gr. $5·90$ and atomic volume $6·71$. Boullay found the specific gravity to be $6·13$; Karsten $6·430$; and Herapath $6·401$. To ascertain whether the difference of results was due to contraction at a high temperature, a portion of the same oxide was exposed to a white heat; 26·62 grains now gave an increase of 4·15, making the sp. gr. 6·414 and the atomic volume 6·17.

Oxide of Cadmium, $CdO = 63·8$.—Karsten states the specific gravity of this oxide at 6·950, which gives 9·18 for the atomic volume.

Oxide of Chromium, $Cr_2O_3 = 80·0$.—The oxide of chromium occurs in two distinct states, commonly known as the brown and green oxides.

The green oxide gave the following result :—24·3 grains gave an increase of 4·95, making sp. gr. 4·909. Wöhler found the sp. gr. of the crystallized oxide 5·210. The mean gives sp. gr. 5·059, and the atomic volume 15·81.

Oxide of Bismuth, $BiO = 79·07$.—26·5 grains of this oxide gave an increase of 3·28, making the sp. gr. 8·079, and the atomic volume 9·79. Karsten found a sp. gr. of 8·173.

Oxide of Tin, $SnO = 65·90$.—According to Herapath, the specific gravity of this oxide is 6·666; its atomic volume will therefore be 9·8.

Peroxide of Tin, $SnO_2 = 73·9$.—Herapath describes the specific gravity of this oxide as 6·640; hence its atomic volume will be 11·1.

Arsenious Acid, $AsO_3 = 99·34$.—This acid has been frequently examined, and the following results are those possessing the greatest uniformity :—

> 3·702   Karsten.
> 3·695   Guibourt.
> 3·690   Leonhard.
> 3·710   Leonhard.
> 3·698   Le Royer and Dumas.

The mean sp. gr., 3·699, yields 26·86 as the atomic volume.

Arsenic Acid, $AsO_5 = 115·34$.—29·53 grains gave in one experiment an increase of 7·34, in another of 7·41, making the sp. gr. respectively 4·023 and 3·985. Karsten states it at 3·734. The mean, 3·914, gives the atomic volume 29·47.

Oxide of Antimony, $SbO_3 = 153·28$.—26·52 grains of this

oxide caused an increase of 5·05, making a sp. gr. of 5·251, and a volume of 29·19.

Antimonious Acid, $SbO_4 = 161·28$.—20·37 grains of this compound, prepared by igniting antimonic acid, gave an increase of 5·0, making the sp. gr. 4·074, and the atomic volume 39·59.

Antimonic Acid, $SbO_5 = 169·28$.—This acid, prepared in the usual way by treating the oxide with nitric acid, and gently igniting the residue, gave the specific gravity 3·779, and atomic volume 45·06.

Oxide of Molybdenum, $MoO_2 = 63·96$.—According to Bucholz, this oxide has a sp. gr. of 5·666, which gives the atomic volume 11·28.

Molybdic Acid, $MoO_3 = 71·96$.—According to Berzelius, the sp. gr. of this acid is 3·49; according to Thomson 3·46. The mean, 3·475, gives the atomic volume 20·71.

Tungstic Oxide, $WO_2 = 110·8$.— According to Karsten, this oxide has a sp. gr. 12·11 ; hence the atomic volume will be 9·14.

Tungstic Acid, $WO_3 = 118·8$.—It is well known that this acid heightens in colour as the heat to which it has been exposed is increased. The specific gravity probably differs in the same proportion ; and this may account for the contradictory results given by chemists with respect to its specific gravity. De Luyart states it at 6·12 ; Herapath at 5·274; and Karsten at 7·139. The mean result, 6·177, gives the atomic volume as 19·23.

Titanic Acid, $TiO_2 = 40·33$.—Titanic acid occurs in various forms, to which especial reference will afterwards be made. In the meantime we assume Brookite as the representative of this oxide. The sp. gr. of that mineral, 4·128, gives 9·77 as the atomic volume.

Uranous Oxide, $UO = 68$.—According to Richter, this oxide has a sp. gr. of 6·94, making the atomic volume 9·8.

Hydrated Uranic Oxide, $U_2O_3 + HO = 153$.—Gmelin describes this oxide as having a sp. gr. of 5·926, making an atomic volume of 25·82.

Uranoso-Uranic Oxide, $UO + U_2O_3 = 212$.—Karsten states

that this oxide enjoys the sp. gr. 7·193, making the atomic volume 29·47.

Suboxide of Lead, $Pb_2O = 215·2$.—This oxide was made by slowly heating the oxalate to redness. 48·47 grains gave increase 4·96, making sp. gr. 9·772, and atomic volume 22·02.

Protoxide of Lead, $PbO = 111·6$.—48·47 grains gave an increase of 5·24, making sp. gr. 9·250. Herapath states it at 9·277; Karsten at 9·209. The mean result, 9·245, gives the volume 12·07.

Peroxide of Lead, $PbO_2 = 119·6$.—The oxide experimented upon was made by projecting the protoxide into melted chlorate of potash. 41·46 grains gave an increase of 4·66, making sp. gr. 8·897; in a second experiment, 32·3 grains gave increase 3·69, making sp. gr. 8·756. Herapath states the sp. gr. at 8·90. The mean of the three results is 8·851, making the atomic volume 13·51.

Minium, $PbO_2 + 2PbO = 342·8$.—Herapath describes the specific gravity of this compound as 9·096; its atomic volume must therefore be 37·7. Boullay states the sp. gr. at 9·190, a result confirmatory of Herapath's experiment.

Suboxide of Mercury, $Hg_2O = 210·8$.—Herapath describes the density of this oxide as 10·69, which gives a volume of 19·70.

Protoxide of Mercury, $HgO = 109·8$.—56·72 grains of this oxide, prepared from the nitrate, gave an increase of 5·0, making the sp. gr. 11·344. Berzelius found the sp. gr. 11·074; Herapath 11·085; Karsten 11·196; Boullay 11·000; and Le Royer and Dumas 11·29. The mean sp. gr., 11·164, gives the atomic volume 9·83.

Oxide of Silver, $AgO = 116·0$—Boullay describes the sp. gr. of this oxide as 7·250; Herapath as 7·140; and we ourselves found it 7·147 in an experiment with 44·10 grains. The mean of these results, 7·179, gives the atomic volume 16·14.

Potash, $KO = 47·15$.—According to Karsten, anhydrous potash has a sp. gr. 2·656; its atomic volume is therefore 17·75.

Soda, $NaO = 31\cdot3$.—Karsten describes the sp. gr. of anhy‑drous soda as $2\cdot805$; hence its atomic volume is $11\cdot1$.

Barytes, $BaO = 76\cdot70$.—$23\cdot86$ grains of anhydrous barytes gave an increase of $4\cdot94$, and in another experiment $4\cdot785$, giving the specific gravity $4\cdot829$ and $4\cdot986$ respectively. Karsten states it at $4\cdot733$. The mean, $4\cdot849$, gives an atomic volume $15\cdot82$.

Peroxide of Barium, $BaO_2 = 84\cdot70$.—$23\cdot86$ grains of this oxide, made by passing an excess of oxygen over barytes heated in a porcelain tube, gave the increase $4\cdot812$, making the sp. gr. $4\cdot958$, and the atomic volume $17\cdot08$.

Strontia, $SrO = 51\cdot8$.—Karsten states the specific gravity of anhydrous strontia at $3\cdot932$; hence its atomic volume will be $13\cdot2$.

TABLE V.—Showing the Atomic Volumes and Specific Gravities of the Metallic Oxides.

Designation.		Specific Gravity and Atomic Volume.					
Name.	Formula.	Atomic weight.	Volume by experiment.	1·225 taken as a standard for comparison.	Volume by theory.	Specific gravity by experiment.	Specific gravity by theory.
Protoxide of Manganese ..	MnO	35·7	6·63	5½	6·73	5·380	5·304
Sesquioxide of Manganese.	Mn_2O_3..........	79·4	17·29	14	17·15	4·593	4·629
Red Oxide of Manganese..	$MnO + Mn_2O_3$ ..	115·1	24·49	20	24·5	4·700	4·697
Peroxide of Manganese ..	MnO_2..........	43·7	9·09	7½	9·18	4·810	4·760
Varvicite	{ $2(MnO_2) +$ $(Mn_2O_3) + HO$ }	175·8	38·8	32	39·20	4·531	4·484
Peroxide of Iron	Fe_2O_3..........	80·0	17·09	14	17·15	4·679	4·664
Ditto, strongly heated	Fe_2O_3..........	80·0	15·57	13	15·93	5·135	5·031
Magnetic Oxide of Iron ..	$FeO + Fe_2O_3$	116·0	22·44	18	22·05	5·168	5·260
Oxide of Cobalt	CoO	37·5	6·70	5½	6·73	5·597	5·572
Peroxide of Cobalt	Co_2O_3..........	83·0	17·24	14	17·15	4·814	4·830
Ditto, strongly heated	Co_2O_3..........	83·0	15·6	13	15·93	5·322	5·220
Oxide of Nickel..........	NiO	37·5	6·70	5½	6·73	5·597	5·572
Peroxide of Nickel	Ni_2O_3..........	83·0	17·18	14	17·15	4·830	4·839
Oxide of Zinc	ZnO	40·3	7·21	6	7·35	5·588	5·482
Suboxide of Copper......	Cu_2O	71·2	12·38	10	12·25	5·749	5·812
Oxide of Copper	CuO	39·6	6·71	5½	6·73	5·90	5·884
Ditto, strongly heated	CuO	39·6	6·17	5	6·12	6·414	6·470

Table (*continued*).

Name.	Formula.	Atomic weight	Volume by experiment.	1·225 taken as a standard for comparison.	Volume by theory.	Specific gravity by experiment.	Specific gravity by theory.
Oxide of Cadmium	CdO	63·8	9·18	7½	9·18	6·95	6·95
Water (ice)	HO	9·0	9·8	8	9·8	0·918	0·918
Oxide of Chromium, strongly heated	Cr₂O₃	80·0	15·81	13	15·93	5·059	5·031
Oxide of Bismuth	BiO	79·07	9·79	8	9·8	8·079	8·068
Oxide of Tin	SnO	65·9	9·8	8	9·8	6·666	6·666
Peroxide of Tin	SnO₂	73·9	11·1	9	11·02	6·640	6·705
Arsenious Acid	AsO₃	99·34	26·86	22	26·95	3·699	3·680
Arsenic Acid	AsO₅	115·34	29·47	24	29·4	3·914	3·923
Oxide of Antimony	SbO₃	153·28	29·17	24	29·4	5·251	5·213
Antimonious Acid	SbO₄	161·28	39·59	32	39·2	4·074	4·114
Antimonic Acid	SbO₅	169·28	45·06	36	44·1	3·779	3·838
Molybdic Oxide	MoO₂	63·96	11·28	9	11·02	5·666	5·803
Molybdic Acid	MoO₃	71·96	20·71	17	20·82	3·475	3·456
Tungstic Oxide	WO₂	110·8	9·14	7½	9·18	12·11	12·070
Tungstic Acid	WO₃	118·8	19·23	16	19·60	6·177	6·061
Titanic Acid (Brookite)	TiO₂	40·33	9·77	8	9·8	4·128	4·115
Uranous Oxide	UO	68·0	9·8	8	9·8	6·94	6·938
Hydrated Uranic Oxide	U₂O₃+HO	153·0	25·82	21	25·72	5·926	5·971
Uranoso-Uranic Oxide	UO+U₂O₃	212·0	29·47	24	29·4	7·193	7·210
Suboxide of Lead	Pb₂O	215·2	22·02	18	22·05	9·772	9·759
Protoxide of Lead	PbO	111·6	12·07	10	12·25	9·246	9·110
Peroxide of Lead	PbO₂	119·6	13·51	11	13·4	8·851	8·878
Minium	PbO₂+2PbO	342·8	37·7	31	37·97	9·096	9·028
Suboxide of Mercury	Hg₂O	210·8	19·7	16	19·6	10·69	10·755
Peroxide of Mercury	HgO	109·8	9·83	8	9·8	11·164	11·204
Oxide of Silver	AgO	116·0	16·14	13	15·9	7·179	7·295
Potash	KO	47·15	17·75	14	17·15	2·656	2·749
Soda	NaO	31·3	11·10	9	11·02	2·805	2·843
Barytes	BaO	76·7	15·82	13	15·90	4·849	4·824
Peroxide of Barium	BaO₂	84·7	17·08	14	17·15	4·958	4·938
Strontia	SrO	51·8	13·2	11	13·47	3·932	3·845
Magnesia	MgO	20·7	6·6	5½	6·73	3·135	3·075
Lime	CaO	28·5	9·13	7½	9·18	3·12	3·104

Magnesia, MgO = 20·7.—Richter states the sp. gr. of this earth at 3·07; Karsten at 3·20. The mean, 3·135, gives 6·60 as the atomic volume.

Lime, CaO = 28·5.—Le Royer and Dumas describe the sp. gr.

of anhydrous lime as 3·08; Karsten states it at 3·16. The mean gives 3·12 as the sp. gr., 9·13 as the atomic volume.

In the preceding list of metallic oxides, we have given the specific gravity of all those the composition of which has been well ascertained. By tabulating the results, we have decisive proof that the number 1·225, or $\frac{1·225}{2}$, is the primitive volume of the metallic oxides.

There can no longer, therefore, be a doubt that either 1·225 or $\frac{1·225}{2}$ forms a submultiple for this large class of compounds. But it is not this fact, in its general enunciation, that is the principal point of interest in the table, for we learn from it at the same time that, in most of the oxides, 1·225 × 2 forms the combining volume of oxygen. ·In the class of magnesian metals we find sometimes the volume $\frac{1·225}{2}$, but only when the oxide has not been strongly heated. Are we then to adopt ·6125 as the standard volume? We still prefer using for the present 1·225 as the standard for comparison, especially as we conceive that the equivalents of the magnesian metals should be doubled. We know various hydrates of the magnesian oxides and their salts in which *two* equivalents of the oxide or salt combine with *one* equivalent of water. Instances of this are found in hydrated peroxide of manganese, and in Johnston's sulphate of lime. We know also that we require to use *two* equivalents of a magnesian oxide to substitute for *one* of an oxide of the potash family. From these reasons alone we have almost a right to demand an increase in the equivalents of the magnesian metals. But add to this the simplicity which would ensue were this done, and our right to demand the increase becomes much greater. This class of metals favours us with the following series of oxides :—

$$R_2O \text{——} RO \text{——} R_2O_3 \text{——} RO_2.$$

But the above series, besides being unnaturally complex, is not complete, as it wants the element R_3O_2. It would therefore be more rational, as well as more simple, to double the equivalents of the magnesian metals, in which case the series would be one of arithmetical progression :—

$$RO\text{——}RO_2\text{——}RO_3\text{——}RO_4.$$

Admitting this view, the division of 1·225 ceases to be necessary.

At present we cannot discuss with propriety the difference of volumes of an oxide before and after being heated ; but we are prepared to give a satisfactory account of this difference in the latter part of the paper. We must content ourselves at present with pointing out the general facts exhibited by the table under consideration.

In the first place, we are shown by it, that, in general, the oxygen in the oxides possesses the volume 1·225 × 2. Thus the protoxides of the magnesian metals in their densest state possess the volume 1·225 × 5. We have already seen that the magnesian radical has a volume 1·225 × 3 ; hence oxygen in the protoxides enjoys two unit volumes. Representing 1·225 by V, the following rule prevails for the protoxide :—

$$RO, V \times 5 = R, V \times 3 + O, V \times 2.$$

The same rule holds generally for all protoxides having the formula RO, the value of the oxide being the corresponding RV added to OV × 2, cadmium and tin being exceptions although in all probability CdO would cease to be an exception, if that oxide were examined after strong ignition. The RV must in all cases be taken in the state when it exhibits itself in its natural power uncontrolled by cohesion. The exception of StO, in which O is made to possess a unit vol., is obviously connected with the dimorphous relations of the metal. But although O in the oxides RO generally possesses two unit volumes, it does not do so in all the oxides : thus the second equivalent O in BaO_2, SnO_2, PbO_2 possesses only one unit volume. In the case of Cu_2O, Ni_2O_3, Mn_2O_3, &c., it is highly probable that the metal assumes a volume of V × 4 as in zinc. If so, the O in these compounds will still enjoy a volume of V × 2. Again, if in the case of ice we assume the volume of the O to be V × 2, the volume of H will be V × 6, which would exactly coincide with the volumes of the other magnesian metals, if, according to what we have already said, their equivalents were doubled. There are other points of interest in the table, the consideration of which we defer to another place.

Section III.

Sulphurets.

Sulphuret of Manganese, $MnS = 43\cdot8$.—The following estimations are given of manganese-glance :—

$$3\cdot95 \quad \text{Leonhard.}$$
$$4\cdot01 \quad \text{Leonhard.}$$
$$4\cdot014 \quad \text{Mohs.}$$

Mean . . . $3\cdot991$

This makes the atomic volume $10\cdot98$.

Subsulphuret of Iron, $Fe_2S = 72\cdot1$.—This sulphuret was made by passing hydrogen over FeO, SO_3, at a red heat. $19\cdot2$ grains gave an increase of $3\cdot31$; hence the sp. gr. is $5\cdot80$, and the atomic volume $12\cdot43$.

Sulphuret of Iron, $FeS = 44\cdot1$.—A specimen made by holding sulphur in contact with iron at a white heat, yielded the following result :—$38\cdot3$ grains gave an increase of $7\cdot59$, making sp. gr. $5\cdot046$. Another specimen, prepared by fusing iron with sulphur, was examined in like manner. $38\cdot3$ grains gave an increase of $7\cdot62$, making sp. gr. $5\cdot026$. The mean result, $5\cdot035$, gives an atomic volume $8\cdot75$.

Sesquisulphuret of Iron, $Fe_2S_3 = 104\cdot3$.— This sulphuret was made by passing HS over Fe_2O_3 with the usual precautions. $15\cdot16$ grains gave an increase in one experiment of $3\cdot62$, in another $3\cdot53$. The mean, $3\cdot57$, gives the sp. gr. $4\cdot246$, and the atomic volume $24\cdot56$.

Bisulphuret of Iron, $FeS_2 = 60\cdot2$.—Karsten describes the sp. gr. of this compound as 4·9, from which we deduce the volume 12·28.

Subsulphuret of Copper, $Cu_2S = 79·3$.—Mohs describes the sp. gr. of copper-glance as 5·695, which makes the atomic volume 13·93.

Sulphuret of Copper, $CuS = 47·7$. — Walchner states the sp. gr. of this sulphuret at 3·8 ; Karsten at 4·163; the mean gives sp. gr. 3·981, and atomic volume 11·98.

Sulphuret of Zinc, $ZnS = 48·3$.—Karsten has determined the sp. gr. of this sulphuret at 3·923, which gives 12·3 for the atomic volume.

Subsulphuret of Nickel, $Ni_2S = 74·1$.—This sulphuret was made by passing H over NiO, SO_3 at a red heat. 12·1 grains gave an increase of 2·0, making a sp. gr. 6·05, and atomic volume 12·24.

Bisulphuret of Cobalt, $CoS_2 = 61·7$.—This sulphuret, made by heating the oxide of cobalt with an excess of sulphur, gave the following results :—22·8 grains gave increase 5·31 in one experiment, and 5·37 in another; the mean, 5·34, gives sp. gr. 4·269, and atomic volume 14·45.

Sesquisulphuret of Chromium, $Cr_2S_3 = 104$.—This sulphuret was made by passing dry HS over Cr_2O_3 at an elevated temperature. 16·37 grains gave increase 4·0, making sp. gr. 4·092, and volume 25·41.

Sulphuret of Bismuth, $BiS = 87·07$.—According to Wehrle the sp. gr. is 7·807 ; to Herapath 7·591 ; to Karsten 7·000. The mean, 7·466, gives the atomic volume 11·66.

Sulphuret of Cadmium, $CdS = 71·8$.—The sp. gr. of Greinocket was found by Breithaupt to be 4·90; its atomic volume is therefore 14·65.

Sulphuret of Lead, $PbS = 119·7$. Karsten found its sp. gr. 7·50 ; Mohs 7·568 ; Leonhard 7·40 and 7·60 ; Brisson 7·587 ; and Musschenbroek 7·22. The mean, 7·479, gives the atomic volume 16·00.

Sesquisulphuret of Lead, $Pb_2S_3 = 255·4$.—This sulphuret was prepared by passing HS over Pb_2O_3 heated in the waterbath, the oxide being prepared as recommended by Winkelblech, by pouring chloride of soda into a solution of oxide of lead in potash. No separation of sulphur took place in the preparation. 16·6 grains gave increase 2·62, giving the sp. gr. 6·335, and atomic volume 40·32.

Sulphuret of Platinum, $PtS = 114·9$.—Bottger describes the sp. gr. of this sulphuret as 8·84, making the atomic volume 12·88.

Bisulphuret of Platinum, $PtS_2 = 131·0$.—Bottger's determination of 7·224 for the sp. gr. gives a volume of 18·1.

Sulphuret of Silver, $AgS = 124·$—Karsten's sp. gr. of 6·85 gives the volume 18·1.

Sulphuret of Tin, $SnS = 74·3$.—Boullay states the sp. gr. at 5·267; Karsten at 4·852. The mean, 5·059, gives a volume 14·67.

Bisulphuret of Tin, $SnS_2 = 90·0$.—According to Boullay, the sp. gr. is 4·415; according to Karsten 4·600. The mean, 4·507, gives the atomic volume of 19·97.

Sulphuret of Molybdenum, $MoS = 64$.—According to Mohs, the sp. gr. of this mineral is 4·59, which gives an atomic volume of 13·9.

Sulphuret of Arsenic, $As_2S_3 = 123·7$.—This compound has, according to Karsten, a sp. gr. of 3·459, making the atomic volume 35·7.

Sulphuret of Antimony, $Sb_2S_3 = 177·5$. — According to Mohs, this substance possesses the sp. gr. 4·62, and hence the volume must be 3·84.

The Table opposite, besides affording additional evidence of the law of multiple proportions, embraces other points of interest, to which we may briefly refer. We have already stated the volume of sulphur to be 7V. Testing therefore the results of the table by the hypothesis of the sulphur retaining that volume in its combinations, we find :—

1st. A class of sulphurets in which the volumes of both metal and sulphur remain the same as they are in the uncombined state. The following examples of this class are furnished by the table :—

Sulphuret of Copper. . . . . Cu, $3V + S$, $7V = CuS$, $10V$.
Sulphuret of Cadmium . . . Cd, $5V + S$, $7V = CdS$, $12V$.
Sulphuret of Platinum. . . . Pt, $4V + S$, $7V = PtS$, $11V$.
Sulphuret of Molybdenum . Mo,$4\frac{1}{2}V + S$,$7V = MoS$, $11\frac{1}{2}V$.

TABLE VI.—Showing the Volumes of certain Sulphurets.

Designation.			Volume.				
Name.	Formula.	Atomic weight.	Volume by experiment.	No. of 1·225 vols.	Volume by theory.	Sp. gr. by experiment.	Sp. gr. by theory.
Sulphuret of Manganese ..	MnS ..	43·8	10·98	9	11·02	3·991	3·974
Subsulphuret of Iron	Fe_2S ..	72·1	12·43	10	12·25	5·800	5·885
Sulphuret of Iron........	FeS ..	44·1	8·75	7	8·57	5·035	5·145
Sesquisulphuret of Iron ..	Fe_2S_3 ..	104·3	24·56	20	24·50	4·246	4·257
Bisulphuret of Iron	FeS_2 ..	60·2	12·28	10	12·25	4·900	4·914
Subsulphuret of Copper ..	Cu_2S ..	79·3	13·93	11	13·47	5·695	5·887
Sulphuret of Copper......	CuS ..	47·7	11·98	10	12·25	3·981	3·894
Sulphuret of Zinc........	ZnS ..	48·3	12·3	10	12·25	3·923	3·943
Subsulphuret of Nickel ..	Ni_2S ..	74·10	12·24	10	12·25	6·050	6·050
Bisulphuret of Cobalt	CoS_2 ..	61·7	14·45	12	14·7	4·269	4·197
Sesquisulphuret of Chromium	Cr_2S_3 ..	104	25·41	21	25·72	4·092	4·044
Sulphuret of Bismuth	BiS....	87·07	11·66	9½	11·63	7·466	7·485
Sulphuret of Cadmium .	CdS ..	71·8	14·65	12	14·7	4·900	4·884
Sulphuret of Lead	PbS ..	119·7	16·0	13	15·92	7·479	7·519
Sesquisulphuret of Lead ..	Pb_2S_3 ..	255·4	40·31	33	40·42	6·335	6·318
Sulphuret of Platinum....	PtS....	114·9	12·9	11	13·47	8·840	8·530
Bisulphuret of Platinum ..	PtS_2 ..	131	18·1	15	18·37	7·224	7·131
Sulphuret of Silver	AgS ..	124	18·1	15	18·37	6·850	6·750
Sulphuret of Tin	SnS....	74·3	14·67	12	14·7	5·059	5·054
Bisulphuret of Tin	SnS_2 ..	90	19·97	16	19·6	4·507	4·597
Sulphuret of Molybdenum.	MoS ..	64	13·9	11½	14·08	4·590	4·545
Sulphuret of Arsenic	AsS_3 ..	123·7	35·7	29	35·52	3·459	3·483
Sulphuret of Antimony ..	SbS_3 ..	177·5	38·4	31	37·97	4·620	4·674

2nd. A class in which the volumes of one or more of the atoms of metal or sulphur disappear, whilst the remaining atoms continue to possess the same volumes as when uncombined. Of this very remarkable class we have several examples, viz. :—

Subsulphuret of Iron . Fe, 3V × Fe,0V + S,7V = Fe_2S,10V.

Sulphuret of Iron . . . Fe, 0V + S, 7V = FeS, 7V.

Sesquisulphuret of Iron. Fe_2,6V + S,0V + S_2,14V = Fe_2S_3,20V.

Bisulphuret of Iron. . . Fe, 3V + S, 0V + S, 7V = FeS_2, 10V.

Magnetic Iron Pyrites. Fe_7, 0V + S_8, 56V = Fe_7S_8, 56V.

Subsulphuret of Nickel Ni, 3V + Ni, 0V + S, 7V = Ni_2S, 10V.

Sesquisulphuret of Chromium, Cr_2,0V + S_3,21V = Cr_2S_3,21V.

3rd. A class in which sulphur assumes the volume 6V. In his group we have included the bisulphuret of cobalt, in which the volume of the metal is merged.

Sulphuret of Manganese . Mn, $3V + S, 6V = MnS, 9V$.
Sulphuret of Zinc . . . . Zn, $4V + S, 6V = ZnS, 10V$.
Sulphuret of Silver. . . . Ag, $9V + S, 6V = AgS, 15V$.
Bisulphuret of Cobalt . . Co, $0V + S_2, 12V = CoS_2, 12V$.

There are several compounds included in the table which do not come under the above heads, and the uncertainty whether their observed volumes are owing to the reduction in the volume of the metal, or in that of the sulphur, prevents us from advancing anything very positive respecting them. There are, however, two substances, viz. the sulphuret and the sesquisulphuret of lead, in which Pb evidently assumes a volume of 6V, thus :—

$$Pb, 6V + S, 7V = PbS, 13V.$$
$$Pb_2 12V + S_3, 21V = Pb_2S_3, 33V.$$

Section IV.

Non-metallic Elements and their Oxides, and the Oxides of Metals of Unknown Specific Gravity.

Sulphur and Carbon.—For convenience these elements will be discussed in the following section.

Selenium, Se $= 39\cdot6$.—According to Berzelius, the sp. gr. varies from $4\cdot30$ to $4\cdot32$; Boullay found it exactly the mean, $4\cdot31$, which therefore we adopt as the specific gravity. This makes the atomic volume $9\cdot188$.

Phosphorus, P $= 31\cdot44$.—The specific gravity of phosphorus is $1\cdot77$, according to Berzelius; this gives the atomic volume $17\cdot76$.

Boron, B $= 10\cdot9$.—The sp. gr. of this element is generally stated as twice that of water, in which case its volume $= 5\cdot45$.

Iodine, I $= 126\cdot39$. — Gay-Lussac and Thenard describe the sp. gr. of this element as $4\cdot948$, in which case its atomic volume is $25\cdot54$.

Taking $2\cdot0$ as the sp. gr. of charcoal, and $2\cdot01$ as that for crystallized sulphur, for reasons afterwards to be stated, we have the following table of non-metallic elements.

TABLE VII.—Showing the Volumes of certain Non-metallic Elements.

Name.	Atomic weight.	Volume by experiment.	1·225 as standard.	Volume by theory.	Sp. gr. by experiment.	Sp. gr. by theory.
Carbon	12	6	5	6·12	2	1·96
Sulphur	16·03	7·97	6½	7·96	2·010	2·013
Selenium	39·6	9·188	7½	9·187	4·31	4·31
Phosphorus ..	31·44	17·76	14½	17·76	1·77	1·77
Boron	10·9	5·45	4½	5·51	2·00	1·978
Iodine	126·39	25·54	21	25·72	4·948	4·914

In the above table we have given the volumes of the elements without the reference to the power of cohesion which we made in a former part of the paper. But nevertheless the coincidence between the theoretical and experimental specific gravities is such that we are forced to admit either that the non-metallic elements are entirely free from the influence of cohesion, or that, as we have before hinted, the contraction thus occasioned exhibits itself in submultiples of the volume 1·225.

We have now to examine the volumes of certain oxides which were neglected in the previous sections, because we were ignorant of the specific gravity of their radicals.

Boracic Acid, $BO_3 = 34·9$.—The following estimations are given of the sp. gr. of vitreous boracic acid :—1·83 Berzelius, 1·803 Davy, 1·75 Breithaupt. Mean, 1·794, which gives the atomic volume 19·45.

Phosphoric Acid, $PO_5 = 71·44$.—This acid, in its vitreous state, has, according to Brisson, 2·387 sp. gr., and 29·74 atomic volume.

Silica, $SiO_3 = 46·22$.—Quartz has a sp. gr. of 2·66, according to the mean of many determinations. This gives the volume 17·37.

Alumina, $Al_2O_3 = 51·4$.—The ruby and sapphire possess sp. gr. 3·531, equal to an atomic volume of 14·56.

Oxide of Thorium, $ThO = 67·6$.—The sp. gr. of this oxide

is 9·402 according to Berzelius; hence its atomic volume will be 7·19.

Glucina, $G_2O_3 = 77$.—The sp. gr. of this earth is stated to be 3·0, in which case its volume is 25·66.

Yttria, $YO = 40·2$.—Ekeberg found the sp. gr. of this earth 4·842, which gives the volume 8·3.

Zirconia, $Zr_2O_3 = 91·4$.—This earth, according to Klaproth, has a sp. gr. 4·3, which is equivalent to an atomic volume of 21·2.

It must be borne in mind that considerable uncertainty exists with regard to the atomic weights of the last four bodies.

TABLE VIII.—Showing the Volume of Oxides of Elements of Unknown Specific Gravity.

Designation.			Volume.				
Name.	Formula.	Atomic weight.	Volume by experiment.	1·225 as standard.	Volume by theory.	Specific gravity by experiment.	Specific gravity by theory.
Boracic Acid ..	BO_3	34·9	19·45	16	19·6	1·794	1·781
Phosphoric Acid.	PO_5	71·44	29·74	24½	30·01	2·387	2·381
Silica..........	SiO_3	46·22	17·37	14	17·15	2·660	2·695
Alumina	Al_2O_3 ..	51·4	14·56	12	14·7	3·531	3·496
Thoria	ThO	67·6	7·19	6	7·35	9·402	9·197
Glucina........	G_2O_3	77	25·66	21	25·72	3·000	2·993
Yttria	YO	40·2	8·3	7	8·57	4·842	4·691
Zirconia.	Zr_2O_3	91·4	21·2	18	22·05	4·300	4·145

SECTION V.

Dimorphism and Polymorphism.

In Section II. we pointed out distinct differences in the volumes of oxides before and after being heated, and we showed that these differences were either 1·225 or $\frac{1\cdot225}{2}$. The question therefore naturally arose—Is there any such difference between the volumes of bodies of the same composition but of unlike forms? If any such difference could be established, a flood of light would be thrown upon this obscure subject, and the alteration in properties would become at once comprehensible. Let us take some of the most noted examples of polymorphism and dimorphism to test this view.

Example I.—Carbon offers three very unlike forms and conditions—the diamond, graphite, and charcoal. The sp. gr. of diamond varies from 3·4 to 3·5, but as an average may be taken at 3·45. Graphite enjoys a sp. gr.·of 2·5, according to Berzelius *. Karsten states it at 2·328; the mean gives 2·414. The purest charcoal obtained from alcohol has a sp. gr. of only 2·0.

Now, assuming from the considerations deduced by Regnault from his experiments on specific heat, that the equivalent of carbon, 6·12, should be doubled to 12·24, we have the following simple relations :—

| | No. of Vols. | | | | |
	Vol. by experiment.	1·225 as unity.	Vol. by theory.	Sp. gr. by experiment.	Sp. gr. by theory.
Alcohol-charcoal	6·00	5	6·12	2·0	1·96
Graphite	4·97	4	4·90	2·414	2·448
Diamond	3·48	3	3·67	3·25	3·27

The result exhibited in this table is very striking, as it shows that the three forms of carbon differ from each other by one unit in volume, viz. 3, 4, 5, so that we may conceive the diamond to pass into the graphite by expanding one unit volume, and graphite into charcoal by a similar expansion, or, *vice versâ*, by condensation.

Example II.—As a second example we may take calc spar and arragonite, dimorphous forms of carbonate of lime, of which the following table gives the recorded specific gravities :—

* Ann. der Chemie, Bd. xlix. S. 250.

Calc spar.		Arragonite.	
2·718	Brisson.	2·995	Breithaupt.
2·778	Baumgartner.	2·931	Mohs.
2·727	,,	2·946	Beudant.
2·6987	Karsten.		
2·7064	,,	Mean . . 2·957	
2·7230	Beudant.		

Mean . . 2·725

Calculating the volumes on the equivalent 50·5, we have the following result :—

	Vol. by experiment.	1·225 as unity.	Vol. by theory.	Sp. gr. by experiment.	Sp. gr. by theory.
Calc spar . . .	18·53	15	18·37	2·725	2·749
Arragonite . .	17·08	14	17·15	2·957	2·945

Thus the same alteration of volume is exhibited here as in the previous example, the difference in this case also being one unit volume, or 1·225. The experimental and theoretical results are sufficiently near when calculated on the mean specific gravity, but actually accord if we assume that usually adopted for calc spar, viz. 2·75 according to Neumann, and 2·946 for arragonite according to Beudant.

Example III.—A notable instance of dimorphism is exhibited in iron pyrites, viz. in cubic iron pyrites, and in white pyrites or cockscomb spar.

According to Karsten, cubic iron pyrites has the sp. gr. 4·90, and Mohs states that of cockscomb pyrites to be 4·678. The volume calculated on the atomic weight is as follows :—

	Vol. by exp.	No. of vols.	Vol. by theory.
Cubic pyrites	12·28	10	12·25
Cockscomb pyrites .	12·87	10½	12·86

The difference in this case is therefore $\frac{1·225}{2}$; but if our view be correct, that the magnesian equivalents should be doubled, the number of volumes will be actually 20 for iron pyrites and 21 for cockscomb pyrites, making the difference of one unit volume or 1·225, as in the previous cases.

Example IV.—The sulphurets offer examples of unlike forms and allotropic conditions of the same substance. Black sulphuret of mercury has, according to our own experiments, a sp. gr. of 7·331; whilst, according to Musschenbroek and others, cinnabar enjoys a sp. gr. of 8·00. The volume calculated on the atomic weight, 117, is as follows : —

	Volume by experiment.	No. of unit volumes.	Vol. by theory.
Red sulphuret of mercury . .	14·6	12	14·7
Black sulphuret of mercury .	15·95	13	15·9

The difference in this case also is one unit volume, or 1·225.

A relation is also shown between the native sesquisulphuret of antimony, with a sp. gr. of 4·62 according to Mohs, and of amorphous kermes, with sp. gr. 4·15 according to Gmelin.

	Vol. by experiment.	No. of unit volumes.	Volume by theory.
Native Sb_2S_3 . . .	$\frac{1·775}{4·62} = 38·42$	$31\frac{1}{2}$	38·59
Kermes „ . . .	$\frac{177·5}{4·15} = 42·77$	35	42·87

the difference in this case being $3\frac{1}{2}$ volumes.

Example V.—The researches of Rose have thrown much light on the difference in specific gravity between the varieties of titanic acid Thus we have :—

	Sp. gr.	
Artificial TiO_2, . .	3·66 Rose.	
Anatase	3·912 Rose.	⎫
	3·927 Rose.	⎪
	3·829 Mohs.	⎬ 3·85
	3·759 Breithaupt.	⎪
	3·826 Mohs.	⎭
Brookite	4·128 Rose.	
Rutil	4·256 Rose.	

These differences are very notable; and when calculated on

M 2

the generally received atom of titanic acid, 40·33, viewing it as TiO_2, we obtain the following relations :—

	Vol. by exp.	No. of vols.	Vol. by theory.
Artificial TiO_2	11·01	9	11·02
Anatase	10·48	$8\frac{1}{2}$	10·41
Brookite	9·77	8	9·80
Rutile	9·47	$7\frac{3}{4}$	9·49

The relations here are very striking, but at the same time give rise to new questions. We have previously seen re-peated examples of a volume represented by $\frac{1·225}{2}$, but this is the first instance of $\frac{1·225}{4}$. Are we to admit the subdivision of the volume 1·225 ? It may be necessary to do so, but not on an isolated case like the present. May we not rather suppose that the contraction of volumes takes place on associated atoms ? Thus :—

1 eq. TiO_2 contracting 1 unit volume forms Brookite.
4   „   „   1 unit volume   „   Rutile.
2   „   „   5 unit volumes „   Anatase.

Which of these hypotheses is right it is impossible to say in the present state of our knowledge; but we do not feel warranted in admitting a further division of 1·225 on a single case, although we by no means deny that such division may afterwards become necessary, especially as far as regards $\frac{1·225}{2}$.

Example VI.—*Quartz and Opal.*—Silica presents two well-defined forms in quartz and anhydrous opal, the sp. gr. of the former being 2·66, while the anhydrous opal analyzed by Klaproth possesses the sp. gr. 2·072.

	Vol. by exp.	1·225 as unity.	Theory.
Quartz . .	$\frac{46·22}{2·66} = 17·3$	14	17·15
Opal . . .	$\frac{46·22}{2·072} = 22·3$	18	22·05

The difference in this case is 4 .

Example VII.—*Alumina* forms a similar instance :—

	Vol. by exp.	1·225 as unity.	Theory.
Ruby	$\frac{51\cdot4}{3\cdot531}=14\cdot56$	12	14·70
Ignited alumina	$\frac{51\cdot4}{4\cdot15}=12\cdot38$	10	12·25

The difference is 1·225 × 2. Numerous other instances of this uniformity of difference might be cited if necessary.

Example VIII.—*Elementary Allotropism.*—In the metals we have several well-marked instances of elementary allotropism. Thus Berzelius, in his memoir on allotropism, has pointed out that iridium artificially prepared has a sp. gr. not exceeding 16·0, while the native metal enjoys a sp. gr. approaching to that of platinum.

The most uniform results for native metallic iridium are as follows :—

$$19\cdot5 \quad \text{Mohs.}$$
$$18\cdot68 \quad \text{Children.}$$
$$22\cdot1 \quad \text{Breithaupt.}$$

Mean . . . 20·09

Hence we have for native iridium $\frac{98\cdot84}{20\cdot09}=4\cdot92$; and for the metal as obtained by reduction, $\frac{98\cdot84}{16\cdot0}=6\cdot18$.

	Exp.	No. of vols.	Theory.
Native iridium	4·92	4	4·90
Reduced iridium	6·18	5	6·12

In this case, therefore, we have the variation of one unit volume between the two different states of the metal.

Example IX.—*Osmium* presents a similar uniformity in the difference between its three states. Thus, according to Thenard, wrought osmium has a sp. gr. of 19·5, while that reduced from the oxide has a sp. gr. of only 10·0 according to Berzelius, and when obtained by heating its amalgam, 7·0.

	Exp.	No. of vols.	Theory.
$\frac{99\cdot97}{19\cdot5}=$	5·12	4	4·9
$\frac{99\cdot97}{7\cdot0}=$	14·28	12	14·7
$\frac{99\cdot97}{10\cdot0}=$	9·99	8	9·8

Example X.—*Platinum* exhibits several very decided allo-tropic differences. Thus in its densest state it has sp. gr. 23·5 (Cloud) ; after fusion the sp. gr. is 20·3 (Brisson) ; the powder obtained by heating the oxide enjoys sp. gr. 17·76 (P. and J.), while platinum black has a sp. gr. 16·557 (Liebig).

	Exp.	No. of vols.	Theory.
Platinum black . . .	$\frac{98\cdot8}{16\cdot557}=5\cdot97$	5	6·12
Platinum powder . .	$\frac{98\cdot8}{17\cdot76}=5\cdot56$	$4\frac{1}{2}$	5·51
Platinum after fusion	$\frac{98\cdot8}{20\cdot3}=4\cdot86$	4	4·9
Platinum hammered.	$\frac{98\cdot8}{23\cdot5}=4\cdot24$	$3\frac{1}{2}$	4·28

These numbers, and those deduced from the allotropic conditions of osmium and iridium, offer much confirmation of the view which we are inclined to entertain, that the attraction of cohesion is itself governed by the volume 1·225. Another instance of this contraction is seen in palladium, which in its usual state possesses sp. gr. 10·923, but when hammered attains sp. gr. 12·14.

	Exp.	No. of vols.	Theory.
	$\frac{53\cdot3}{10\cdot923}=4\cdot88$	4	4·9
	$\frac{53\cdot3}{12\cdot14}=4\cdot39$	$3\frac{1}{2}$	4·29

The differences between the theoretical and experimental numbers are not greater than can be accounted for by the uncertain atomic weight. The very numerous instances of a change of volume consequent on the combination of the elements may all be brought to reinforce this part of our argument. For instance, in the case of lead uncombined we find a volume 8V ; but when combined with sulphur a volume of 6V. Nay more, it appears from our former as well as the present paper, that the volume of an element is frequently actually merged. Such facts lead us to hope for the dis-covery of far more wonderful examples of allotropism than those we have already given.

Berzelius has thrown out the suggestion that the presence of the different forms of the radical may account for allotropism in their compounds, and our researches go far to prove this sagacious view. We had intended to have devoted ourselves to further proofs of this theory, but our memoir has already reached an unreasonable length and we must defer the subject.

We have, however, one other notable instance of allotropism to which we must refer. SULPHUR exists in several distinct states : in the first state, that of native sulphur, we have the following determinations :—

2·033   Brisson.
2·050   Karsten.
1·989   Karsten.
1·898   Fontenelle.
2·072   Mohs.
2·027   Osann.
2·066   Marchand and Scheerer.

Mean . . . 2·019

This result agrees very closely with our own determination, 2·010 (see Section 1).

The specific gravity of waxy sulphur, procured by pouring viscid melted sulphur into water, is 1·959 according to Marchand and Scheerer; according to our own experiments 1·921; the mean being 1·940. The other estimation by ourselves we have described in a previous section, and we now find the following relations :—

Description.	Volume by experiment.	No. of 1·225 vols.	Volume by theory.	Specific gravity by experiment.
Sulphur in viscid melted state ..	9·17	7½	9·18	1·748
Sulphur in clear amber liquid ..	8·87	7¼	8·88	1·807
Sulphur in flowers	8·38	7	8·57	1·913
Sulphur in soft state...........	8·26	6¾	8·27	1·940
Sulphur in crystals from fusion ..	7·97	6	7·96	2·010

The results of the above table afford the most powerful argument for the assumption of a volume $\frac{1·225}{4}$ that we have yet seen. But still we adhere to our primitive volume 1·225, believing that a contraction of whole volumes of 1·225 may take place on associated atoms, as we alleged in the case of titanic acid.

TABLE IX.—Showing the Simple Relation in Volumes between Unlike Forms of the Same Body.

Name.	Formula.	Atomic weight.	Volume by experiment.	1·225 taken as standard.	Volume by theory.	Specific gravity by experiment.	Specific gravity by theory.
Charcoal			6·12	5	6·12	2·0	2·0
Graphite	C_2	12·24	4·90	4	4·90	2·5	2·5
Diamond			3·50	3	3·67	3·5	3·33
Calc spar	CaO, CO_2	50·5	18·53	15	18·37	2·725	2·749
Arragonite			17·08	14	17·15	2·957	2·945
Cockscomb Pyrites	FeS_2	60·6	12·87	10½	12·86	4·678	4·664
Cubic Pyrites			12·28	10	12·25	4·90	4·897
Black Sulph. Mercury	HgS	117	15·95	13	15·9	7·331	7·358
Cinnabar			14·6	12	14·7	8·001	7·959
Artificial Kermes	Sb_2S_3	177·5	42·77	35	42·87	4·150	4·140
Native Kermes			38·42	31½	38·59	4·620	4·600
Titanic Acid			11·01	9	11·02	3·660	3·659
Anatase	TiO_2	40·33	10·48	8½	10·41	3·850	3·874
Brookite			9·77	8	9·8	4·128	4·115
Rutil			9·47	7¾	9·49	4·256	4·249
Anhydrous Opal	SiO_3	46·22	22·3	18	22·05	2·072	2·096
Quartz			17·3	14	17·15	2·660	2·695
Ruby	Al_2O_3	51·4	14·56	12	14·70	3·531	3·496
Ignited Alumina			12·38	10	12·25	4·150	4·195
Reduced Iridium	Ir	98·84	6·18	5	6·12	16·0	16·17
Native Iridium			4·99	4	4·90	20·09	20·17
Osmium from amalgam.			14·28	12	14·70	7·0	6·721
Reduced Osmium	Os	99·97	9·92	8	9·80	10·0	10·20
Native Osmium			5·12	4	4·90	19·5	20·40
Platinum black			5·97	5	6·12	16·557	16·143
Platinum powder			5·56	4½	5·51	17·760	17·931
Platinum after fusion	Pt	98·8	4·86	4	4·90	20·300	20·163
Platinum hammered			4·20	3½	4·28	23·500	23·084
Sulphur in viscid melted state			9·17	7½	9·18	1·748	1·746
Sulphur in clear amber melted			8·87	7¼	8·88	1·807	1·805
Sulphur in flowers of	S	16·03	8·38	7	8·57	1·913	1·870
Sulphur in soft state			8·26	6¾	8·27	1·940	1·938
Sulphur in crystals from fusion			7·97	6½	7·96	2·010	2·013

CONCLUSION.

We have now to sum up the results of the previous researches, which we announce generally under the following proposition :—

I. *The volumes of solid bodies bear a simple relation to each other, being multiples of a submultiple of the volume of ice, which for convenience is adopted as a standard.*

a. The force of cohesion in the metals prevents the assumption of their natural volumes, which, however, appear when they are placed in a position not to yield to the solicitation of cohesion.

b. It is probable that the differences from the natural volume produced by cohesion can be expressed by a submultiple of the volume of ice.

c. The differences between unlike forms of dimorphous, polymorphous, and allotropic substances are expressed by a submultiple of the volume of ice.

The results to which we were led by our previous researches now become intelligible. We then asserted that the volumes of salts were multiples of 9·8 or of 11. We guarded ourselves against stating that the latter number was absolutely 11, but we averred it to be " a number very nearly approaching the number 11." We now know exactly what that number is, and see its relation to 9·8 or the volume of ice :—

$$9·8 + \frac{9·8}{8} \quad 11·025,$$

a number very nearly approaching 11. We have shown in this memoir that 9·8 is composed of 1·225 × 8, and in the same way 11 is 1·225 × 9. That we should have assumed 11 as the unit volume for salts when we were ignorant of its root, is not surprising, considering its frequent occurrence in salts.

Thus if RV be the volume of any radical, OV that of oxygen, and AV of an acid, the general formula RV + OV + AV = XV will apply to a large class of salts divisible by the number 11, although the latter number itself is a multiple of 1·225. The magnesian sulphates have a volume of 22 according to

our former paper. By the results of this paper, the volume
of sulphur in its solid state, as flowers of sulphur, is 8·57.
Oxygen we found in most cases to have a volume of 1·225 × 2;
hence SO_3 will be $SV + OV × 3 = 15·92$. The magnesian
metals were shown to have a volume of 1·225 × 3; hence
their oxides, as shown also by experiment, have a volume
$= RV + OV = 6·12$.

$$RO \ldots \ldots 6·12$$
$$SO_3 \ldots \ldots 15·92$$
$$\overline{}$$
$$RO, SO_3 . 22·04$$

So that we are conducted by entirely different considerations
to the same result which we gave in Table VII. of our former
memoir. We stated at Table VI. of that series of researches
that KO, SO_3 had the volume of 33·05. By Table V. of the
present paper KO is shown to have a volume of 17·15. SO_3
we have just shown to have 15·92.

$$KO \ldots \ldots 17·15$$
$$SO_3 \ldots \ldots 15·92$$
$$\overline{}$$
$$KO, SO_3. . 33·07$$

a result almost identical with that given, and warranting the
numbers 11 × 3. We pointed out at the same time that
NaO, SO_3 differed from this law, but the apparent exception
is now explained. Our sp. gr. for this salt, 2·597, gave the
volume 27·5; Karsten's result then referred to, 2·631, gave
the volume of 27·14. Now by Table V.,

$$NaO \ldots \ldots 11·025$$
$$SO_3 \ldots \ldots 15·920$$
$$\overline{}$$
$$26·945$$

The curious class of chromates find now a complete
elucidation. Chromium possesses an atomic volume 4·9
(Table I.), while O in the acids occasionally has a volume
of 4·9,

$$CrV + OV × 3 = 19·6,$$

the exact number which we found by experiment for CrO_3 in our former paper. KO, CrO_3 ought to be as follows :—

$$KO \ . \ . \ . \ . \ . \ 17\cdot15$$
$$CrO_3 \ . \ . \ . \ . \ 19\cdot60$$
$$\overline{}$$
$$KO, CrO_3 \ . \ . \ 36\cdot75$$

the number which we found being 37·1, a result we then allowed as difficult to be explained. On $KO, 2CrO_3$ we find an increase of 1·225 above the second atom of CrO_3, due perhaps to the dimorphous relations of chromium. The results as then found, and those calculated according to our advanced knowledge in the present paper, are as follow * :—

	Vol. found.	Theory.
KO, CrO_3	37·1	36·75
$KO, 2CrO_3$ . . .	57·8	57·57
$KO, 3CrO_3$ . . .	76·8	77·17

Finally, let us take the singular results obtained for the carbonates as a test of this view. We found a sp. gr. for KO, CO_2 of 2·103, which is too low a number when compared with the only other recorded result, that of Gmelin, 2·264. The latter gives the atomic volume $\frac{69\cdot4}{2\cdot264} = 30\cdot65$. Deducting 17·15 for KO, we have a volume of 13·5 for CO_2. The volume of $KO, CO_2 + HO, CO_2$, according to Gmelin's results and our own, was 49·0, from which 30·65 must be deducted for KO, CO_2, and 4·9, as in the case of many hydrates, for combined water. The difference, 13·45, is the volume of CO_2 in bicarbonate of potash. Carbonate of soda we found to possess a sp. gr. 2·427, which agrees very well with Karsten's result, 2·465. From the resulting volume, 22·0V, must be deducted 11V for the volume of soda, leaving 11V for CO_2. $NaO, CO_2 + HO, CO_2$ had a volume of 38·6, from

---

* We avail ourselves of this opportunity to correct an error into which we fell in our former paper. We there stated that $KO, 2CrO_3$ was not reduced to KO, CrO_3 by litharge, but formed a compound $2KO, 3CrO_3$. By continued action, however, the bichromate is wholly reduced to yellow chromate, and we are therefore satisfied that what we described as $2KO, 3CrO_3$, was merely a mixture of KO, CrO_3 and $KO, 2CrO_3$.

which must be deducted $22.05 + 4.9$, for reasons above stated, leaving 11.65 for CO_2.   Here, then, we have,

> CO_2 in the carbonate of potash $= 13.45$.
> CO_2 in the carbonate of soda $= 11.65$.

These differences are therefore connected with the character of the base itself.   Are the differences due to the allotropic condition of carbon?   Does CO_2 in the potash-salts contain charcoal carbon, and in the soda-salts graphite carbon? Does the denser form cause the volume of oxygen itself to become condensed? And may we thus explain the differences of calc spar and arragonite?   These are questions which shall be answered in a future memoir.

At present the examples above given may suffice for illustrations of the accuracy of the standards which we took in our former paper.   We reserve for a distinct memoir the important considerations which flow from our present results with regard to the constitution of salts, and the behaviour of the water entering into their constitution.

It would be easy to give a classification of the metals from the views developed in the preceding sections, were it not that their allotropic conditions associate the groups by a distinct chain of connection.   We intend to follow up this paper with another, in which we shall endeavour to show that the elements, not only in their natural states, but even in their unusual forms, may be brought under one simple mathematical law, and thus clearly proving that sufficient grounds exist for the acceptance of the law which we have tried to establish in the present memoir,—THAT THE ATOMIC VOLUMES OF BODIES STAND IN A SIMPLE MULTIPLE RELATION TO EACH OTHER.

Researches on Atomic Volume and Specific Gravity.
By JAMES P. JOULE, *Esq., and* Dr. LYON PLAYFAIR.

['Memoirs of the Chemical Society,' vol. iii. p. 199.]

SERIES III.—*On the Maximum Density of Water.*

IN all researches on specific gravity the selection of a proper standard of comparison is a matter of great importance. For obvious reasons water has been universally selected as this standard; but a diversity of opinion has been entertained as to the temperature at which the gravity of water should be called unity. Hence, whilst our continental neighbours have adopted the freezing temperature, our own countrymen have generally chosen the temperature of 60° Fahr. Water at 60° cannot be a desirable standard on account of its high rate of expansion at that temperature; and for the same and other reasons, the temperature of 32° is not at all more convenient. We conceive that it would be much more philosophical, and lead to many practical advantages, if water at its maximum density were taken as the unit *. In that case, calculations would often be greatly facilitated; and if the temperature of the water should happen to be a degree under or over the maximum point in any experiment, the extreme slowness of the expansion would avert the possibility of a grave error.

Intending to give still greater accuracy to our future experiments, we deem the present a favourable opportunity for fixing upon a good unit of comparison. We propose therefore to occupy the present series with a brief discussion as to the point of temperature at which water arrives at its state of greatest density.

Although a variety of methods have been employed in the investigation of this subject, they may be classified under two general heads. The first of these embraces all the methods involving the necessity of an accurate acquaintance

* Since this paper was communicated to the Society we have seen the *Annuaire du Bureau des Longitudes* for 1845, in which, under the high authority of Arago, water at its point of maximum density is assumed as the unit for comparison.

with the rate of expansion of some solid body by heat. The inquiry, when made in this way, is one of great difficulty, and has occupied the attention of many of our most accurate experimenters. The Florentine Academicians, Croune, Deluc, Dalton, and others, compared the indications of a thermometer filled with water with one filled with mercury; Lefevre, Gineau, and Hallström weighed a solid body in water at different temperatures; and Blagden and Gilpin measured the variation of volume by the variation of the weight of water contained in the same vessel at different temperatures.

The other general method does not require a previous acquaintance with the expansion of a solid by heat: it consists virtually in weighing water in water—the heavier water descending, while the lighter ascends to replace it. This principle was introduced by Dr. Hope, who applied it in the following elegant manner :—He filled with water at different temperatures tall glass jars having thermometers at top, middle, and bottom. In this way he observed that when water was cooled down to 40° at the surface, it sank to the bottom; and when cooled below 40° at the bottom, it rose again to the surface. Trallés, Count Rumford, Ekstrand, and Despretz have repeated Hope's experiment with a similar apparatus.

Believing that the second general method is susceptible of a far greater degree of accuracy than the first, we at once determined to employ it in our own experiments. The particular apparatus of Dr. Hope did not, however, appear to us to present the method in a form calculated to give results of great accuracy; and hence we have found it necessary to devise a new instrument combining all the theoretical advantages with the requisite facilities for exact observation.

Our instrument is represented in fig. 6, drawn to a scale of $\frac{1}{24}$ the real size. $a\ a$ are two upright vessels of tinned iron, each $4\frac{1}{2}$ feet high and 6 inches in diameter : they are connected at the bottom by means of a brass pipe b furnished with an accurately wrought stopcock. This pipe is altogether 6 inches long, and enters 1 inch within each vessel. When the stopcock is opened, a clear passage of 1 inch

diameter throughout forms a communication
between the vessels. A rectangular trough
of tinned iron, c, 6 inches long, 1 inch broad,
and 1 inch deep, forms a communication
between the tops of the vessels. In the
middle of this trough there is a slide, by
means of which the motion of a current
along the trough can be stopped when requisite.

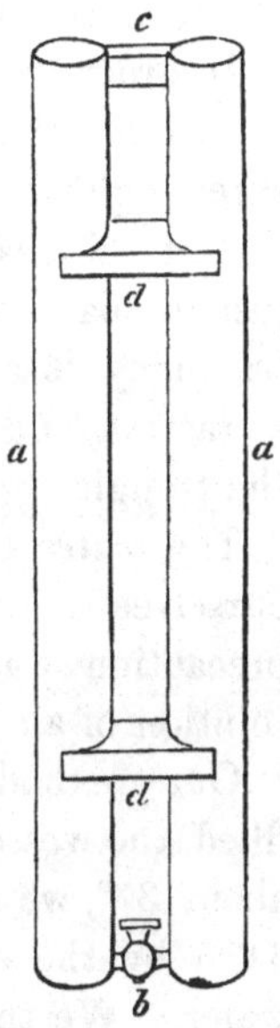

The vessels were supported in two places
by means of the wooden brackets d, d; and
in order to prevent the greater than desired
effect of the atmosphere in raising or de-
pressing their temperature, they were completely covered with hay-bands. During the
experiments the instrument was placed upon
a tripod stool resting upon a support quite
independent of the floor of the laboratory,
in order to keep it entirely free from vibra-
tion.

Now if the two vessels be filled with water and made to
communicate with one another by opening the stopcock and
removing the slide, it is evident that a current will tend to
flow through the trough connecting the tops of the vessels,
if the density of the water in one of the vessels be in the
least degree greater than that of the water in the other
vessel. Although the changes in density are very minute
near the maximum point, the extreme mobility of fluids led
us to expect that we might in this way arrive at an exact and
incontrovertible result.

The *thermometers* employed by us were of extreme accuracy,
having been calibrated throughout their whole length, and
their delicacy was such as to indicate a change of temperature
considerably less than $\frac{1}{100}$ of a degree Fahrenheit. The
freezing-points of the thermometers were carefully determined
within a few hours of the experiments. Each vessel was
furnished with a *stirrer*, consisting of a disk of tinned iron
4 inches in diameter attached to the end of a slender rod of

Fig. 6.

iron, by means of which the water was thoroughly stirred before each determination of temperature.

In order to measure the motion of the water in the trough connecting the tops of the vessels, a hollow glass ball of about three eighths of an inch diameter was placed in it. The weight of this glass ball was carefully adjusted so as only just to float : a matter of great importance, as the slightest buoyancy is accompanied by a certain degree of capillary attraction, and makes the ball liable to adhere to the sides of the trough.

The *water* employed in the experiments was distilled by ourselves in clean vessels of tinned iron; and the additional precaution was taken to prevent, as far as possible, the solution of air.

Our method of experimenting was as follows :—Having filled the vessels with distilled water at a temperature of about 37°, we increased the temperature of one of them to 41°·5 by the addition of a small quantity of hot distilled water. We then placed two of the delicate thermometers upon a proper stand, so that their bulbs dipped in the water to the depth of 6 inches. Having then closed the stopcock and adjusted the slide, we stirred the water in each vessel thoroughly, and noted the temperatures indicated by the thermometers. The stopcock was then opened, and the slide carefully removed from the trough. After waiting three minutes the glass ball was put into the trough, and its motion watched for two or three minutes with the help of a graduated rule placed at the top of the trough. In conclusion, the stopcock was again turned, the slide readjusted, the water stirred, and the temperatures again noted. The mean of the temperatures thus observed before and after each trial of the velocity of the current was taken as the temperature of the observation.

The following table contains the results of a series of observations taken in the above manner. The temperature of the laboratory being about 38°, the water in the warmer vessel cooled down more rapidly than the water in the other vessel increased in temperature; and therefore after two or three hours had elapsed, the water in the cooler vessel was found to have acquired greater buoyancy than that in the warmer vessel, although at the commencement of the experiments the current indicated a greater degree of buoyancy in the warmer water.

SERIES 1.

Temperature of the water in the warmer vessel.	Temperature of the water in the cooler vessel.	Mean of the temperatures of the two vessels.	Velocity of the current in inches per hour.
41·183	37·348	39·265	280 from the warmer vessel.
41·129	37·368	39·248	240 „ „
40·959	37·363	39·161	20 „ „
40·905	37·368	39·136	8 „ „
40·711	37·317	39·014	40 from the cooler vessel.

We could hardly have anticipated more satisfactory results than those of the above table. They show clearly that while water at a temperature of 40°·905 is lighter than water at a temperature of 37°·368, water at 40°·711 is heavier than water at 37°·317 ; in other words, that 39°·136 is above, whilst 39°·014 is below the maximum point. By drawing a curve from the results, we find that the exact point of maximum density indicated by the above series of observations is 39°·102.

During the next series of observations the temperature of the laboratory was about 41°, which occasioned a gradual increase of the temperature of both vessels.

SERIES 2.

Temperature of the water in the warmer vessel.	Temperature of the water in the cooler vessel.	Mean of the temperatures of the two vessels.	Velocity of the current in inches per hour.
40·742	37·368	39·055	22 from the cooler vessel.
40·758	37·420	39·089	8 from the warmer vessel.
40·773	37·470	39·121	60 „ „

The point of maximum density indicated by this second series of observations is 39°·078.

SERIES 3.

Temperature of the water in the warmer vessel.	Temperature of the water in the cooler vessel.	Mean of the temperatures of the two vessels.	Velocity of the current in inches per hour.
40°·332	37°·633	38°·982	70 from the cooler vessel.
40·402	37·682	39·042	80 ,, ,,
40·425	37·709	39·067	60 ,, ,,
40·440	37·745	39·092	8 ,, ,,
40·448	37·791	39·120	30 ,, ,,
40·467	37·837	39·152	12 from the warmer vessel.
40·483	37·873	39·178	30 ,, ,,

The point of maximum density indicated by the above series will be situated at about 39°·134.

We now proceeded to apply a severer test to our method. In the next series of experiments we arranged matters so that the temperature of the water in one vessel was only a degree and a half higher than that of the other vessel. The expansion of water increasing as the square of the temperature from that of the maximum density, it was obvious that the current in the trough would be much more feeble than in the former experiments. We therefore allowed the vessels to be in perfect repose for six minutes before we introduced the glass ball, and we afterwards watched its motion for 4 or 5 minutes.

SERIES 4.

Temperature of the water in the warmer vessel.	Temperature of the water in the cooler vessel.	Mean of the temperatures of the two vessels.	Velocity of the current in inches per hour.
39°·921	38°·382	39°·151	30 from the warmer vessel.
39·864	38·398	39·131	0.
39·821	38·362	39·091	0.
39·782	38·332	39·057	2½ from the cooler vessel.

The position of the point of maximum density according to the above fourth series of experiments will be at 39°·091.

I. Point of maximum density of pure water . . 39°·102
II.   ,,   ,,   ,,   . . 39°·078
III.   ,,   ,,   ,,   . . 39°·134
IV.   ,,   ,,   ,,   . . 39°·091

Mean . . . 39°·101

Although in the different series of observations there are several irregular results, there is on the whole sufficient consistency among them to enable us to receive 39°·1, the mean of the four sets of observations, as the actual point of maximum density. We think it highly probable that this temperature is within one-hundredth of a degree of the truth : it certainly cannot be more than one-twentieth of a degree in error. We were prevented by the mildness of the season from extending the experiments further, but we doubt not that by repeating them more frequently we should be able to bring the determination of the point to any required degree of accuracy. The result arrived at by Despretz from a very extensive series of experiments, with an apparatus similar to that employed by Hope, is 39°·176 *, which agrees very well with our determination. But other results, such as those of Hallström 39°·38, Blagden and Gilpin 39°, Hope 39°·5, Deluc 41°, Lefebvre Gineau 40°, Dalton 38°, Rumford 38°·8, Muncke 38°·804, Stampfer 38°·75, &c., show by their discordance with one another, and their disagreement with our result, the little dependence which can in general be placed on the results of former methods.

We believe that our new method may be applied with great advantage to a variety of interesting problems. One of the most important of these applications is the determination of the dilatation of glass bulbs by heat, which, though formerly presenting great practical difficulties, can now be accomplished in the most simple and decisive manner. The bulb has only to be filled with pure water and reduced successively to two

* 'Annales de Chimie,' 1839, t. lxx. p. 45.

temperatures, one as much above as the other is below the
point of maximum density : the rise of the liquid in the stem
of course indicates the contraction of the glass in passing
from the higher to the lower temperature. The expansion
of the glass bulbs being thus accurately ascertained, they
may be advantageously applied in determining the dilatations
of solutions and other liquids.

---

*Researches on Atomic Volume and Specific Gravity.
By* JAMES P. JOULE, *Esq., Corresponding Member of
the Royal Academy of Sciences, Turin, and* LYON
PLAYFAIR, *F.R.S., of the Museum of Practical
Geology.*

['Journal of the Chemical Society,' vol. i. p. 121.]

SERIES IV —*Expansion by Heat of Salts in the Solid State.*

IN pursuing our researches on atomic volume and specific
gravity, we have thought it desirable, as has been already
intimated in a previous memoir, to ascertain the expansion
of the salts by heat, as well as their volume at a given
temperature.   In this way we hope to arrive ultimately at
the solution of the apparent discrepancies between theory
and experiment.   We are not aware that any experiments of
this kind have hitherto been made.   Brunner has indeed
determined the expansion of ice; and we are led to expect
that Pierre, who has already given valuable results on the
expansion of liquids, will extend his labours to solid salts.
We hope, therefore, that our own results, as detailed in the
present memoir, will be speedily confirmed. The expansion
of solids by heat is a subject which, although little cultivated
hitherto, is of very great importance to science, and will
require, in all probability, the labours of many experimenters
for its complete development.   It will not, therefore, be
expected that we shall be able to include the expansions of

the whole range of known inorganic salts in one memoir; our object being simply to examine a sufficient number of them, in order to throw light on the causes which produce variations in the sp. gr. of bodies, and thus enable us to confirm or correct our views on atomic volumes.

It was only after much consideration and some preliminary trials, that we were enabled to select what appeared to us an unexceptionable method of conducting the experiments. The first form of apparatus which suggested itself to us, consisted of a glass volumenometer (fig. 7), in which *a* is a bulb of two

Fig. 7.   ⅓ actual size.

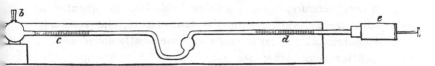

and a half cubic inches capacity, having a neck fitted with a perforated glass stopper *b*. A graduated tube *c d*, of small diameter, was attached to this bulb, having at the centre a smaller bulb, of one cubic inch capacity. A syringe *e* was attached tightly to the extremity, with the exception of which the whole was immersed in a large vessel containing water. The method of experimenting with this apparatus was as follows :—The bulb *a* was filled with turpentine or any liquid incapable of dissolving the salt, and the exact quantity was noted on the graduated stem at *c*. The piston of the syringe was then gradually withdrawn, so as to draw the liquid nearly to the end of the tube *d*, entirely filling the intermediate bulb. A known quantity of the salt was then thrown into the bulb *a*, and the stopper being readjusted, the liquid was driven back by the syringe till it ascended into the perforation of the stopper. By examining the position of the liquid in the graduated tube, the space occupied by the salt was rendered evident. By conducting experiments with a salt in the above manner, at different temperatures, its expansion could be made evident, regard being of course paid to the expansion of the glass tube.

Several trials were made with the above-described apparatus, the use of which appeared to offer a great advantage in not requiring an acquaintance with the expansion of the liquid employed. However, it was speedily found that there were grave inconveniences attending it, in consequence of the difficulty of drying the tube thoroughly after each experiment, as well as the danger of losing a portion of the salt whilst introducing it through the narrow neck of the bulb. But these objections might have given way to time and patience. Our chief and, with this apparatus, insurmountable difficulty, arose from the fact that a quantity of air always remained attached to the salt after immersion. Since, therefore, it was highly probable that the quantity of adhering air varied with the temperature, the apparatus appeared liable to error, and was consequently abandoned.

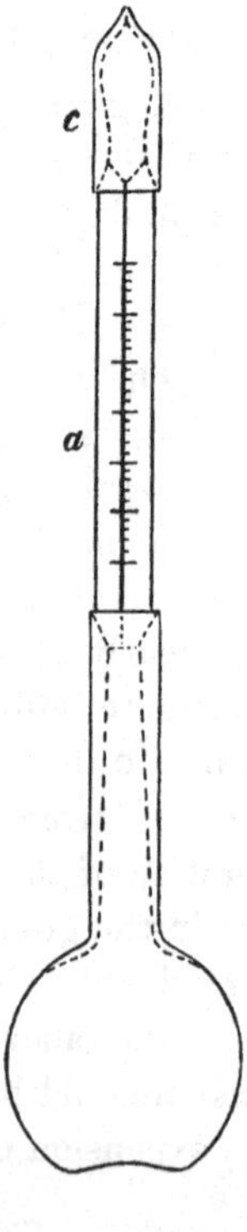

After some other attempts, we at last fixed upon a plan, which appeared to combine the advantages of accuracy with great practical facility. We procured four sp. gravity bottles, made of the same sample of glass. Two of them, marked No. 1 and No. 3, were capable of containing somewhat more than 500 grs. of water, whilst the other two, No. 2 and No. 4, had a capacity for about 270 grs. of water, the latter being destined for ascertaining the volumes and expansions of weighty articles in small quantities. We have given a half-size representation of one of the larger bulbs in fig. 8, where *a* is a stopper formed from a piece of thermometer-tube of narrow bore, terminated at the upper end by a small conical cavity, over which a cap *c* is accurately fitted by grinding, so that any liquid which may have ascended through the capillary is confined in

Fig. 8.

the cap without loss. There is a fine perforation at the summit of the cap, for allowing the egress of the air displaced by the liquid.

In order to ascertain the expansion of the glass of these volumenometers, the capillary tubes which served for their stoppers were carefully calibrated and graduated. They were then filled with distilled water, and immersed for half an hour in a bath containing a large quantity of water, kept constantly at 38°·92 Fahr. The exact height of the water in the capillary tubes being then noted, the water in the bath was from time to time increased in temperature. At every successive increase, the position of the water in the stems was observed. It first descended, and then of course ascended, until at the temperature of 45°·82 the water stood at exactly the same point as at 38°·92. Hence it was evident that the glass had expanded exactly as much as the water, through the interval between 38°·92 and 45°·82. According to the table given by Despretz, it appears that the expansion of water through this interval is 0·0001069; so that, supposing the expansion of glass to go on at the same rate through 180°, the expansion of bulbs of the volumenometers will be 0·002788 between the freezing- and boiling-points of water.

Another experiment of the same kind showed that the expansions of the glass and water were the same through the interval between 38°·478 and 46°·218, which placed the expansion of the volumenometers at 0·002798.

The expansion of the glass bulbs being thus known, it was easy to see how they might be employed in ascertaining the expansion of turpentine and salts. For by weighing them filled with turpentine, at different temperatures, we could obtain the expansion of that fluid. And then by weighing the bulbs filled with turpentine, and a given weight of salt at different temperatures, we could readily obtain the volume and expansion of the salt.

It need hardly be observed, that experiments of this nature require to be performed with very great accuracy. We therefore employed an excellent calibrated thermometer, in which each division was equal to $\frac{1}{12\cdot92}$ of a degree of

Fahrenheit's scale, and the freezing-point, which had remained nearly stationary * for half a year, stood at 16·1 divisions. It was easy to read off temperatures to within $\frac{1}{150}$th of a degree by this instrument. We also employed a very sensible balance by Dancer, which would turn with the addition of $\frac{1}{1000}$th of a grain, when each scale was loaded with 1000 grains. The barometric pressure was always noted, in order to correct the observed weights, and, before weighing, the bulbs were reduced nearly to the temperature of the apartment, to prevent the inaccuracy arising from the currents of air they would otherwise occasion in the balance case. In taking temperatures, regard was had to the temperature of that part of the stem not immersed in the bath; and, whenever necessary, the rule of the Committee of the Royal Society was employed in order to supply the requisite correction.

In the first place the weights of the bulbs, both empty and filled with water, were ascertained, in order to find their exact capacity. The water employed for this purpose was distilled, and had been recently boiled. The stoppers were placed in the necks of the bulbs, causing the water to ascend through the capillary tubes into the small cavities at the tops of the stoppers. A noose of string was now placed on the neck of each of the bulbs, which were then immersed, to within an inch of the tops of the stoppers, in the bath of water. The thermometer in the bath stood at the 113th division, and kept as nearly as possible at that point for about three quarters of an hour, during the whole of which time the water was repeatedly agitated by a stirrer. Experience had already shown that three quarters of an hour was more than sufficient to reduce the temperature of the bulbs to the exact temperature of the bath; the water remaining in the cavities at the tops of the stoppers was therefore now removed by means of bibulous paper, and the caps were replaced. The bulbs were then removed in succession from the bath, reduced to the temperature of the

* The actual rise of the freezing-point in half a year would be about $\frac{1}{15}$th of a degree Fahr. See Vol. i. p. 558.—NOTE, 1885.

room, dried with a soft silk handkerchief, and weighed. The
results, corrected for barometrical pressure, were as follows :—

	No. 1 Volu-menometer.	No. 2 Volu-menometer.	No. 3 Volu-menometer.
Bulbs and water at 113	623·572	882·042	922·522
Bulbs alone	349·546	377·036	397·649
Capacity in grs. of water at 113 .	274·026	505·006	524·873

The bulbs being now thoroughly dried, were filled with the
turpentine destined for the experiments. The range of
temperature fixed upon was between about 100 and 750 of
the thermometer, corresponding to about 38°·5 and 88°·8
Fahr. The low temperature was attained by dissolving
carbonate of soda in the water of the bath, adding a small
quantity from time to time, to keep the temperature uniform.
The high temperature was maintained by a lamp, burning
under the bath. When, after half an hour's immersion, the
turpentine had attained the exact temperature of the bath,
the bulbs were successively weighed, as in the case of the
trial with water already described. The bulbs were then
again immersed in the bath for half an hour, and re-weighed,
in order to preclude the possibility of an accidental error.

Volume-nometer.	Temp.	Weight.	Mean temp.	Mean weight.	Weight of empty bulb.	Weight of tur-pentine only.
No. 1.	120	589·648	123·5	589·616	349·546	240·070
	127	589·584				
	745·2	583·854	747·8	583·828	349·546	234·280
	750·4	583·802				
No. 2.	107·6	819·779	111·8	819·710	377·036	442·674
	116·0	819·640				
	744·5	808·874	747·45	808·831	377·036	431·795
	750·4	·788				
No. 3.	106	857·852	110·5	857·773	397·649	460·124
	115	857·694				
	744·9	846·456	747·55	846·411	397·649	448·762
	750·2	846·366				

The above results were obtained on the 11th of August,
1846, and immediately afterwards the experiments on the
expansion of salts were commenced, and continued until the
20th of August, when the turpentine (which was kept in a
large glass bottle, holding about a gallon) was again tested,

in order to ascertain whether any change had taken place in
its density. Such was found to have happened, in consequence of the absorption of atmospheric air, as will be seen
by the results given below.

Volume-nometer.	Temp.	Weight.	Mean temp.	Mean weight.	Weight of empty bulb.	Weight of turpentine alone.
No. 1.	93	590·052	93·5	590·052	349·546	240·506
	94	590·052				
	399·9	587·214	399·7	587·220	349·546	237·674
	399·5	587·227				
	754·4	583·906	752·5	583·926	349·546	234·380
	750·6	583·946				
No. 2.	93·9	820·271	95·1	820·255	377·036	443·219
	96·4	820·239				
	399·6	815·044	399·3	815·050	377·036	438·014
	399·0	815·056				
	752·8	808·934	751·3	808·962	377·036	431·926
	749·9	808·991				
No. 3.	93·8	858·358	94·2	858·353	397·649	460·704
	94·7	858·349				
	399·9	852·905	399·7	852·907	397·649	455·258
	399·5	852·909				
	753·5	846·554	751·8	846·582	397·649	448·933
	750·2	846·610				

In the above table we have recorded observations taken at
a temperature intermediate between the two extremes. The
same was done in the case of a large proportion of the salts
tried, with a view to ascertain whether their expansion was
uniform. Such appeared to be invariably the case; at least,
if any discrepancy occurred, it was within the limits of
experimental errors. We have not, however, thought it right
to extend our paper, already too voluminous, by these details.
It will be seen from the observations on the volume of
turpentine at the three temperatures above given, that that
fluid is remarkably uniform in its expansion, a circumstance
which eminently adapts it for experiments on the expansion
of salts.

Owing to the slight change which had occurred in the
density of the turpentine, we calculated the expansion of the
salts first tried by the first table of results on the expansion
of turpentine, and that of the salts last tried by the last
table, employing the mean of the two tables for the intermediate experiments.

In trying a salt, the bulb, partially filled with turpentine, was accurately weighed. A quantity of salt having been introduced, the bulb was again weighed, the increase of weight giving the exact quantity of salt, attention being paid to the correction for barometrical pressure. The bulb was then placed under the exhausted receiver of an air-pump, until the air adhering to the salt was entirely boiled away. This done, the vacant space in the bulb was filled up with turpentine, and the stopper inserted. The weights of the volumenometers and their contents at different temperatures were then ascertained, as in the case of the experiments with turpentine alone, already described. We may mention in this place that, for convenience sake, two or three bulbs containing different salts were always tried at the same time.

Exp. I.—634·200 grs. of Powdered Red Oxide of Mercury, in No. 3 volumenometer.

Temp.	Weight.	Mean Temp.	Mean Weight.
108°	1442·086 ⎫	109°	1442·072
110	1442·058 ⎭		
747·9	1431·882 ⎫	748·5	1431·863
749·2	1431·844 ⎭		

In calculating the volume and expansion of the red oxide of mercury from the above results, we proceeded as follows. From our first table of results for the expansion of turpentine, it appears that the weights of No. 3 volumenometer, filled with turpentine at the temperatures 110·5 and 747·55, are respectively 857·773 and 846·411 grs. Hence, at the temperatures of the above experiment (109 and 748·5), the weights would have been 857·800 and 846·394. Subtracting the weight of oxide from the weights of the volumenometer at the two temperatures, we have 1442·072−634·200= 807·872, and 1431·863−634·200=797·663. The numbers being subtracted from 857·800 and 846·394, leave 49·928 and 48·731 as the quantities of turpentine displaced by the oxide at the respective temperatures. But from the first table of the expansion of turpentine, it appears that 48·731 grs. at 748·5 are equivalent to 50·008 grs. at 109, regard being had to the expansion of glass. The expansion of the oxide for the interval of temperature between 109 and 748·5 is therefore 0·001596, which gives 0·005802 as the expansion for an interval of 2325 divisions of the thermometer, corresponding to 180° Fahr.

It appears also that the volume of 634·200 grs. of the oxide is equal to the volume of 49·928 grs. of turpentine at 109. But from the relative weights of the volumenometer containing water and turpentine already given, we find that the sp. gravity of turpentine at 109 is to that of water at 107·7 as 0·87669 is to 1. Hence we readily find that the sp. gravity of the oxide at 107·7, the point of maximum density of water, is 11·136.

Exp. II.—540·940 grs. of Sulphuret of Lead in powder, in No. 2 volumenometer.

Temp.	Weight.	Mean Temp.	Mean Weight.
108·2	1292·228 ⎫	111·3	1292·173
114·4	1292·116 ⎭		
748·2	1282·824 ⎫	748·9	1282·807
749·6	1282·791 ⎭		

Therefore expansion for 180° = 0·01045. Sp. gr. at 39°·1 = 6·9238.

Exp. III.—585·500 grs. of Bichromate of Potash in small crystals, in No. 3 volumenometer.

Temp.	Weight.	Mean Temp.	Mean Weight.
111·2	1252·440 ⎫	112·7	1252·421
114·2	1252·402 ⎭		
749·8	1245·290 ⎫	749·3	1245·297
748·9	1245·303 ⎭		

Therefore expansion for 180° = 0·0122   Sp. gr. at 39°·1 = 2·692.

Exp. IV.—146·355 grs. of pounded Muriate of Ammonia, in No. 2 volumenometer.

Temp.	Weight.	Mean Temp.	Mean Weight.
111·8	882·348 ⎫	113·8	882·329
115·8	882·310 ⎭		
750·0	873·149 ⎫	749·5	873·149
748·8	873·150 ⎭		

Therefore expansion for 180° = 0·0191. Sp. gr. at 39°·1 = 1·5333.

Exp. V.—376·520 grs. of Peroxide of Tin in powder ; No. 1 volumenometer.

T.	W.	M. T.	M. W.
109·2	917·088 }	111·1	917·073
113	917·058		
750	912·379 }	749·5	912·381
749	912·383		

Therefore expansion for 180° = 0·00172. Sp. gr. at 39°·1 = 6·7122.

Exp. VI.—405·131 grs. of Sulphate of Iron, prepared by pounding, and pressing between bibulous paper ; in No. 3 volumenometer.

T.	W.	M. T.	M. W.
104	1074·826 }	107·5	1074·792
111	1074·758		
749·4	1067·586 }	750·0	1067·576
750·6	1067·567		

Therefore expansion for 180° = 0·01153. Sp. gr. at 39°·1 = 1·8889.

Exp. VII.—426·672 grs. of Sulphate of Copper, prepared by pressing the pounded salt between folds of bibulous paper ; No. 2 volumenometer.

T.	W.	M. T.	M. W.
104·5	1083·046 }	108·2	1083·004
112·0	1082·963		
749·6	1075·777 }	750·0	1075·776
750·4	1075·775		

Therefore expansion for 180° = 0·009525. Sp. gr. at 39°·1 = 2·2901.

Exp. VIII.—552·605 grs. of Protoxide of Lead in powder ; No. 1 volumenometer.

T.	W.	M. T.	M. W.
103·5	1090·670 }	106·4	1090·649
109·4	1090·629		
749·6	1085·902 }	750·0	1085·9
750·4	1085·898		

Therefore expansion for $180° = 0·00795$. Sp. gr. at $39°·1$ $= 9·3634$.

Exp. IX.—327·760 grs. of Sulphate of Magnesia, prepared by pressing the pounded salt between folds of bibulous paper; No. 3 volumenometer.

T.	W.	M. T.	M. W.
90	1014·948 ⎫	93·3	1014·913
96·6	1014·879 ⎭		
750	1007·181 ⎫	749·5	1007·182
749	1007·182 ⎭		

Therefore expansion for $180° = 0·01019$. Sp. gr. at $39°·1$ $= 1·6829$.

Exp. X.—417·706 grs. of Nitrate of Potash, in No. 2 volumenometer.

T.	W.	M. T.	M. W.
89·4	1063·822 ⎫	93·4	1063·776
97·4	1063·731 ⎭		
750·3	1056·109 ⎫	749·4	1056·114
748·6	1056·119 ⎭		

Therefore expansion for $180° = 0·01967$. Sp. gr. at $39°·1$ $= 2·1078$.

Exp. XI.— 287·080 grs. of Copper, prepared from the oxide by passing H over it at a red heat; in No. 1 volumenometer.

T.	W.	M. T.	M. W.
89	846·962 ⎫	92·3	846·932
95·6	846·903 ⎭		
401·2	844·420 ⎫	400·6	844·426
400	844·432 ⎭		

Therefore expansion for $180° = 0·0055$. Sp. gr. at $39°·1 = 8·367$.

Exp. XII.—621·528 grs. of Yellow Chromate of Potash, in No. 3 volumenometer.

T.	W.	M. T.	M. W.
91·4	1278·445 ⎫	95·2	1278·402
99	1278·360 ⎭		
752·5	1271·323 ⎫	752·8	1271·304
753·2	1271·285 ⎭		

Therefore expansion for $180° = 0.01134$. Sp. gr. at $39°.1$ $= 2.7110$.

Exp. XIII.—531·794 grs. of Nitrate of Soda, in No. 2 volumenometer.

T.	W.	M. T.	M. W.
92	1145·372 ⎫	95·8	1145·323
99·6	1145·274 ⎭		
752·4	1138·724 ⎫	752·7	1138·720
753	1138·715 ⎭		

Therefore expansion for $180° = 0.0128$. Sp. gr. at $39°.1 =$ 2.2606.

Exp. XIV.—133·131 grs. of Red Oxide of Manganese, in No. 1 volumenometer.

T.	W.	M. T.	M. W.
91·2	696·149 ⎫	94·6	696·113
98·0	696·077 ⎭		
753·4	690·654 ⎫	753·2	690·660
753·0	690·666 ⎭		

Therefore expansion for $180° = 0.00522$. Sp. gr. at $39°.1$ $= 4.325$.

Exp. XV —424·298 grs. of Sugar Candy, coarsely pounded ; No. 3 volumenometer.

T.	W.	M. T.	M. W.
102·6	1048·504 ⎫	107·0	1048·472
111·5	1048·440 ⎭		
752·6	1042·251 ⎫	753·8	1042·244
755	1042·238 ⎭		

Therefore expansion for $180° = 0.01116$. Sp. gr. at $39°.1$ $= 1.5927$.

Exp. XVI.—546·662 grs. of Nitrate of Lead, in No. 1 volumenometer.

T.	W.	M. T.	M. W.
99·4	1029·299 ⎫	104·1	1029·280
108·8	1029·261 ⎭		
753·4	1025·773 ⎫	753·9	1025·772
754·4	1025·771 ⎭		

Therefore expansion for $180° = 0\cdot00839$.　Sp. gr. at $39°\cdot1$ $= 4\cdot472$.

The above experiments were, as we have already stated, completed before the commencement of September 1846, and a variety of circumstances prevented our resumption of the research before March in the succeeding year.　However, previously to the commencement of the new series, we provided ourselves with a large quantity of turpentine, which, after having been well mixed, was decanted into small bottles, and preserved over mercury.　In this way, the turpentine was kept without material change, as will be seen from the following series of results obtained with it before the commencement of the experiments, and after their completion.

SERIES I.—*Experiments on the Expansion of Turpentine, on March* 20, 1847.

Volume-nometer.	Tempe-rature.	Weight.	Mean temperature.	Mean weight.	Weight of bulb.	Weight of turpentine.
No. 1.	89·5 87·0	590·119 590·136	88·4	590·127	349·546	240·581
	747·6	584·024	747·6	584·024	349·546	234·478
No. 2.	89·4 87·0	820·472 820·512	88·2	820·492	377·036	443·456
	746·6	809·669	746·6	809·169	377 026	432·133
No. 3.	89 87	858·553 858·584	88	858·568	397·649	460·918
	734·1 745·7	847·030 846·806	739·9	846·918	397·649	449·269

SERIES II.—*Experiments on the Expansion of Turpentine, on January* 21, 1848.

Volume-nometer.	Tempe-rature.	Weight.	Mean temperature.	Mean weight.	Weight of bulb.	Weight of turpentine.
No. 2.	73 70	820·752 820·805	71·5	820·778	377·036	443·742
	754·6 754·6	809·031 809·032	754·6	809·031	377·036	431·995
No. 3.	70·8 73·2 70·2	858·871 858·831 858·886	71·4	858·863	397·649	461·214
	754·6 754·6 754·6	846·645 846·641 846·647	754·6	846·644	397·649	448·995

In calculating the expansion of the salts, the meanof the above two series was employed, viz. :—

Volumeno-meter.	Tempera-ture.	Weight.	Weight of volumenometer.	Weight of turpentine.
No. 1.	{ 88·2	590·127	349·546	240·581
	747·6	584·024	349·546	234·478
No. 2.	{ 79·85	820·635	377·036	443·599
	750·6	809·100	377·036	432·064
No. 3.	{ 79·7	858·715	397·649	461·066
	747·25	846·781	397·649	449·132

Exp. XVII.—466·202 grs. of Nitrate of Potash in large crystals; No. 3 volumenometer.

T.	W.	M. T.	M. W.
75	1129·601 }	76·5	1129·589
78	1129·578 }		
739·3	1121·954 }	742·9	1121·937
746·5	1121·920 }		

Therefore expansion for 180°=0·017237. Sp. gr. at 39°·1 =2·09584.

Exp. XVIII.—654·992 grs. of Sulphate of Potash in small crystals; No. 2 volumenometer.

T.	W.	M. T.	M. W.
74·6	1259·066 }	76·3	1259·060
78·0	1259·054 }		
738·4	1252·752 }	742·5	1252·721
746·6	1252·690 }		

Therefore expansion for 180°=0·010697. Sp. gr. at 39°·1 =2·65606.

Exp. XIX.—302·609 grs. of Copper, prepared from the oxide by passing hydrogen over it at a red heat; No. 1 volumenometer.

T.	W.	M. T.	M. W.
74·0	861·271 }	75·9	861·263
77·8	861·256 }		
740·0	855·884 }	743·7	855·850
747·4	855·816 }		

Therefore expansion for 180°=0·00767. Sp. gr. at 39°·1
=8·41613.

Exp. XX.—501·286 grs. of Nitrate of Potash, pounded
small; No. 3 volumenometer.

T.	W.	M.T.	M.W.
60·0	1151·237 ⎱	61·4	1151·217
62·8	1151·197 ⎰		
736·6	1143·634 ⎱	739·8	1143·596
743·0	1143·558 ⎰		

Therefore expansion for 180°=0·019487   Sp. gr. at 39°·1
=2·10657.

Exp. XXI.—427·790 grs. of Sulphate of Copper and Ammonia, in large crystals; No. 2 volumenometer.

T.	W.	M.T.	M.W.
60·8	1050·197 ⎱	62	1050·182
63·2	1050·167 ⎰		
736·0	1043·559 ⎱	739·5	1043·529
743·0	1043·499 ⎰		

Therefore expansion for 180°=0·0066113.   Sp. gr. at 39°·1
=1·89378.

Exp. XXII.—429·784 grs. of Sulphate of Magnesia and
Ammonia, in good crystals; No. 3 volumenometer.

T.	W.	M.T.	M.W.
76·0	1068·645 ⎱	76·6	1068·644
77·2	1068·643 ⎰		
739·0	1062·177 ⎱	737·3	1062·202
735·6	1062·227 ⎰		

Therefore expansion for 180°=0·007161.   Sp. gr. at 39°·1
=1·71686.

Exp. XXIII.—507·958 grs. of Sulphate of Potash and
Zinc, in good crystals; No. 2 volumenometer.

T.	W.	M.T.	M.W.
75·0	1129·497 ⎱	75·8	1129·480
76·6	1129·463 ⎰		
738·7	1122·899 ⎱	736·6	1122·920
734·5	1122·941 ⎰		

Therefore expansion for 180°=0·008235.   Sp. gr. at 39°·1
=2·24034.

Exp. XXIV.—444·494 grs. of Sulphate of Magnesia and Potash, in good crystals; No. 3 volumenometer.

T.	W.	M. T.	M. W.
75·0	1113·108 ⎫	75·4	1113·101
75·8	1113·095 ⎭		
747·9	1105·678 ⎫	748·0	1105·683
748·1	1105·688 ⎭		

Therefore expansion for 180°=0·009372. Sp. gr. at 39°·1 =2·05319.

Exp. XXV.—456·314 grs. of Sulphate of Copper and Potash, in good crystals; No 2 volumenometer.

T.	W.	M. T.	M. W.
74·9	1091·772 ⎫	75·4	1091·761
75·9	1091·750 ⎭		
748·3	1084·695 ⎫	748·0	1084·698
747·7	1084·701 ⎭		

Therefore expansion for 180°=0·009043. Sp. gr. at 39°·1 =2·16376.

Exp. XXVI.—449·535 grs. of Sulphate of Copper, in small crystals, prepared by stirring the cupreous solution while cooling. This specimen contained 5·112 equivalents of water, or an excess, due to a mechanical admixture of water. No. 3 volumenometer.

T.	W.	M. T.	M. W.
83·7	1132·089 ⎫	83·9	1132·094
84·1	1132·099 ⎭		
746·2	1124·632 ⎫	746·2	1124·653
746·2	1124·674 ⎭		

Therefore expansion for 180°=0·005315. Sp. gr. at 39°·1 =2·2422.

Exp. XXVII.—320·027 grs. of Sulphate of Ammonia, in fine small crystals; No. 3 volumenometer.

T.	W.	M. T.	M. W.
89·4	1019·049 ⎫	85·3	1019·095
81·2	1019·141 ⎭		
746·2	1011·018 ⎫	746·1	1011·011
746·0	1011·005 ⎭		

Therefore expansion for 180°=0·010934. Sp. gr. at 39°·1 =1·76147.

Exp. XXVIII.—377·686 grs. of Sulphate of Chromium and Potash, in good crystals; No. 2 volumenometer.

T.	W	M. T.	M. W.
89·0	1019·481 ⎫	85·1	1019·526
81·2	1019·571 ⎭		
746·2	1012·614 ⎫	746·2	1012·617
746·2	1012·620 ⎭		

Therefore expansion for 180°=0·005242. Sp. gr. at 39°·1 =1·85609.

Exp. XXIX.—517·725 grs. of Sulphate of Copper, pounded and well pressed between folds of bibulous paper; No. 3 volumenometer.

T.	W.	M. T.	M. W.
86·0	1176·752 ⎫	87·0	1176·743
88·0	1176·734 ⎭		
745·3	1169·776 ⎫	743·7	1169·794
742·1	1169·812 ⎭		

Therefore expansion for 180°=0·00812. Sp. gr. at 39°·1 =2·2781.

Exp. XXX.—600·594 grs. of Yellow Chromate of Potash, in fine small crystals; No. 2 volumenometer.

T.	W.	M. T.	M. W.
86·0	1227·440 ⎫	87·0	1227·433
88·0	1227·427 ⎭		
745·3	1220·612 ⎫	743·8	1220·634
742·3	1220·656 ⎭		

Therefore expansion for 180°=0·011005. Sp. gr. at 39°·1 =2·72309.

Exp. XXXI.—467·184 grs. of Potash Alum, in good crystals; No. 2 volumenometer.

T.	W.	M. T.	M. W.
80·0	1053·484 ⎫	80·0	1053·482
80·0	1053·480 ⎭		
748·0	1047·995 ⎫	746·4	1048·018
744·8	1048·042 ⎭		

Therefore expansion for 180°=0·003682. Sp. gr. at 39°·1 =1·75125.

Exp. XXXII.—402·116 grs. of Binoxalate of Potash, in good crystals ; No. 3 volumenometer.

T.	W.	M. T.	M. W.
80·0	1088·032 ⎤	80·0	1088·033
80·0	1088·034 ⎦		
748·2	1080·146 ⎤	746·6	1080·170
745·0	1080·194 ⎦		

Therefore expansion for 180°=0·011338. Sp. gr. at 39°·1 =2·04401.

Exp. XXXIII.—447·312 grs. of Oxalate of Potash, in good crystals ; No. 2 volumenometer.

T.	W.	M. T.	M. W.
61·6	1083·389 ⎤	63·3	1083·374
65·0	1083·359 ⎦		
752·2	1075·978 ⎤	751·6	1075·996
751·0	1076·014 ⎦		

Therefore expansion for 180°=0·01162. Sp. gr. at 39°·1 =2·12657.

Exp. XXXIV.—406·552 grs. of Chloride of Potassium, coarsely pounded; No. 3 volumenometer.

T.	W.	M. T.	M. W.
60·8	1084·916 ⎤	63·2	1084·886
65·6	1084·856 ⎦		
752·5	1076·959 ⎤	751·9	1076·970
751·3	1076·981 ⎦		

Therefore expansion for 180°=0·010944. Sp. gr. at 39°·1 =1·97756.

Exp. XXXV.—292·039 grs. of Oxalate of Ammonia, in fine small crystals ; No. 3 volumenometer.

T.	W.	M. T.	M. W.
79·2	979·728 ⎤	80·8	979·713
82·4	979·698 ⎦		
752·0	971·874 ⎤	750·9	971·890
749·8	971·907 ⎦		

Therefore expansion for 180°=0·00876. Sp. gr. at 39°·1 =1·49985.

Exp. XXXVI.—621·193 grs. of Nitrate of Barytes, in small crystals; No. 2 volumenometer.

T.	W.	M.T.	M.W.
85·8	1269·101 ⎱	87·9	1269·085
90·0	1269·069 ⎰		
745·8	1262·089 ⎱	746·4	1262·086
747·0	1262·083 ⎰		

Therefore expansion for 180° = 0·004523. Sp. gr. at 39°·1 = 3·16052.

Exp. XXXVII.—450·560 grs. of Bisulphate of Potash, pounded; No. 3 volumenometer.

T.	W.	M.T.	M.W.
86·9	1149·461 ⎱	88·5	1149·445
90·1	1149·429 ⎰		
746·0	1141·345 ⎱	746·5	1141·339
747·0	1141·333 ⎰		

Therefore expansion for 180° = 0·012287. Sp. gr. at 39°·1 = 2·47767.

Exp. XXXVIII.—373·783 grs. of Oxalic Acid, in good crystals; No. 2 volumenometer.

T.	W.	M.T.	M.W.
89·2	994·334 ⎱	87·8	994·346
86·4	994·359 ⎰		
744·9	986·783 ⎱	744·8	986·784
744·7	986·785 ⎰		

Therefore expansion for 180° = 0·027476. Sp. gr. at 39°·1 = 1·64138.

Exp. XXXIX.—393·044 grs. of Chlorate of Potash, in small crystals; No. 3 volumenometer.

T.	W.	M.T.	M.W.
89·2	1103·259 ⎱	87·9	1103·269
86·6	1103·279 ⎰		
744·7	1094·718 ⎱	744·7	1094·722
744·7	1094·726 ⎰		

Therefore expansion for 180° = 0·017112. Sp. gr. at 39°·1 = 2·32643.

Exp. XL.—541·833 grs. of Chloride of Barium, in small crystals; No. 2 volumenometer.

T.	W.	M. T.	M. W.
64·8	1206·826 ⎫	67·1	1206·800
69·4	1206·774 ⎭		
755·4	1198·800 ⎫	755·8	1198·797
756·2	1198·794 ⎭		

Therefore expansion for 180°=0·009873.　Sp. gr. at 39°·1 =3·05435.

Exp. XLI.—326·462 grs. of Sugar of Milk, pounded; No. 3 volumenometer.

T.	W.	M. T.	M. W.
65·0	998·393 ⎫	67·2	998·383
69·4	998·373 ⎭		
755·6	990·721 ⎫	755·9	990·720
756·2	990·719 ⎭		

Therefore expansion for 180°=0·009111.　Sp. gr. at 39°·1 =1·53398.

Exp. XLII.—307·866 grs. of Binoxalate of Ammonia in good crystals; No. 2 volumenometer.

T.	W.	M. T.	M. W.
81·8	960·874 ⎫	83·6	960·862
85·4	960·850 ⎭		
750·8	953·203 ⎫	750·9	953·210
751·0	953·217 ⎭		

Therefore expansion for 180°=0·013718.　Sp. gr. at 39°·1 =1·61341.

Exp. XLIII.—234·865 grs. of Bichromate of Chloride of Potassium, in good crystals; No. 3 volumenometer.

T.	W.	M. T.	M. W.
81·6	1010·939 ⎫	83·4	1010·914
85·2	1010·889 ⎭		
750·7	1000·808 ⎫	750·9	1000·816
751·1	1000·824 ⎭		

Therefore expansion for 180°=0·015902.　Sp. gr. at 39°·1 =2·49702.

Exp. XLIV.—392·901 grs. of Quadroxalate of Potash in good crystals; No. 2 volumenometer.

No. of expt.	Name of Salt.	Formula.	Atomic weight.	Expansion for 180°.	Sp. gr. at 39°·1.	Atomic volume.	Atomic volume divided by 1·225.
11.	Copper	Cu	31·66	0·0055	8·367	3·7839	3·0889
19.	Ditto	Ditto	31·66	0·00767	8·4161	3·7618	3·0709
1.	Red Oxide of Mercury	HgO	108·07	0·00580	11·136	9·7046	7·9221
8.	Protoxide of Lead	PbO	111·56	0·00795	9·3634	11·887	9·7037
14.	Red Oxide of Manganese	Mn_3O_4	115·0	0·00522	4·325	26·590	21·706
5.	Peroxide of Tin	SnO_2	74·82	0·00172	6·7122	11·147	9·0995
2.	Sulphuret of Lead	PbS	119·56	0·01045	6·9238	17·268	14·096
34.	Chloride of Potassium	KCl	74·5	0·01094	1·9776	37·673	30·753
40.	Chloride of Barium	$BaCl+2HO$	122·14	0·00987	3·0543	39·989	32·644
4.	Chloride of Ammonium	NH_4Cl	53·5	0·0191	1·5333	34·892	28·483
13.	Nitrate of Soda	NaO,NO_5	85·0	0·0128	2·2606	37·601	30·694
10.	Nitrate of Potash	KO,NO_5	101·0	0·01967	2·1078	47·917	39·116
17.	Ditto	Ditto	101·0	0·01724	2·0958	48·191	39·339
20.	Ditto	Ditto	101·0	0·01949	2·1066	47·945	39·139
16.	Nitrate of Lead	PbO,NO_5	165·56	0·00839	4·472	37·021	30·222
36.	Nitrate of Barytes	BaO,NO_5	130·64	0·00452	3·1605	41·335	33·743
39.	Chlorate of Potash	KO,ClO_5	122·5	0·01711	2·3264	52·656	42·984
12.	Chromate of Potash	KO,CrO_3	99·15	0·01134	2·711	36·573	29·856
30.	Ditto	Ditto	99·15	0·01101	2·7231	36·411	29·723
3.	Bichromate of Potash	$KO,2CrO_3$	151·3	0·0122	2·692	56·204	45·880
43.	Bichromate of Chloride of Potassium	$KCl+2CrO_3$	178·8	0·0159	2·497	71·605	58·463

No.	Name	Formula					
33.	Oxalate of Potash	KO, C_2O_3+HO	92·0	0·01162	2·1266	3·262	35·316
32.	Binoxalate of Potash	$KO, 2C_2O_3+3HO$	146·0	0·01134	2·0440	71·428	58·309
44.	Quadroxalate of Potash	$KO, 4C_2O_3+7HO$	254·0	0·01592	1·8488	137·384	112·150
35.	Oxalate of Ammonia	NH_4O, C_2O_3+HO	71·0	0·00876	1·49985	47·338	38·643
42.	Binoxalate of Ammonia	$NH_4O, 2C_2O_3+3HO$	125·0	0·01372	1·61341	77·476	63·245
45.	Quadroxalate of Ammonia	$NH_4O, 4C_2O_3+7HO$	233·0	0·01435	1·65194	141·046	115·140
18.	Sulphate of Potash	KO, SO_3	87·0	0·01070	2·65606	32·755	26·739
37.	Bisulphate of Potash	KO, SO_3+HO, SO_3	136·0	0·01229	2·47767	54·890	44·808
27.	Sulphate of Ammonia	NH_4O, SO_3+HO	75·0	0·01093	1·76147	42·578	34·758
7.	Sulphate of Copper	CuO, SO_3+5HO	124·66	0·00952	2·2901	54·434	44·436
26.	Ditto	Ditto	124·66	0·00531	2·2422	55·597	45·385
29.	Ditto	Ditto	124·66	0·00812	2·2781	54·721	44·670
6.	Sulphate of Iron	FeO, SO_3+7HO	139·0	0·01153	1·8889	73·588	60·072
9.	Sulphate of Magnesia	MgO, SO_3+7HO	123·67	0·01019	1·6829	73·486	59·989
21.	Sulphate of Copper and Ammonia	$CuO, SO_3+NH_4O, SO_3+6HO$	199·66	0·00661	1·8938	105·430	86·065
25.	Sulphate of Copper and Potash	CuO, SO_3+KO, SO_3+6HO	220·66	0·00904	2·1638	101·980	83·249
24.	Sulphate of Magnesia and Potash	MgO, SO_3+KO, SO_3+6HO	201·67	0·00937	2·0532	98·223	80·182
28.	Sulphate of Chromium and Potash	$Cr_2O_3, 3SO_3+KO, SO_3+24HO$	503·3	0·00524	1·8561	271·161	221·356
31.	Potash Alum	$Al_2O_3, 3SO_3+KO, SO_3+24HO$	476·38	0·00368	1·75125	272·023	222·060
23.	Sulphate of Potash and Zinc	ZnO, SO_3+KO, SO_3+6HO	221·0	0·00823	2·2403	98·646	80·527
22.	Sulphate of Magnesia and Ammonia	$MgO, SO_3+NH_4O, SO_3+6HO$	180·67	0·00716	1·71686	105·233	85·904
15.	Cane-Sugar	$C_{12}H_{11}O_{11}$	171·0	0·01116	1·5927	107·365	87·645
41.	Sugar of Milk	$C_{12}H_{12}O_{12}$	180·0	0·00911	1·5340	117·342	95·789

T.	W.	M. T.	M. W.
74·5	1026·945 ⎱	75·3	1026·940
76·1	1026·935 ⎰		
751·0	1019·509 ⎱	752·2	1019·502
753·4	1019·494 ⎰		

Therefore expansion for 180° = 0·015916.   Sp. gr. at 39°·1
= 1·84883.

Exp. XLV.—396·646 grs. of Quadroxalate of Ammonia in
good crystals; No. 3 volumenometer.

T.	W.	M. T.	M. W.
74·9	1044·521 ⎱	75·6	1044·518
76·3	1044·515 ⎰		
751·0	1037·270 ⎱	752·2	1037·263
753·4	1037·256 ⎰		

Therefore expansion for 180° = 0·014347.   Sp. gr. at 39°·1
= 1·65194.

The foregoing results are collated in the foregoing Table,
pp. 200, 201.

On a cursory inspection of the above table, it will be observed that considerable discrepancy occurs in the results
obtained with salts of the same name.   This must not throw
doubt on the accuracy of the experiments, as it arose from
the state in which the salts were tried.   The specific gravity
of copper in the 19th experiment is somewhat greater than
that in the 11th experiment, the reason for which is, that
the copper in the former case was exposed to hydrogen at a
red heat for a considerable time after the oxygen had been
removed from the oxide.   In the case of nitrate of potash,
the 17th experiment was made with large crystals averaging
$\frac{3}{16}$ of an inch in diameter, whilst in experiment 20 we employed another portion of the same salt pounded very small,
so as to free it from mechanical water, and in experiment 10
a different specimen was used in small crystals.   It will be
observed that a very considerable difference exists between
the results of the three trials.   The large crystals were lighter
and in a more coerced state, as shown by their smaller
expansion.   Of the three specimens of sulphate of copper
tried, that employed in experiment 26 had, as we have stated,

a larger quantity of water than 5 equivalents attached to it, and hence we observed a great difference in its expansion and specific gravity, the decrease of gravity being, as before, accompanied by a decrease of expansion. With regard to the expansions, we may call attention to the close approximation of the highly hydrated salts to 0·01125, or the expansion of ice as determined by Brunner. The expansion of oxalic acid is remarkable as being greater, and that of peroxide of tin as being less than those of any solid bodies on record. In all cases of hydrated salts, it is extremely difficult to obtain them in their exact normal state of hydration, and therefore the results of experiments with different specimens are subject to variation.

Researches on Atomic Volume and Specific Gravity.
By LYON PLAYFAIR, *F.R.S., and* J. P. JOULE.

[Journal of the Chemical Society, vol. i. p. 139.]

SERIES V.—*On the Disappearance of the Volume of the Acid, and in some cases of the Volume of the Base, in the Crystals of highly Hydrated Salts.*

WE have had the honour to present to the Society several Memoirs on the Atomic Volumes of Salts. In the first of these, we pointed out three distinct facts with reference to hydrated salts, which increased experience has only the more confirmed us in believing to be strictly true.

1st. That the anhydrous portion of highly hydrated salts, when dissolved in water, occupies no space, if allowance be made for the increased expansibility of the solution.

2nd. That the hydrated salts of one class take up, in their solid state, the space which would be occupied by their water frozen into ice, the volume of their anhydrous portion ceasing to be recognized.

3rd. That hydrated salts of a second class take up the space represented by the number of their equivalents of water multiplied by the empirical number 11, instead of 9·8, which latter number represents the volume of ice.

In our first Memoir, we fully admitted that the empirical number 11 was merely an approximation to the truth; and in our second communication, we showed that it was more strictly $9\cdot8 + \frac{9\cdot8}{8} = 11\cdot025$. The number $9\cdot8$, assumed for the volume of ice, is derived from that denoting the specific gravity of that body, which our experiments showed to be $0\cdot9184$, a result almost identical with the number $0\cdot918$ as determined by Brunner.

Two classes of objections have been made to our researches. The first, by Berzelius and others, that the methods which we employed for obtaining the specific gravity were not likely to be very accurate. Although fully confident of their general accuracy, we considered it due to criticism, generally so just as that of this philosopher, to repeat all our experiments with the greatest care, not with new instruments such as those we had previously used, but by processes universally recognized. In the previous paper we have done this with much labour, and at the same time elicited the influence of temperature on specific gravity, but without obtaining results in any degree materially different, or nearer to theory than were obtained with the apparatus formerly used. Our new experiments are contained in the Memoir just read to the Society.

A second class of objectors urged that no rule was given when the volume $9\cdot8$ should be taken, and when the volume $11\cdot0$. This objection was perfectly just, and was always entertained by ourselves. We stated in our first Memoir*, "That for the present we announce the number 11 empirically," as being an expression of experiment, not as being its rational exposition. Future research alone could explain the cause of this number, and to its elucidation we have devoted ourselves. In our second Memoir, we endeavoured to show that this number 11 was itself a multiple of a number, $\frac{9\cdot8}{8} = 1\cdot225$, at least in the case of a large number of bodies; and we now proceed to point out how it is that in the hydrated salts this number so frequently appears as the expression of

* 'Memoirs of the Chemical Society,' vol. ii. p. 477.

experiment. We take one class of salts at a time, because the deliberate study of one is better calculated to lead us to truth, and enable us to abandon error. In pursuance of our plan this Memoir is confined to hydrated salts only.

In the first place, we call attention to the fact already noticed in our first Memoir, that there are certain hydrated salts in which the water alone takes up space in the solid state. The specific gravities of these had already been frequently determined, our results agreeing closely with those of former experimentalists. We therefore adduce the table given on a former occasion, no further experiments having been made.

TABLE I.

Designation.		Volume in Salt.					
Name.	Formula.	Atomic weight.	Volume of salt by experiment.	9·8, or volume of ice, as unity.	Volume by theory.	Specific gravity by theory.	Specific gravity by experiment.
rbonate of Soda ..	NaO, CO$_2$+10HO	143·4	98·6	10	98·0	1·463	1·454
osphate of Soda ..	2NaO, HO, PO$_5$ +24HO	359·1	235·5	24	235·2	1·527	1·525
bphosphate of Soda.	3NaO, PO$_5$+24HO	381·6	235·2	24	235·2	1·622	1·622
seniate of Soda....	HO, 2NaO, AsO$_5$ +24HO	402·9	232·0	24	235·2	1·713	1·736
barseniate of Soda.	3NaO, AsO$_5$+24HO	425·2	235·6	24	235·2	1·808	1·804

Cane-sugar is a remarkable example of this class, as showing that the hydrogen and oxygen are present *quasi* water, the carbon occupying no space. It was interesting to test this with milk-sugar, the specific gravity of which we had carefully determined in the preceding paper. The following table proves that in the two cases of cane-sugar and sugar of milk, the carbon ceases to occupy space, and the water takes up the exact volume which it should do if it were frozen into ice.

TABLE II.

Name.	Formula.	Atomic weight.	Volume by theory, supposing the HO to take the space of ice.	Specific gravity by theory.	Volume by experiment.	Specific gravity by experiment.
Cane Sugar..	$C_{12}H_{11}O_{11}$ ..	171	107·8	1·586	107·4	1·593
Milk Sugar..	$C_{24}H_{24}O_{24}$ ..	360	235·2	1·531	234·7	1·534

The specific gravity of cane-sugar according to other experimentalists, is 1·606 by Musschenbroeck, 1·600 and 1·585
by Schübler and Renz; and of milk-sugar 1·548 by Schübler.

The coincidence between the observed and calculated
numbers, in the above two tables, is so striking as to leave
no doubt that both the acid and base of the salts and the
carbon of the sugars * cease to occupy space, having merged
into the volume of the water. In these tables there is
nothing new beyond the addition of milk-sugar to the series,
the general fact having been communicated in our first
Memoir.

We now direct attention to the hydrated salts possessing
a larger volume than that due to the water contained in
them; that is, to those salts in which the factor, representing
the number of atoms of water, had to be multiplied by 11
instead of by 9·8.

Table III. is constructed upon two assumptions : first, that
the water, as in other hydrated salts, takes up the space of
ice, and, second, that the base takes up the same space in the
hydrated as it does in the anhydrous salt. These two
assumptions being admitted, the sum of the volumes of the
solid water and of the base is exactly equal to the total volume
of the salt, *nothing being left for the acid, which therefore has*

---

* We have not estimated the sp. gr. of grape-sugar. Guérin states it
to be 1·386, which would appear to make it belong to a different class, as
the sp. gr. necessary to give it the volume due to its solid water only
would be 1·442.

Table III.

Name.	Formula.	Atomic weight.	Space occupied by solid water.	Space occupied by base.	Sum of spaces occupied by base and water.	Theoretical sp. gravity.	Volume of salt by experiment.	Sp. gravity by experiment.	Specific gravity by various Authorities.
Sulphate of Copper	CuO, SO_3+5HO	124·66	49·00	6·12	55·12	2·261	54·7	2·279	2·284, P. and J. / 2·274, Kopp.
Sulphate of Zinc	ZnO, SO_3+7HO	143·43	68·6	6·12	74·72	1·919	74·2	1·931	1·90, Musschenbroeck. / 1·912, Hassenfratz.
Sulphate of Magnesia	MgO, SO_3+7HO	123·67	68·6	6·12	74·72	1·655	73·5	1·683	1·660, Hassenfratz. / 1·674, Kopp.
Sulphate of Iron	FeO, SO_3+7HO	139·0	68·6	6·12	74·72	1·860	73·58	1·888	1·660, P. and J. / 1·84, Hassenfratz.
Sulphate of Nickel	NiO, SO_3+6HO	131·74	58·8	6·12	64·92	2·029	64·6	2·037	1·58, Musschenbroeck. / 2·037, Kopp.
Sulphate of Soda	NaO, SO_3+10HO	161·48	98·0	12·25	110·25	1·464	109·9	1·469	1·470, Leonhard. / 1·481, Mohs. / 1·380, Watson.
Sulphate of Alumina	$Al_2O_3, 3SO_3+18HO$	333·7	176·4	22·05	198·45	1·681	199·6	1·671	1·720, Musschenbroeck. / 1·723, Hassenfratz.
Biborate of Soda	$NaO, 2BO_3+10HO$	191·23	98·0	12·25	110·25	1·734	110·5	1·730	1·740, Kirwan. / 1·716, Mohs. / 1·740, Accum.
Pyrophosphate of Soda	$2NaO, PO_5+10HO$	224·15	98·0	24·50	122·50	1·829	122·0	1·836	1·905, Tunnermann.

ceased to occupy space. This result being very remarkable, it is necessary to examine closely into the evidence on which it rests. It has been shown in our second Memoir, that the volumes of the protoxides of copper, zinc, magnesia, nickel, and iron, in combination, are 6·12, which is also the volume of the oxides in their state of greatest density when separated. The experimental volume of the sulphates of these oxides is as a mean 22·17 *, which is very near the theoretical number 22·05 †. The volume of sulphuric acid being 15·92, the difference 6·12 leads to the same theoretical result as that obtained for the oxides in their densest state. If, instead of this theoretical number, that procured by experiment were substituted, the difference would only be in the decimal place, and not affect the general accordance of the above table.

The volume of soda in its salts is derived from the mean specific gravity 2·562 for sulphate of soda, as determined by various experimentalists (2·46 Mohs, 2·631 Karsten, 2·597 P. and J.). This mean result gives 27·9 for the volume of sulphate of soda, from which must be deducted 15·9, the known volume of sulphuric acid, leaving 12 as the volume of soda; the theoretical result is 12·25 ‡.

The volumes of all the oxides in the table, except alumina, are therefore derived from experiment, but in the latter case we are obliged to make an assumption. The specific gravity of ignited alumina is known, but not of alumina as it exists in its salts. In order to arrive at its volume, it is necessary to compare it with what we know of other oxides. Now we are aware that in the sesquioxides, oxygen frequently takes up the volume 1·225 × 3, or 3·675. Wohler gives 2·5 as the sp. gr. of aluminium, and 2·67 as the result of another estimation. The first estimation, in giving the same volume to aluminium as to chromium in its least dense state, has analogy in its favour, and the volume calculated from it would be 5·47, which is not far from theory, the latter affording 5·51. The volume of alumina in its least dense form would,

---

* Chem. Mem. vol. i. p. 433.   † Idem, vol. ii. p. 101.

‡ The volume of soda when separate, so far as we may rely on one experiment, that of Karsten, is 11·0, instead of 12·2.

therefore, be $5·51 \times 2 + 3·675 \times 3 = 22·04$. This number, then though strengthened by analogy, is only an assumption, and, must be distinguished as such from the experimental facts above recorded.

It is now important to extend our evidence to another class of highly hydrated salts, and in these it will also be seen that the acid ceases to occupy appreciable space. The alums offer a good series of salts for testing this view; but from our ignorance of the specific gravity of their bases before calcination, certain assumptions have to be made with regard to their volumes. It is well known that alumina, oxide of chromium, and peroxide of iron contract amazingly when heated, so that the bulk of the calcined oxides by no means denotes that of the base in combination.

In the class of the alums, we assume, for reasons formerly given, that alumina has the volume 22·05. Chromium in its least dense form has, according to Thomson, the sp. gr. 5·10, or a volume of 5·5, the same as that found for aluminium; its corresponding oxide will therefore have a volume in its salts of 22·05. With respect to iron-alum, the assumption is made, which is only justified by its isomorphism, that the volume of peroxide of iron in its salts is equal to that of the oxide of chromium and that of alumina in their respective salts; it must be remembered, however, that in the isolated state we do not know this oxide to have a higher volume than 17·15. Potash has a volume of 17·75, according to the sp. gr. of Karsten (2·756), which does not differ much from the volume deduced from the sulphate, viz. 17·1. The volume of anhydrous sulphate of ammonia is 39·2 by experiment *, from which 15·9 being deducted for sulphuric acid, 23·3 is left as the volume of the oxide of ammonium in the sulphate. The assumption in this table is in the case of peroxide of iron; but this being admitted, no other conclusion can be drawn from it than that, in the class of alums, *the water takes up the space which it would do if frozen into ice, the bases assume their own proper volumes, the 4 equivalents of acid occupying no appreciable space.* If we refuse

* Vol. i. Chem. Mem. p. 42).

TABLE IV.

Name.	Formula.	Atomic weight.	Volume of solid water.	Volume of proper metallic base.	Volume of alkaline base.	Sum of vols. of water and base.	Theoretical sp. gravity.	Volume by experiment.	Sp. gr. by mean experiment.	Various authorities.
Potash-Alum	$Al_2O_3, 3SO_3 + KO, SO_3 + 24HO$	476·38	235·2	22·05	17·15	274·4	1·736	274·0	1·731	1·724, Kopp. / 1·74, Mohs. / 1·714, Muschenbroeck. / 1·726 & 1·751, P. & J.
Ammonia-Alum	$Al_2O_3, 3SO_3 + NH_4O, SO_3 + 24HO$	455·38	235·2	22·05	23·27	280·52	1·623	280·2	1·625	1·626, Kopp. / 1·625, P. and J.
Chrome-Alum	$Cr_2O_3, 3SO_3 + KO, SO_3 + 24HO$	503·3	235·2	22·05	17·15	274·4	1·834	273·1	1·848	1·848, Kopp. / 1·826 & 1·856, P. & J.
Ammonia-Iron Alum	$Fe_2O_3, 3SO_3 + NH_4O, SO_3 + 24HO$	481·03	235·2	22·05	23·27	280·52	1·714	280·5	1·715	1·712, Kopp. / 1·718, P. and J.

all assumption in this class of alums, and suppose the
metallic sesquioxides in combination to have the volume of
their ignited oxides (17·15), there would still be only 4·89
left for 4 equivalents of sulphuric acid, which ought to take
up the volume of 63·6; it is therefore certain that at least
3¾ equivalents of acid have disappeared.

It is unnecessary to tabulate the volumes of the double
sulphates of copper, zinc, magnesia, iron, and nickel, with
the sulphates of potash and ammonia, the following list
giving sufficient information for our present purpose.

		Atomic weight.	Sp. gr. by expt.	Atomic vol. by expt.
Sulphate of Copper and Potash	......	220·66	2·163*	102·0
,,	,, Ammonia	199·66	1·893	105·4
,,	Zinc and Potash	221·0	2·240	98·6
,,	,, Ammonia	200·0	1·897	105·4
,,	Magnesia and Potash	201·67	2·053	98·2
,,	,, Ammonia	180·67	1·717	105·2
,,	Iron and Potash	216·73	2·202	98·4
,,	,, Ammonia	195·5	1·848	105·8

The average volume of double potash salts, as deduced
from the above table, is 99·3, including the .copper salt,
which appears to be an exception to the rule. The other
salts, excluding this, have a volume of 98·4. The average
volume of the double ammonia salts is 105·4. The general
formula of the potash salts is $RO, SO_3 + KO, SO_3 + 6HO$.
We have already shown that the volume best representing
that of the magnesian oxides in combination is 6·125, and
the volume of sulphate of potash by experiment is 33·075.

$$6 \text{ eq. Water} = 9\cdot8 \times 6 = 58\cdot8$$
$$1 \text{ eq. Magnesian oxide} = 6\cdot125$$
$$1 \text{ eq. Sulphate of Potash} = 33\cdot075$$
$$\overline{ 98\cdot000}$$

If, therefore, we except sulphate of copper and potash,

* We here give the result obtained with great care as described in the
preceding paper, although it differs considerably from that formerly
obtained, 2·244.

which perhaps from containing mechanical water may have given a specific gravity lower than the truth, we find that the double salts are made up of the volumes of their constituents, the sulphuric acid of the magnesian salt having ceased to occupy space.

The ammoniacal sulphates are made up in the same way.

$$
\begin{array}{ll}
\text{6 eq. Solid water} & =58\cdot8 \\
\text{1 eq. Magnesian oxide} & = 6\cdot125 \\
\text{1 eq. Sulphate of Ammonia} & =39\cdot2 \\
\hline
& 104\cdot12
\end{array}
$$

There are probably other hydrated salts, which might be included among those treated in this paper, but we are anxious at present not to complicate the question, having selected those whose sp. gr. may be considered well established. We now sum up the preceding observations as follows :—

1. The water in highly hydrated salts always occupies the volume of ice.

a. In the class of hydrated arseniates and phosphates with 24 atoms of water, neither acid nor base occupies space, the volume of the solid water alone accounting for that of the salt.

b. In cane- and milk-sugar the carbon ceases to occupy space.

2. Another class of salts, including all the hydrated magnesian sulphates, sulphate of alumina, borax, pyrophosphate of soda, and the alums, possess a volume made up of that of their bases and of their solid water, their acids ceasing to be recognizable in space.

Attention has now to be drawn to some points, which at present we introduce incidentally, without giving them the character of a substantive law, rather looking upon them in the light of coincidences.

We have already viewed the volumes of bodies as being the multiples of $\frac{9\cdot8}{8}$ or of $1\cdot225$. We gave reasons for taking this as a standard unit, and not merely as an arbitrary number for comparison. The consideration of the previous tables now leads to the remarkable result, that, speaking generally, there is one atom of constitutional* water added for every unit volume existing in the base. Thus the constitutional water of the magnesian sulphates amounts to five atoms, the two additional atoms present in certain cases being lost in dry air. The volume of the base of these oxides is 1·225 × 5. Again, there are 18 atoms of water in sulphate of alumina, the oxide itself having a volume of 1·225 × 18. Sulphate of soda affects 10 atoms of water, and soda enjoys a volume of 1·225 × 10. Biborate of soda also possesses 10 atoms of water. It is this circumstance, be it accidental or otherwise, which led to the result, that in many cases the volume of a hydrated salt is the number of its atoms of water multiplied by 11; the number 9·8, the true volume of solid water, being so generally associated with the unit volume 1·225 of the base, and the sum of these giving the number 11·025. There are exceptions to this, which prevent it being announced as general, even in this limited class of salts. The pyrophosphate of soda, containing 2 atoms of soda, possesses only 10 atoms of water, although it might have had 20. The potash of the sulphate of potash in alum has not the number of atoms of water added to it corresponding to its unit volumes, although its acid ceases to occupy space.

In the first class of hydrated salts there is one atom of water added for every unit volume of their *acid*. Thus phosphoric acid and arsenic acid both affect a volume of 1·225 × 24, and the atoms of water attached to their salts are also 24 in number; but in this case the volumes, both of the base and the acids, are merged.

This, however, cannot be looked upon in any other light than as fortuitous; because in the carbonate of soda belonging to this series, carbonic acid affects only 9 of their unit volumes, though in the carbonates of potash it has 110 unit

---

* The term *constitutional water* applied here refers to the water of crystallization, and not to that to which the same name is applied by Graham.

volumes.   Another fact, which, on account of the limited
class of salts examined, may be altogether accidental, is, that
the number of atoms of water attached to the salts in Table I.
is exactly double the number of ultimate atoms in the
anhydrous salt.   Thus there are 12 anhydrous atoms in the
phosphates and arseniates, and 24 atoms of water; in carbonate of soda, there are 5 anhydrous atoms and 10 atoms of
water.   It is impossible to avoid the speculation that the
attachment of atoms of water to the magnesian sulphates, in
some such manner as in the class of phosphates, may be the
cause of the volume of their bases disappearing when these
salts are dissolved in water.

The undoubted result of the disappearance of the volume
of the acid is of much importance in considering the constitution of hydrated salts.   By this disappearance we should
be led to suppose, that the water and the acid in hydrated
salts are in more intimate union than the acid and base; in
other words, that the salt was rather water and acid + base,
than acid and base + water.   Looking back to the results
in this point of view, the constitutional water added to the
sulphuric acid of all the sulphates examined, except sulphate
of soda, amounts to 6 atoms, allowing that the magnesian
oxide plays the part of water, as it so often does.   Thus then,
we have $RO, 5HO, SO_3$ in the sulphates of the magnesian
class; in the alums we have $4 (6HO, SO_3) + Al_2O_3 + KO.$

In the case of the sulphates of zinc, magnesia, and iron,
there is a disposition to affect seven atoms of water, and we
find their radicals also inclined under certain circumstances
to affect larger volumes than others.   Thus magnesium has
a volume of 5·5; iron in the alums has also a similar volume,
and zinc usually possesses a volume of 4·9 instead of 3·12.

It may be merely a coincidence that the number of unit
volumes in the base corresponds to that of the number of
atoms of the water, for we possess few other salts upon
which we can test this view; but on the doctrine of proba-
bilities the wager is a high one against it being merely
chance.   It accounts, however, sufficiently for our having

supposed that water in combination possessed the volume 11. We now withdraw that opinion, having convinced ourselves by the previous study, *that in all the cases mentioned in this paper, the water attached to salts has exactly the volume of ice.*

NOTE, 1885.—It was in the year 1843 that I read a paper "On the Calorific Effects of Magneto-Electricity and the Mechanical Value of Heat" to the Chemical Section of the British Association assembled at Cork. With the exception of some eminent men, among whom I recollect with pride Dr. Apjohn, the president of the Section, the Earl of Rosse, Mr. Eaton Hodgkinson, and others, the subject did not excite much general attention; so that when I brought it forward again at the meeting in 1847, the chairman suggested that, as the business of the section pressed, I should not read my paper, but confine myself to a short verbal description of my experiments. This I endeavoured to do, and discussion not being invited, the communication would have passed without comment if a young man had not risen in the section, and by his intelligent observations created a lively interest in the new theory. The young man was William Thomson, who had two years previously passed the University of Cambridge with the highest honour, and is now probably the foremost scientific authority of the age. My work with Thomson was chiefly experimental, performed in Manchester and the neighbourhood. We pursued the discussion of the mtheral effects of fluids in motion until the experiments were interrupted by the action of the owners of the adjacent property, who, on the strength of an obsolete clause in the deeds of conveyance, threatened legal proceedings, the cost of which I did not feel disposed to incur.

On the Thermal Effects experienced by Air in rushing through small Apertures.   By J. P. JOULE *and* W. THOMSON*.

[Phil. Mag. 4th Series, Suppl. vol. iv. p. 481.]

THE hypothesis that the heat evolved from air compressed and kept at a constant temperature, is mechanically equivalent to the work spent in effecting the compression, assumed by Mayer as the foundation for an estimate of the numerical relation between quantities of heat and mechanical work, and adopted by Holtzmann, Clausius, and other writers, was made the subject of an experimental research by Mr. Joule †, and verified as at least approximately true for air at ordinary atmospheric temperatures.   A theoretical investigation founded on a conclusion of Carnot's ‡, which requires no modification § in the dynamical theory of heat, also leads to a verification of Mayer's hypothesis within limits of accuracy as close as those which can be attributed to Mr. Joule's experimental tests.   But the same investigation establishes the conclusion, that that hypothesis cannot be rigorously true except for one definite temperature within the range of Regnault's experiments on the pressure and latent heat of saturated aqueous vapour, unless the density of the vapour both differs considerably at the temperature 100° Cent. from what it is usually supposed to be, and for other temperatures and pressures presents great discrepancies from the gaseous laws.   No experiments, however, which have yet been published on the density of saturated aqueous

---

* Read to the British Association, September 3, 1852.   The experiments were made by us in one of the cellars of the house No. 1 Acton Square, Salford.

† Phil. Mag. (3rd Series) vol. **xxvi.** May 1845, p. 369.   " On the Changes of Temperature produced by the Rarefaction and Condensation of Air."

‡ Transactions of the Royal Society of Edinburgh, vol. **xvi.** pt. 5 (April 1849).   Appendix to " Account of Carnot's Theory," §§ 46–51.

§ Ibid. vol. **xx.** pt. 2 (March 1851); also Phil. Mag. (Series 4) vol. iv. " On the Dynamical Theory of Heat," § 30.

vapour are of sufficient accuracy to admit of an uncon-
ditional statement of the indications of theory regarding the
truth of Mayer's hypothesis, which cannot therefore be
considered to have been hitherto sufficiently tested either
experimentally or theoretically. The experiments described
in the present communication were commenced by the
authors jointly in Manchester last May. The results which
have been already obtained, although they appear to establish
beyond doubt a very considerable discrepancy from Mayer's
hypothesis for temperatures from 40° to 170° Fahr., are far
from satisfactory ; but as the authors are convinced that,
without apparatus on a much larger scale, and a much more
ample source of mechanical work than has hitherto been
available to them, they could not get as complete and
accurate results as are to be desired, they think it right at
present to publish an account of the progress they have made
in the inquiry.

The following brief statement of the proposed method, and
the principles on which it is founded, is drawn from §§ 77,
78 of Part 4 of the series of articles on the Dynamical
Theory of Heat, republished in the Phil. Mag. from the
'Transactions of the Royal Society of Edinburgh' in 1851*
(vol. xx. part 2, pp. 296, 297).

Let air be forced continuously and as uniformly as possible,
by means of a forcing-pump, through a long tube, open to
the atmosphere at the far end, and nearly stopped in one
place so as to leave, for a short space, only an extremely
narrow passage, on each side of which, and in every other
part of the tube, the passage is comparatively very wide ;
and let us suppose, first, that the air in rushing through the
narrow passage is not allowed to gain any heat from, nor
(if it had any tendency to do so) to part with any to, the
surrounding matter. Then, if Mayer's hypothesis were true,
the air after leaving the narrow passage would have exactly
the same temperature as it had before reaching it. If, on
the contrary, the air experiences either a cooling or a heating

* See also "Dynamical Theory of Heat," pt. 5: Trans. Roy. Soc.
Edin. 1852.

effect in the circumstances, we may infer that the heat
produced by the fluid friction in the rapids, or, which is the
same, the thermal equivalent of the work done by the air in
expanding from its state of high pressure on one side of the
narrow passage to the state of atmospheric pressure which
it has after passing the rapids, is in one case less, and in the
other more, than sufficient to compensate the cold due to
the expansion; and the hypothesis in question would be
disproved.

The apparatus consisted principally of a forcing-pump of
$10\frac{1}{2}$ inches stroke and $1\frac{3}{8}$ internal diameter, worked by a
hand-lever, and adapted to pump air, through a strong copper
vessel * of 136 cubic inches capacity (used for the purpose
of equalizing the pressure of the air) into one end of a spiral
leaden pipe 24 feet long and $\frac{5}{16}$ of an inch in diameter,
provided with a stop-cock at its other end. The spiral was
in all the experiments kept immersed in a large water-bath.

In the first series of experiments, the temperature of the
bath was kept as nearly as possible the same as that of the
surrounding atmosphere; and the stop-cock, which was kept
just above the surface of the water, had a vulcanized india-
rubber tube tied to its mouth. The forcing-pump was
worked uniformly, and the stop-cock was kept so nearly
closed as to sustain a pressure of from two to five atmospheres
within the spiral. A thermometer placed in the vulcanized
india-rubber tube, with its bulb near the stop-cock, always
showed a somewhat lower temperature than another placed
in the water-bath †; and it was concluded that the air had
experienced a cooling effect in passing through the stop-
cock.

---

* This and the forcing-pump are parts of the apparatus used by Mr.
Joule in his original experiments on air. See Phil. Mag. (Series 3) vol.
xxvi. p. 370 (1845).

† When the forcing-pump is worked so as to keep up a uniform
pressure in the spiral, and the water of the bath is stirred so as to be at
a uniform temperature throughout, this temperature will be, with almost
perfect accuracy, the temperature of the air as it approaches the stop-cock.
It is to be remarked, however, that when, by altering the aperture of the

To diminish the effects which might be anticipated from the conduction of heat through the solid matter round the narrow passage, a strong vulcanized india-rubber tube, a few inches long, and of considerably less diameter than the former, was tied on the mouth of the stop-cock in place of that one which was removed, and tied over the mouth of the narrower. The stop-cock was now kept wide open, and the narrow passage was obtained by squeezing the double india-rubber tube by means of a pair of wooden pincers applied to compress the inner tube very near its end, through the other surrounding it. The two thermometers were placed, one, as before, in the bath, and the other in the wide india-rubber tube, with its bulb let down so as to be close to the end of the narrower one within. It was still found that, the forcing-pump being worked as before, when the pincers were applied so as to keep up a steady pressure of two atmospheres or more in the spiral, the thermometer placed in the current of air flowing from the narrow passage showed a lower temperature than that of the air in the spiral, as shown by the other. Sometimes the whole of the narrow india-rubber tube, the wooden pincers, and several inches of the wider tube containing the thermometer, were kept below the surface of the bath, and still the cooling effect was observed; and this even when hot water, at a temperature of about 150° Fahr., was used, although in this case the observed cooling effect was less than when the temperature of the bath was lower.

As it was considered possible that the cooling effects observed in these experiments might be due wholly or partly to the air reaching the thermometer-bulb before it had lost all the *vis viva* produced by the expansion in the narrow

stop-cock, or the rate of working the pump, the pressure within the spiral is altered, even although not very suddenly, the air throughout the spiral, up to the narrow passage, alters in temperature on account of the expansion or condensation which it is experiencing, and there is an immediate corresponding alteration in the temperature of the stream of air flowing from the *rapids*, which produces often a most sensible effect on the thermometer in the issuing stream.

passage, and consequently before the full equivalent of heat had been produced by the friction, and as some influence (although this might be expected to diminish the cooling effect) must have been produced by the conduction of heat through the solid matter round the air, especially about the narrow passage, an attempt was made to determine the whole thermal effect by means of a calorimetrical apparatus applied externally. For this purpose the india-rubber tubes were removed, and the stop-cock was again had recourse to for producing the narrow passage. A piece of small block-tin tube, about 10 inches long, was attached to the mouth of the stop-cock and was bent into a spiral as close to the round the stop-cock as it could be conveniently arranged. A portion of the block-tin pipe was unbent from the principal spiral, and was bent down so as to allow the stop-cock to be removed from the water-bath, and to be immersed with the exit spiral in a small glass jar filled with water. The forcing-pump was now worked at a uniform rate, with the stop-cock nearly closed, for a quarter of an hour, and then nearly open for a quarter of an hour, and so on for several alternations. The temperature of the water in the large bath and the glass jar were observed at frequent stated intervals during these experiments; but instead of there being any cooling effect discovered when the stop-cock was nearly closed, there was found to be a slight elevation of temperature during every period of the experiments, averaging nominally $0°·06525$ F. for four periods of a quarter of an hour when the stop-cock was nearly closed, and $0°·06533$ when it was wide open, or, within the limits of accuracy of the observations, $0°·065$ in each case ; a rise due no doubt to the rising temperature of the surrounding atmosphere during the series of experiments. Hence the results appear at first sight only negative; but it is to be remarked that, the temperature of the bath having been on an average $3\frac{1}{2}°$ F. lower than that of the water in the glass jar, the natural rise of temperature in the glass jar must have been somewhat checked by the air coming from the principal spiral; and had there been no cooling effect due to the rushing through the stop-cock when it was nearly

closed, would have been more checked when the stop-cock
was wide open than when it was nearly closed, as the same
number of strokes of the pump must have sent considerably
more air through the apparatus in one case than in the
other. A cooling effect on the whole, due to the rushing
through the nearly closed stop-cock, is thus indicated, if not
satisfactorily proved.

Other calorimetric experiments were made with the stop-
cock immersed in water in one glass jar, and the air from it
conducted by a vulcanized india-rubber tube, to flow through
a small spiral of block-tin pipe immersed in a second glass
jar of equal capacity ; and it was found that the water in the
jar round the stop-cock was cooled, while that in the other,
containing the exit spiral, was heated, during the working of
the pump, with the stop-cock nearly closed, and a pressure
of about three atmospheres in the principal spiral. The
explanation of this curious result is clearly, that the water
round the stop-cock supplied a little heat to the air in the
first part of the rapids, where it has been cooled by expansion and has not yet received all the heat of the friction ; and
that the heat so obtained, along with the heat produced by
friction throughout the rapids, raises the temperature of the
air a little above what it would have had if no heat had been
gained from without ; so that about the end of the rapids the
air has a temperature a little above that of the surrounding
water, and is led, under the protection of the india-rubber
tube, to the exit spiral with a slightly elevated temperature.
This is what would *necessarily* happen in any case of an
arrangement such as that described, if Mayer's hypothesis
were strictly true ; but then the quantity of heat emitted to
the water in the second glass jar, from the air in passing
through the exit spiral, would be exactly equal to that taken
by conduction through the stop-cock from the water in the
first. In reality, according to the discrepancy from Mayer's
hypothesis, which the other experiments described in this
communication appear to establish, there must have been
somewhat more heat taken in by conduction through the
stop-cock than was emitted by it in flowing through the

exit spiral; but the experiments were not of sufficient
accuracy, and were affected by too many disturbing circumstances, to allow this difference to be tested.

To obtain a decisive test of the discrepancy from Mayer's
hypothesis indicated by the experiments which have been
described, and to obtain either comparative or absolute
determinations of its amount for different temperatures,
some alterations in the apparatus, especially with regard to
the narrow passage and the thermometer for the temperature
of the air flowing from it, were found to be necessary by
Mr. Joule, who continued the research alone, and made the
experiments described in what follows.

A piece of brass piping (see *a*, fig. 9, which is drawn half

Fig. 9.

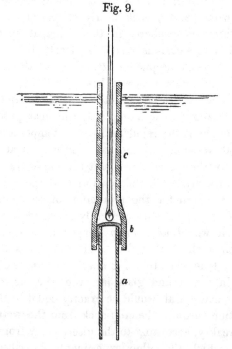

the actual size) was soldered to the termination of the leaden
spiral, and a bit of calf-skin leather, *b*, having been tightly
bound over its end, it was found that the natural pores of

the leather were sufficient to allow of a uniform and conveniently rapid flow of air from the receiver.   By protecting
the end over which the leather diaphragm was bound, with a
piece of vulcanized india-rubber tube c, the former could be
immersed to the depth of two inches in the bath of water.
A small thermometer*, having a spherical bulb $\frac{1}{6}$ of an inch
in diameter, was placed within the india-rubber tube, the
bulb being allowed to rest on the central part of the leather
diaphragm†.

In making the experiments, the pump was worked at a
uniform rate until the pressure of the air in the spiral and
the temperature of the thermometer had become sensibly
constant.   The water of the bath was at the same time
constantly stirred, and by various devices kept as uniform as
possible during each series of experiments.   The temperature
of the stream of air having been observed, the same
thermometer was immediately plunged into the bath to
ascertain its temperature, the difference between the two
readings giving of course the cooling effect of the rushing
air.

According to theory ‡, the cooling effect for a given
temperature would be independent of the kind of aperture

* We had two of these thermometers, one of which had Fahrenheit's,
the other an arbitrary scale.

† The bulb was kept in this position for convenience sake, but it was
ascertained that the effects were not sensibly diminished when it was
raised $\frac{1}{4}$ of an inch above the diaphragm.

‡ See Account of Carnot's Theory, Appendix II. Trans. Royal Soc.
Edin. vol. xvi. p. 566; and Dynamical Theory, § 75, Trans. Royal Soc.
Edin. vol. xx. p. 296; or Phil. Mag. vol. iv. p. 431.   The numbers shown
in the table of § 51 of the former paper being used in the formula of § 75 of
the latter, and 1390 being used for J, we find (according to the numerical
data used formerly for deriving numerical results from the theory) how
much heat would have to be added to each pound of the issuing stream of
air to bring it back to the temperature it had when approaching the narrow
passage; and this number, divided by 0·24, the specific heat of air under
constant pressure, would be the depression of temperature (in Centigrade
degrees) actually experienced by the air when no heat is communicated
to it in or after the rapids.

and of the copiousness of the stream, and would be simply proportional to the logarithm of the pressure, if the insulation of the current against gain or loss of heat from the surrounding matter were perfect, and if the thermometer be so placed in the issuing stream as to be quite out of the *rapids*. On this account the values of the cooling effect divided by the logarithm of the pressure were calculated, and are shown in the last columns of the tables of results given below. When this was done for the first two series of experiments, the discrepancies (see columns 5 of the first two of the tables given below) were found to be so great, and—especially among the results of the different experiments for the higher temperature of 160° F., all made with the pressure and other circumstances as nearly as possible the same—so irregular, that great uncertainty was felt as to the numerical results, which must obviously have been much affected by purely accidental circumstances. At the same time it was noticed, that in the case of Series 1, in which the temperature of the bath was always as nearly as possible that of the atmosphere, and different pressures were used, the discrepancies showed a somewhat regular tendency of the value of the cooling effect divided by the logarithm of the pressure to increase with the pressure; which was probably owing to the circumstance that the stream was more copious, and that less of the cooling effect was lost (as some probably was in every case) by the conduction of heat from without, the higher the pressure under which the air approached the narrow passage. Hence in all the subsequent experiments the quantity of air pumped through per second was noted.

The following Tables show the results obtained from ten series of experiments conducted in the manner described:—

SERIES 1.

Column 1. Quantity of air pumped, in cubic inches per second. A	Column 2. Temperature of bath. T	Column 3. Pressure of air in atmospheres. P	Column 4. Cooling effect. D	Column 5 *. Cooling effect divided by logarithm of pressure. $\frac{D}{\log P}$
Not noted	$\overset{\circ}{6}1$	1·79	$\overset{\circ}{0}$·5	$\overset{\circ}{1}$·98
Not noted	61	2·64	0·94	2·13
Not noted	61	2·9	0·7	1·51
Not noted	61	3·22	1·5	2·95
Not noted	61	3·4	1·4	2·64
Not noted	61	3·61	1·4	2·51
Not noted	61	3·61	1·3	2·33
Not noted	61	3·61	1·4	2·51
Not noted	61	3·84	1·5	2·57
Not noted	61	4·11	1·7	2·77
Mean....	61	....	....	2·39

SERIES 2.

Not noted	$\overset{\circ}{1}60$	2·64	$\overset{\circ}{0}$·264	$\overset{\circ}{0}$·62
Not noted	160	2·64	0·396	0·94
Not noted	160	2·64	0·66	1·56
Not noted	160	2·64	0·528	1·25
Not noted	160	2·64	0·66	1·56
Mean....	160	2·64	0·502	1·18

SERIES 3.

5·6	$\overset{\circ}{1}70·8$	3·61	$\overset{\circ}{0}$·396	$\overset{\circ}{0}$·71
5·6	170·8	4·11	0·528	0·86
5·6	170·8	4·11	0·66	1·08
5·6	170·8	4·11	0·726	1·18
5·6	170·8	4·26	0·66	1·05
8·4	170·8	4·78	0·858	1·26
8·4	170·8	4·98	0·858	1·23
Mean 6·4	170·8	4·28	0·67	1·05

* The true value of $\frac{D}{\log P}$ for any particular temperature would be the depression of temperature that would be experienced by air approaching the narrow passage at that temperature and under ten atmospheres of pressure, since P is measured in atmospheres, and the *common* logarithm is taken.

Q

Series 4.

Column 1. Quantity · f air pumped, in cubic inches per second. A	Column 2. Temperature of bath. T	Column 3. Pressure of air in at-mospheres. P	Column 4. Cooling effect. D	Column 5. Cooling effect divided by logarithm of pressure. $\dfrac{D}{\log P}$
5·6	37·8	3·4	0·8	1·51
5·6	38·8	3·4	1·1	2·07
5·6	37·9	3·61	0·6	1·08
5·6	44·4	3·04	1·1	2·28
5·6	45·3	3·04	0·9	1·86
5·6	46·3	3·04	1·0	2·07
Mean 5·6	41·75	....	....	1·81

Series 5.

8·4	46·8	3·84	1·2	2·06
8·4	38·7	4·11	1·8	2·93
8·4	39·3	4·11	1·8	2·93
Mean 8·4	41·6	....	....	2·64

Series 6.

11·2	39·7	4·4	1·7	2·64
11·2	40·9	4·4	1·9	2·95
11·2	41·9	4·4	1·5	2·33
11·2	43·0	4·4	1·5	2·33
Mean 11·2	41·38	4·4	1·65	2·56

Series 7.

1·4	64·1	1·9	0·3	1·08
1·4	64·2	1·87	0·45	1·65
1·4	64·0	1·9	0·4	1·43
1·4	64·2	1·9	0·5	1·79
1·4	64·3	1·9	0·45	1·61
Mean 1·4	64·16	1·894	0·42	1·51

SERIES 8.

Column 1. Quantity of air pumped, in cubic inches per second. A	Column 2. Temperature of bath. T	Column 3. Pressure of air in atmospheres. P	Column 4. Cooling effect. D	Column 5. Cooling effect divided by logarithm of pressure. $\dfrac{D}{\log P}$
2·8	64·2	2·41	0·5	1·31
2·8	64·3	2·41	0·5	1·31
2·8	64·5	2·41	0·5	1·31
2·8	64·7	2·41	0·7	1·83
2·8	64·7	2·41	0·6	1·57
Mean 2·8	64·48	2·41	0·56	1·46

SERIES 9.

5·6	64·6	2·9	0·8	1·73
5·6	64·7	2·9	0·8	1·73
5·6	64·8	3·04	0·8	1·66
5·6	65·0	2·97	0·7	1·48
Mean 5·6	64·775	....	....	1·65

SERIES 10.

11·2	65·0	4·11	1·2	1·95
11·2	65·1	4·11	1·3	2·12
11·2	65·1	4·11	1·4	2·28
Mean 11·2	65·06	4·11	1·3	2·12

The numbers in the last column of any one of these tables show, by their discrepancies, how much uncertainty there must be in the results on account of purely accidental circumstances.

The following table is arranged, with double argument of temperature and of quantity of air passing per second, to show a comparison of the means of the different Series (Series 3 being divided into two, one consisting of the first five experiments, and the other of the remaining two).

Table of the Mean Values of $\frac{D}{\log P}$ in different Series of Experiments.

Temperature of bath.	Quantity of air passing per second.				
	1·4	2·8	5·6	8·4	11·2
41½°..	....	....	1·81	2·64	2·56
64½°..	1·51	1·46	1·65	....	2·12
171°..	....	....	0·98	1·25	....

The general increase of the numbers from left to right in this table shows that very much of the cooling effect must be lost on account of the insufficiency of the current of air. This loss might possibly be diminished by improving the thermal insulation of the current in and after the rapids; but it appears probable that it could be reduced sufficiently to admit of satisfactory observations being made, only by using a much more copious current of air than could be obtained with the apparatus hitherto employed.

The decrease of the numbers from the upper to the lower spaces, especially in the one complete vertical column (that under the argument 5·6) shows that the cooling effect is less to a remarkable degree for the higher than for the lower temperatures. Even from 41° to 65° F. the diminution is most sensible; and at 171° the cooling effect appears to be only about half as much as at 41°.

The best results for the different temperatures are probably those shown under the arguments 8·4 and 11·2, being those obtained from the most copious currents; but it is probable that they all fall considerably short of the true values of $\frac{D}{\log P}$ for the actual temperatures; and we may consider it as perfectly established by the experiments described above, that *there is a final cooling effect produced by air rushing through a small aperture at any temperature up to* 170° F., and that *the amount of this cooling effect decreases as the*

temperature is augmented.   Now according to the theoretical
views on this subject brought forward in the papers on
" Carnot's theory " and " On the Dynamical Theory of
Heat," already referred to, a cooling effect was expected for
low temperatures ; and the amount of this effect was expected
to be the *less* the *higher* the temperature ; expectations which
have therefore been perfectly confirmed by experiment.   But
since the excess of the heat of compression above the thermal
equivalent of the work was, in the theoretical investigation, found to diminish to zero * as the temperature is raised
to about 33° Cent., or 92° Fahr., and to be negative at all
higher temperatures, a *heating* instead of a *cooling* effect
would be found for such a temperature as 171° F. if the data
regarding saturated steam used in obtaining numerical results
from the theory were correct.   All of these data except the
density had been obtained from Regnault's very exact experimental determinations ; and we may consequently consider
it as nearly certain, that the true values of the density of
saturated aqueous vapour differ considerably from those
which were assumed.   Thus, if the error is to be accounted
for by the *density* alone, the fact of there being any cooling
effect in the air-experiments at 171° Fahr. (77° Cent.) shows
that the density of saturated aqueous vapour at that temperature must be greater than it was assumed to be, in the
ratio of something more than 1416 to 1390, or must be more
than 1·019 of what it was assumed to be ; and since the
experiments render it almost, if not absolutely, certain, that
even at 100° Cent. air rushing through a small aperture
would produce a final cooling effect, it is probable that the
density of steam at the ordinary boiling-point, instead of
being about $\frac{1}{1693·5}$, as it is generally supposed to be, must be
something more than $\frac{1430·6}{1390}$ of this ; that is, must exceed
$\frac{1}{1645}$.

* See the table in § 51 of the Account of Carnot's Theory, from which
it appears that the element tabulated would have the value 1390, or that
of the mechanical equivalent of the thermal unit, at about 33° Cent.

With a view to ascertain what effect would be produced in the case of air rushing violently against the thermometer-bulb, the leather diaphragm was now perforated with a fine needle, and the bulb so placed on the orifice as to cause the air to rush between the leather and the sides of the bulb. With this arrangement the following results were obtained:—

SERIES 11.

A	T	P	D	$\frac{D}{\log P}$
11·2	64	3·22	3·5	6·90
11·2	64	3·31	3·5	6·73
11·2	64	3·61	3·8	6·82
11·2	64	2·30	4·0	11·05
11·2	64	3·31	6·1	11·73
11·2	64	2·58	4·7	11·41
11·2	64	4·78	5·3	7·8
11·2	64	1·9	4·0	14·34
Mean 11·2	64	....	....	9·60

The great irregularities in the last column of the above table are owing to the difficulty of keeping the bulb of the thermometer in exactly the same place over the orifice. The least variation would occasion an immediate and considerable change of temperature; and when the bulb was removed to only ¼ of an inch above the orifice, the cooling effects were reduced to the amount observed when the natural pores alone of the leather were employed. There can be no doubt but that the reason why the cooling effects experienced by the thermometer-bulb were greater in these experiments than in the former is, that in these it was exposed to the current of air in localities in which a sensible portion of the mechanical effect of the work done by the expansion had not been converted into heat by friction, but still existed in the form of *vis viva* of fluid motion. Hence this series of experiments confirms the theoretical anticipations formerly published * regarding the condition of the air in *the rapids* caused by flowing through a small aperture.

* See Dynamical Theory, § 77, Trans. Roy. Soc. Edin. vol. xx. p. 296; or Phil. Mag. Dec. 1852.

On the Thermal Effects of Fluids in Motion. By
WILLIAM THOMSON, M.A., F.R.S., F.R.S.E., &c.,
Professor of Natural Philosophy in the University
of Glasgow, For. Memb. of the Royal Swedish
Academy of Sciences; and J. P. JOULE, F.R.S.,
F.C.S., Corr. Memb. R. A. Turin, Vice-President
of the Literary and Philosophical Society of
Manchester, &c.

[' Philosophical Transactions,' 1853, p. 357.]

IN a paper communicated to the Royal Society, June 20
1844, " On the Changes of Temperature produced by the
Rarefaction and Condensation of Air," * Mr. Joule pointed
out the dynamical cause of the principal phenomena, and
described the experiments upon which his conclusions were
founded.   Subsequently Professor Thomson pointed out that
the accordance discovered in that investigation between the
work spent and the mechanical equivalent of the heat evolved
in the compression of air may be only approximate; and in a
paper communicated to the Royal Society of Edinburgh in
April 1851, " On a Method of discovering experimentally
the Relation between the Mechanical Work spent and the
Heat produced by the Compression of a Gaseous Fluid †,
proposed the method of experimenting adopted in the present
investigation, by means of which we have already arrived at
partial results ‡.   This method consists in forcing the compressed elastic fluid through a mass of porous non-conducting
material, and observing the consequent change of temperature
in the elastic fluid.   The porous plug was adopted instead
of a single orifice, in order that the work done by the
expanding fluid may be immediately spent in friction,

* Philosophical Magazine, ser. 3, vol. xxvi. p. 369.
† Transactions of the Royal Society, Edinburgh, vol. xx. part 2.
‡ Philosophical Magazine, ser. 4, vol. iv. p. 481.

without any appreciable portion of it being even temporarily employed to generate ordinary *vis viva*, or being devoted to produce sound. The non-conducting material was chosen to diminish as much as possible all loss of thermal effect by conduction, either from the air on one side to the air on the other side of the plug, or between the plug and the surrounding matter.

A principal object of the researches is to determine the value of μ, Carnot's function. If the gas fulfilled perfectly the laws of compression and expansion ordinarily assumed, we should have *

$$\frac{1}{\mu} = \frac{\frac{1}{E} + t}{J} + \frac{K\delta}{E p_0 u_0 \log P},$$

where J is the mechanical equivalent of the thermal unit; $p_0 u_0$ the product of the pressure in pounds on the square foot into the volume in cubic feet of a pound of the gas at $0°$ Cent.; P is the ratio of the pressure on the high-pressure side to that on the other side of the plug; δ is the observed cooling effect; t the temperature Cent. of the bath; and K the thermal capacity of a pound of the gas under constant pressure equal to that on the low-pressure side of the gas. To establish this equation it is only necessary to remark that $K\delta$ is the heat that would have to be added to each pound of the exit stream of air to bring it to the temperature of the bath, and is the same (according to the general principle of mechanical energy) as would have to be added to it in passing through the plug to make it leave the plug with its temperature unaltered. We have therefore $K\delta = -H$, in terms of the notation used in the passage referred to.

On the above hypothesis (that the gas fulfils the laws of compression and expansion ordinarily assumed) $\frac{\delta}{\log P}$ would be the same for all values of P; but Regnault has shown that the hypothesis is not rigorously true for atmospheric air,

* Dynamical Theory of Heat, equation (7), § 80, Transactions of the Royal Society of Edinburgh, vol. xx. p. 207.

and our experiments show that $\dfrac{\delta}{\log \mathrm{P}}$ increases with P.   Hence
in reducing the experiments, a correction must be first applied
to take into account the deviations, as far as they are known,
of the fluid used, from the gaseous laws, and then the value
of μ may be determined.   The formula by which this is to be
done is the following (Dynamical Theory of Heat, equation
(f), § 74, or equation (17), § 95, and (8), § 89)—

$$\frac{1}{\mu} = \frac{\frac{1}{3}\{w-(p'u'-pu)\}+\mathrm{K}\delta}{\dfrac{dw}{dt}},$$

where $w=\int_{u'}^{w} p\,dv,$

u and u' denoting the volumes of a pound of the gas at the
high pressure and low pressure respectively, and at the same
temperature (that of the bath), and v the volume of a pound
of it at that temperature, when at any intermediate pressure
p.   An expression for w for any temperature may be derived
from an empirical formula for the compressibility of air at
that temperature, and between the limits of pressure in the
experiment*.

The apparatus, which we have been enabled to provide by
the assistance of a grant from the Royal Society, consists
mainly of a pump, by which air may be forced into a series
of tubes acting at once as a receiver of the elastic fluid and
as a means of communicating to it any required temperature; nozzles, and plugs of porous material being employed
to discharge the air against the bulb of a thermometer.

The pump a, fig. 10, consists of a cast-iron cylinder of 6
inches internal diameter, in which a piston, fig. 11, fitted
with spiral metallic packing (of antifriction metal), works
by the direct action of the beam of a steam-engine through a
stroke of 22 inches.   The pump is single-acting, the air
entering at the base of the cylinder during the up-stroke,
and being expelled thence into the receiving tubes by the
down-stroke.   The governor of the steam-engine limits the
number of complete strokes of the pump to 27 per minute.

* The experiments were made at the Salford Brewery, New Bailey St.

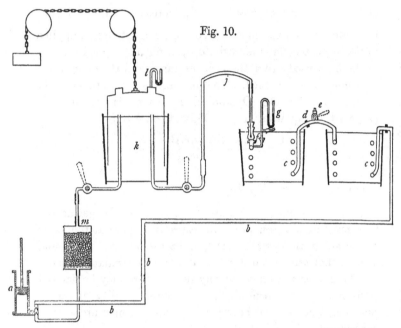

Fig. 10.

The valves, fig. 12, consist of loose spheres of brass 0·6 of an inch in diameter, which fall by their own gravity over orifices 0·45 of an inch diameter. The cylinder and valves in connection with it are immersed in water to prevent the wear and tear which might arise from a variable or too elevated temperature.

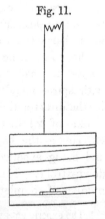

Fig. 11.

Wrought-iron tubing, *b b*, fig. 10, of 2 inches internal diameter, conducts the compressed air horizontally a distance of 6 feet, thence vertically to an elevation of 18 feet, where another length of 23 feet conveys it to the copper tubing, *c c*; the junction being effected by means of a coupling-joint. The copper tubing, which is of 2 inches internal diameter and 74 feet in length, is arranged in two coils, each being immersed in a wooden vessel of 4 feet diameter, from the bottom and

Fig. 12.

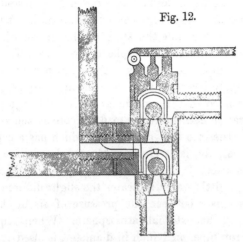

sides of which it is kept at a distance of 6 inches. The coils
are connected by means of a coupling-joint d, near which a
stop-cock, e, is placed, in order to let a portion of air escape
when it is wanted to reduce the pressure. The terminal coil

Fig. 13.

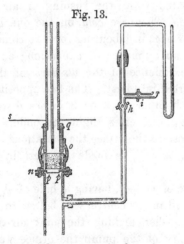

has a flange, f, to which any required nozle may be attached
by means of screw-bolts. Near the flange, a small pipe, g,
is screwed, at the termination of which a calibrated glass
tube, bent (as shown in fig. 13) and partly filled with

mercury, is tightly secured. A stop-cock at h, and another in a small branch-pipe at i, permit the air at any time to be let off, so as to examine the state of the gauge when uninfluenced by any except atmospheric pressure. The branch-pipe is also employed in collecting a small portion of air for chemical analysis during each experiment. A pipe, j, fig. 10, is so suspended that, by means of india-rubber junctions, a communication can readily be made to convey the air issuing from the nozle into the gas-meter, k, which has a capacity of 40 cubic feet, and is carefully graduated by calibration. A bent glass tube, l, inserted in the top of the meter, and containing a little water, indicates the slight difference which sometimes exists between the pressure of air in the meter and that of the external atmosphere. When required, a wrought-iron pipe, m, 1 inch in diameter, is used to convey the elastic fluid from the meter to the desiccating apparatus, and thence to the pump so as to circulate through the entire apparatus.

We have already pointed out the different thermal effects to be anticipated from the rushing of air from a single narrow orifice. They are *cold*, on the one hand, from the expenditure of heat in labouring force to communicate rapid motion to the air by means of expansion; and *heat*, on the other, in consequence of the *vis viva* of the rushing air being reconverted into heat. The two opposite effects nearly neutralize each other at 2 or 3 inches distance from the orifice, leaving, however, a slight preponderance of cooling effect; but close to the orifice the variations of temperature are excessive, as will be made manifest by the following experiments.

A thin plate of copper, having a hole of $\frac{1}{20}$th of an inch diameter, drilled in the centre, was bolted to the flange, an india-rubber washer making the joint air-tight. At the ordinary velocity of the pump the orifice was sufficient to discharge the whole quantity of air when its pressure arrived at 124 lbs. on the square inch. When, however, lower pressures were tried, the stop-cock e was kept partially open. The thermometer used was one with a spherical bulb 0·15 of an inch in diameter. Holding it as close to the orifice as possible without touching the metal, the following observations were made at various pressures, the temperature of the water in which the coils were immersed being 22° Cent. The air was dried and deprived of carbonic acid by passing it, previous to entering the pump, through a vessel $4\frac{1}{2}$ feet long and 20 inches diameter, filled with quicklime.

Total pressure of the air in lbs. on the square inch.	Temperature Centigrade.	Depression below temperature of bath.
124	8·58	13·42
72	11·65	10·35
31	16·25	5·75

The heating effect was exhibited as follows :—The bulb of the thermometer was inserted into a piece of conical gutta-percha pipe is such a manner that an extremely narrow passage was allowed between the interior surface of the pipe and the bulb. Thus armed, the thermometer was held, as represented by fig. 14, at half an inch distance from the orifice, when the following results were obtained :—

Fig. 14.

Total pressure of the air in lbs. on the square inch.	Temperature Centigrade.	Elevation above temperature of bath.
124	45·75	23·75
71	39·23	17·23
31	26·2	4·20

It must be remarked, that the above-recorded thermal effects are not to be taken as representing the maximum results to be derived from the rushing air at the pressures named. The determination of these, in the form of experiment above given, is prevented by several circumstances. In particular it must be observed, that the cooling effects must have been reduced in consequence of the heat evolved by the friction of the rushing air against the bulb of the thermometer. The heating effects, resulting as they do from the absorption and conversion into heat of the *vis viva* of the rushing air, depend very much upon the narrowness of the space between the thermometer and gutta-percha pipe. We intend further on to return to this subject, but in the mean time will mention three forms of experiment whereby the heating effect is very strikingly and instructively exhibited.

Experiment 1.—The finger and thumb are brought over the orifice, as represented in fig. 15, so that by gradually closing them the stream of air is pinched. It is found that the effort to close the finger and thumb is opposed by considerable force, which increases with the pressure applied. At

Fig. 15.

the same time a strong tremulous motion is felt and a shrill noise is heard, whilst the heat produced in five or six seconds necessitates the termination of the experiment.

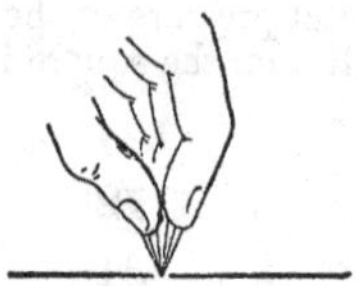

Experiment 2.—Fig. 16. The finger is placed over the orifice and pressed until a thin stratum of air escapes between the copper plate and the finger. In this case the burning heat of the rushing air is equally remarkable in spite of the proximity of the finger to the cold metal.

Fig. 16.

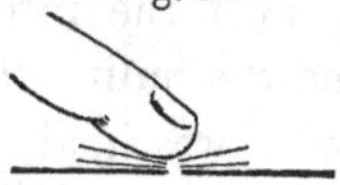

Experiment 3.—Fig. 17. A piece of thick india-rubber is pressed by the finger over the narrow orifice so as to allow a thin stream of air to rush between the india-rubber and the plate of copper. In this case the india-rubber is speedily raised to a temperature which prevents its being handled comfortably.

Fig. 17.

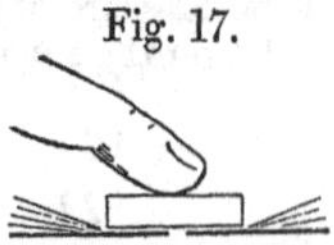

We have now adduced enough to illustrate the immense and sudden changes of temperature which exist in the "rapids" of a current of air, changes which point out the necessity of employing a porous plug, in order that when the air arrives at the thermometer its state may be reduced to a uniform condition. Figs. 13 and 18 represent our first arrangement for the porous plug, where n is a brass casting with flange to bolt to the copper tube. It has eight studs, o, and eight holes, pp,

Fig. 18.

drilled into the inner part of the flange. The studs and holes furnish the means of securing the porous material (in the present instance of cotton wool) in its place, by binding it down tightly with twine. Immediate contact between the cotton and metal is prevented by the insertion of a piece of india-rubber tubing; $q\,q\,q$ are three pieces of india-rubber tube inserted within each other, the inner one communicating with a glass tube r, through which the divisions of the thermometer may be seen, and which serves to convey the air to the meter. In the experiment about to be given, the thermometer was in immediate contact with the cotton plug as represented in the figure, and the nozzle was immersed in the bath up to the line s. The weight of the cotton wool in a dry state was 251 grs., its specific gravity 1·404, and being compressed into a space of 1½ inch in diameter and 1·9 inch long, the opening left for the passage of air must have been equal in volume to a pipe of 1·33 of an inch diameter.

A Liebig tube containing sulphuric acid, specific gravity 1·8, gained 0·03 of a grain by passing through it, during the experiment, 100 cubic inches of air.

The observations tabulated (p. 240) were made at intervals of two or three minutes. It will be observed that the cooling effect appeared to be greater at the commencement than at the termination of the series. This may be attributed in a great measure to the drying of the cotton, which was found to contain at least 5 per cent. of moisture after exposure to the atmosphere. There was also another source of interference with the accuracy of the results, owing to a considerable oscillation of pressure arising from the action of the pump. We had remarked that when the number of strokes of the engine was suddenly reduced from twenty-seven to twenty-five per minute, a depression of the thermometer equal to some hundredths of a degree Cent. took place, a circumstance evidently owing to the entire mass of air in the coils and cotton plugs suffering dilatation without allowing time for the escape of the consequent thermal effect. Hence it was found absolutely essential to keep the pump working at a perfectly uniform rate. For a similar reason

First Series of Experiments. Atmospheric air dried and deprived of carbonic acid by quicklime.
Gauge 73·6 ; barom. 30·04=14·695 lbs. pressure per square inch.

Gauge.	Total pressure in lbs. per square inch.	Cubic inches of air passed per minute reduced to atmospheric pressure.	Temperature of bath* ascertained by Thermometer No. 1, in Centigrade degrees.	Temperature of the issuing air, ascertained by Thermometer No. 2.	Cooling effect.
37·5 37·5 37·8 38 } 37·7	35·854	12703	445 445·5 445·9 446 } 445·6 =18°·2676	414 414 414·6 414·8 415·4 } 414·35=17°·8298	0°·4378
38 37·8 38 } 37·9	35·647	12703	446·1 446·6 446·8 447·1 } 446·65=18°·3128	416 416·8 417·6 } 416·45=17°·9295	0°·3833
37·75 37·5 37·5 } 37·60	35·866	12703	447·2 447·5 447·8 448 } 447·62=18°·3545	418 418·2 418·4 418 } 418·15=18°·0110	0°·3435

* By varying the temperature of the water in which the coils were immersed, it was found that the temperature of the water surrounding the first coil exercised no perceptible influence, the temperature of the rushing air being entirely regulated by that of the terminal coil. However, the precaution was taken of keeping both coils at nearly the same temperature.

it was also most important to prevent the oscillations of pressure due to the action of the pump, particularly as it appeared obvious that the heat evolved by the sudden increase of pressure, on the admission of a fresh supply of air from the pump, would arrive at the thermometer in a larger proportion than the cold produced by the subsequent gradual dilatation. In fact, on making an experiment in which the air was kept at a low pressure, by opening a stop-cock provided for the purpose, the oscillations of pressure amounting to $\frac{1}{20}$th of the whole, it was found that an apparent heating effect, equal to $0°\cdot2$ Cent., was produced instead of a small cooling effect.

It became therefore necessary to obviate the above source of error, and the method first employed with that view was to place a diaphragm of copper with a hole in its centre $\frac{1}{7}$th of an inch in diameter at the junction between the iron and copper pipes. The oscillation being thus reduced, so as to be hardly perceptible, we made the following observations (Second Series, p. 242).

Suspecting that particles of the sperm oil employed for lubricating the pump were carried mechanically to the cotton plug and interfered with the results, we now substituted a box with perforated caps, filled with cotton wool, for the diaphragm used in the last series. With this arrangement the pressure was kept as uniform as with the other, and all solid and liquid particles were kept back by filtration. (Third Series, p. 243.)

Second Series of Experiments. Atmospheric air dried and deprived of carbonic acid by quicklime. Gauge 73·75; barometer 30·162=14·755 lbs. pressure per square inch; thermometer 19°·3 Cent.

Gauge.	Total pressure in lbs. per square inch.	Cubic inches of air issuing per minute at atmospheric pressure.	Temperature of bath by Thermometer No. 1, degrees Centigrade.	Temperature of the issuing air by Thermometer No. 2, degrees Centigrade.	Cooling effect.
39 38·6 38·5 38·5 } 38·65	36·069	11796	467 467 467 467·1 } 467·02 = 19·186	434·6 435 435 } 434·9 = 18°·810	0°·377
38·5 38·8 38·75 38·8 } 38·79	35·912	11796	467·1 467·2 467·2 467·3 } 467·2 = 19·194	435·1 435·4 435·6 435·4 } 435·37 = 18·832	0·362
38·8 38·8 38·8 } 38·8	35·900	11796	467·3 467·4 467·4 467·4 } 467·37 = 19·202	435·6 435·8 435·9 436 } 435·82 = 18·854	0·348

Third Series of Experiments. Atmospheric air dried and deprived of carbonic acid by quicklime * , and filtered through cotton. Gauge 73°·7; thermometer 21°·7 Cent.; barometer 30·10=14·71 lbs. on the square inch.

Time of observation.	Gauge.		Total pressure in lbs. per square inch.	Cubic inches of air issuing per minute at atmospheric pressure.	Temperature of bath by Thermometer No. 1, in degrees Centigrade.		Temperature of the issuing air by Thermometer No. 2, in degrees Centigrade.		Cooling effect.
m 3	39				357·7		337·35		
6	39·1	39·2	34·410	11784	357·8	357·92 =14·506	337·8	337·89 =14·183	0·323
9	39·5				358		338		
12	39·2				358·2		338·4		
15	39·1				358·7		338·8		
16	39·35	39·19	34·418	11784	358·9	358·97 =14·552	338·7	338·85 =14·230	0·322
18	39·1				359·1		339		
21	39·2				359·2		338·9		
23	39·2				359·4		339·25		
25	39·1	39·18	34·426	11784	359·7	359·72 =14·584	339·8	339·69 =14·270	0·314
28	39·2				359·8		339·7		
30	39·2				360		340		
32	39·5				360·1		340		
34	39·3	39·34	34·279	11784	360·2	360·27 =14·607	340·2	340·25 =14·226	0·311
36	39·25				360·4		340·4		
38	39·3				360·4		340·4		

* The use of quicklime as a desiccating agent was suggested to us by Mr. Thomas Ransome. It answered its purpose admirably after it had fallen a little by use, so as to be finely subdivided. The perfection of its action was shown by the desiccating cylinder remaining, after having been used two hours, cold at the lower part, while the upper part for about 9 inches was made very hot. The analysis of the air passed during the third series of experiments showed that one of the Liebig tubes had gained no weight whatever; and in one instance we have observed that the sulphuric acid of 1·8 specific gravity actually lost weight, apparently indicating that the air dried by quicklime was able to remove water from acid of that density.

The stop-cock for reducing pressure being now partially opened, the observations were continued as follows:—

Time of observation.	Gauge.	Total pressure in lbs. per square inch.	Temperature of bath by Thermometer No. 1, in degrees Centigrade.	Temperature of the issuing air by Thermometer No. 2, in degrees Centigrade.	Cooling effect.
h m					
50	55·1		361·7	344	
52	55·1		361·9	344·8	
54	55·1		361·9	345·3	
55	55·1		361·9	345·8	
57	55·1	22·876	362·1	346·0	0°·114
59	55·1		362·3	346·4	
1 1	55·1		362·4	346·9	
3	55·1	55·12	362·7	347·2	
5	55·1		362·7	347·6	
7	55·3		363	347·9	
11	54·3		363·3	348·9	
13	54·4		363·3	348·9	
15	54·4		363·5	349·2	
17	54·7		363·7	349·4	
19	54·5	23·217	363·9	350	0·011
20	54·5		364·1	350	
22	54·6	54·51	364·2	350·3	
24	54·6		364·2	350·4	
26	54·6		364·2	350·6	
30	54·6		375	356·4	
32	54·6		375·4	358·2	
33	54·2		375·4	359·4	
35	54·3	23·277	375·5	359·8	0·032
37	54·4		375·8	360	
39	54·6	54·88	375·7	360·1	
40	54·3		375·8	360·3	
42	54·5		376	360·4	

Group averages — Bath: 362·26 = 14·693; 363·82 = 14·760; 375·7 = 15·270. Air: 346·19 = 14·579; 349·74 = 14·749; 360 = 15·238.

During the above experiment 100 cubic inches of the air was slowly passed through two Liebig tubes containing sulphuric acid, specific gravity 1·8. The first tube gained 0·006 of a grain, the second remained at exactly the same weight.

P.S. Oct. 14, 1853.—The apparently anomalous results contained in the last Table have been fully explained, and shown to depend on the alteration of pressure which took place towards the beginning of the interval of time from 42^m to 50^m, by subsequent researches which we hope soon to lay before the Royal Society.

On the Thermal Effects of Elastic Fluids. By Professor WILLIAM THOMSON, *F.R.S.*, *and* J. P. JOULE, *Esq.*, *F.R.S.*

[Abstract of the preceding paper. Proceedings Royal Society, June 16, 1853.]

THE authors had already proved, by experiments conducted on a small scale, that when dry atmospheric air, exposed to pressure, is made to percolate a plug of non-conducting porous material, a depression of temperature takes place, increasing in some proportion with the pressure of the air in the receiver. The numerous sources of error which were to be apprehended in experiments of this kind conducted on a small scale, induced the authors to apply for the means of executing them on a larger scale; and the present paper contains the introductory part of their researches with apparatus furnished by the Royal Society, comprising a force-pump worked by a steam-engine and capable of propelling 250 cubic inches of air per second, and a series of tubes by which the elastic fluid is conveyed through a bath of water, by which its temperature is regulated, a flange at the terminal permitting the attachment of any nozzle which is desired.

Preliminary experiments were made in order to illustrate the thermal phenomena which result from the rush of air through a single aperture. Two effects were anticipated—one of heat, arising from the *vis viva* of air in rapid motion; the other of cold, arising from dilatation of the gas and the consequent conversion of heat into mechanical effect. The latter was exhibited by placing the bulb of a very small thermometer close to a small orifice through which dry atmospheric air, confined under a pressure of 8 atmospheres, was permitted to escape. In this case the thermometer was depressed 13° Cent. below the temperature of the bath. The former effect was exhibited by causing the stream of air as it issued from the orifice to pass in a very narrow stream between the bulb of the thermometer and a piece of guttapercha tube in which the latter was enclosed. In this experiment, with a pressure of 8 atmospheres, an elevation of temperature equal to 23° Cent. was observed. The same phenomenon was even more strikingly exhibited by pinching the rushing stream with the finger and thumb, the heat resulting therefrom being insupportable.

The varied effects thus exhibited in the "rapids" neutralize one another at a short distance from the orifice, leaving, however, a small cooling effect, to ascertain the law of which and its amount for various gases the present researches have principally been instituted. A plug of cotton wool was employed, for the purpose at once of preventing the escape of thermal effect in the rapids, and of mechanical effect in the shape of sound. With this arrangement a depression of 0°·31 Cent. was observed, the temperature of the dry atmospheric air in the receiver being 14°·5 Cent., and its pressure 34·4 lbs. on the square inch, and the pressure of the atmosphere being 14·7 lbs. per square inch.

Joule. Vol. II.

Plate I.

CHART Nº I.

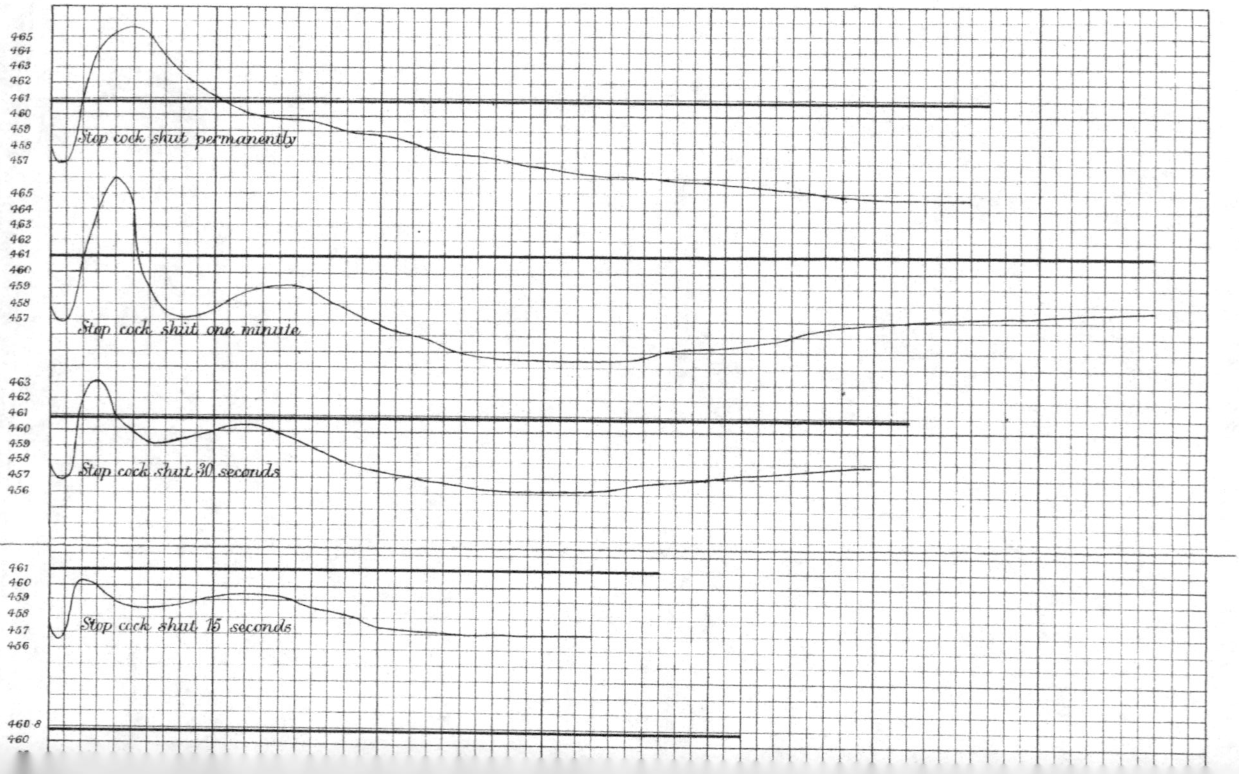

Stop cock shut permanently

Stop cock shut one minute

Stop cock shut 30 seconds

Stop cock shut 15 seconds

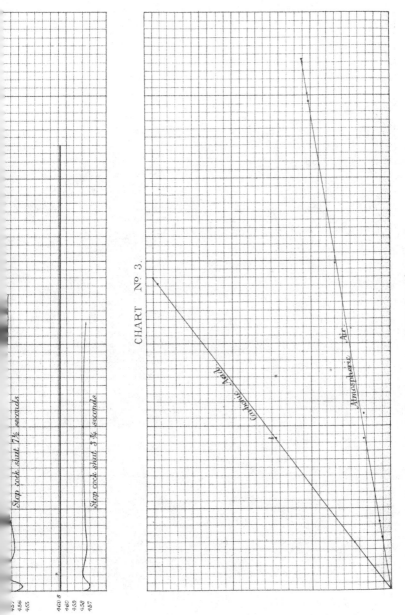

CHART Nº 3.

Carbonic Acid

Atmospheric Air

Stop cock shut 7½ seconds

Stop cock shut 3¾ seconds

4.60.8
4.60
4.59
4.58
4.57

4.56
4.55

Mintern Bros. lith.

On the Thermal Effects of Fluids in Motion.—Part II.
By J. P. Joule, *F.R.S.*, and Professor W. Thomson, *M.A., F.R.S.* *

[Phil. Trans. 1854, vol. cxliv. p. 321.]

[Plates I. & II.]

In the last experiment related in our former paper †, in which a low pressure of air was employed, a considerable variation of the cooling effect was observed, which it was necessary to account for in order to ascertain its influence on the results. We therefore continued the experiments at low pressures, trying the various arrangements which might be supposed to exercise influence over the phenomena. We had already interposed a plug of cotton wool between the iron and copper pipes, which was found to have the very important effect of equalizing the pressure, besides stopping any solid or liquid particles driven from the pump, and which has therefore been retained in all the subsequent experiments. Another improvement was now effected by introducing a nozzle constructed of boxwood, instead of the brass one previously used. This nozzle is represented by fig. 1, Plate I., in which *a a* is a brass casting which bolts upon the terminal flange of the copper piping, *b b* is a turned piece of boxwood screwing into the above, having two ledges for the reception of perforated brass plates, the upper plate being secured in its place by the turned boxwood *c c*, which is screwed into the top of the first piece. The space enclosed by the perforated plates is 2·72 inches long and an inch and a half in diameter, and being filled with cotton, silk, or other material more or less compressed, presents as much resistance to the passage of the air as may be desired. A tin can *d*, filled with cotton wool, and screwing to the brass casting, serves to keep the water of the bath from coming in contact with the boxwood nozzle.

---

* The experiments were made at the Salford Brewery, New Bailey Street.

† Phil. Trans. 1853, part iii.

In the following experiments, made in order to ascertain the variations in the cooling effect above referred to, the nozle was filled with 382 grs. of cotton wool, which was sufficient to keep up a pressure of about 34 lbs. on the inch in the tubes, when the pump was working at the ordinary rate. By opening the stop-cock in the main pipe this pressure could be further reduced to about 22 lbs. by diminishing the quantity of air arriving at the nozle. By shutting and opening the stop-cock we had therefore the means of producing a temporary variation of pressure, and of investigating its effect on the temperature of the air issuing from the nozle. In the first experiments the stop-cock was kept open for a length of time, until the temperature of the rushing air became pretty constant; it was then shut for a period of $3\frac{3}{4}$, $7\frac{1}{2}$, 15, 30, or 60 seconds, then reopened. The oscillations of temperature thus produced are laid down upon the Chart No. 1, in which the ordinates of the curves represent the temperatures according to the scale of thermometer C, each division corresponding to 0·0477 of a degree Centigrade. The divisions of the horizontal lines represent intervals of time equal to a quarter of a minute. The horizontal black lines show the temperature of the bath in each experiment.

The effect upon the pressure of the air produced by shutting the stop-cock during various intervals of time, is given in the following Table :—

Stop-cock shut for	5ˢ.	15ˢ.	30ˢ.	1ᵐ.	2ᵐ.
m s					
Initial pressure	22·35	22·35	22·35	22·35	22·35
Pressure after 0 5	24·92	24·92	24·92	24·92	24·92
Pressure after 0 15	23·07	28·46	28·46	28·46	28·46
Pressure after 0 30	22·43	23·38	30·84	30·84	30·84
Pressure after 0 45	22·35	22·5	24·27	32·03	32·03
Pressure after 1 0	22·35	22·43	22·83	32·79	32·79
Pressure after 1 15	....	22·35	22·45	24·54	33·08
Pressure after 1 30	....	22·35	22·35	22·83	33·25
Pressure after 1 45	....	....	22·35	22·43	33·33
Pressure after 2 0	....	....	....	22·35	33·41
Pressure after 2 15	....	....	....	22·35	24·54
Pressure after 2 30	....	....	....	....	22·54
Pressure after 2 45	....	....	....	....	22·40
Pressure after 3 0	....	....	....	....	22·35

Plate II .

CHART Nº 2 .

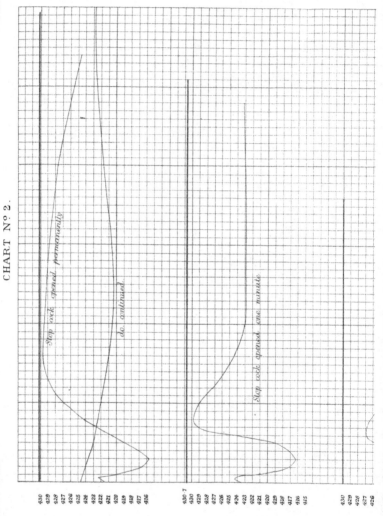

Stop cock opened permanently

do. continued.

Stop cock opened one minute

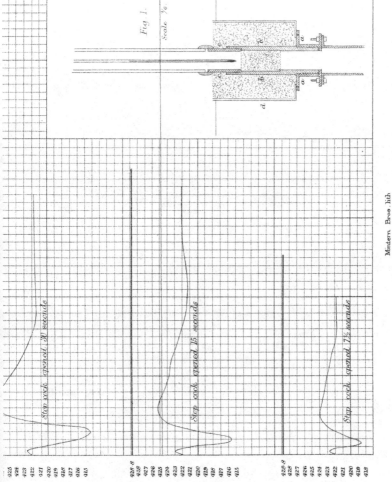

Fig 1.

Scale ⅔

Stop cock opened 30 seconds

Stop cock opened 15 seconds

Stop cock opened 7½ seconds

Mintern Bros. lith

The last column gives also the effect occasioned by the permanent shutting or opening of the stop-cock, 33·41 lbs. being nearly equal to the pressure when the stop-cock has been closed for a long time.

In the next experiments, the opposite effect of opening the stop-cock was tried, the results of which are laid down on Chart No. 2.

The effect upon the pressure of the air produced by opening the stop-cock during the various intervals of time employed in the experiments, is exhibited in the next Table :—

Stop-cock opened for	$3\frac{3}{4}$s.	$7\frac{1}{2}$s.	15s.	30s.	1m.
m s					
Initial pressure	34·37	34·37	34·37	34·37	34·37
Pressure after ...: 0 $3\frac{3}{4}$	29·57	29·57	29·57	29·57	29·57
Pressure after .... 0 $7\frac{1}{2}$	....	27·43	27·43	27·43	27·43
Pressure after 0 15	32·47	30·41	25·15	25·15	25·15
Pressure after 0 30	33·5	32·47	30·41	23·23	23·23
Pressure after 0 45	33·94	33·5	32·4	29·4	22·9
Pressure after 1 0	34·1	34·1	33·5	32·13	22·76
Pressure after 1 15	34·2	34·3	33·94	33·24	28·82
Pressure after 1 30	34·33	34·37	34·14	33·90	31·44
Pressure after 1 45	34·37	34·37	34·30	34·14	32·9
Pressure after 2 0	....	....	34·37	34·33	33·66
Pressure after 2 15	....	...‚	....	34·37	34·06
Pressure after 2 30	....	....	....	....	34·20
Pressure after 2 45	....	....	....	....	34·37

The remarkable fluctuations of temperature in the issuing stream accompanying such changes of pressure, and continuing to be very perceptible in the different cases for periods of from 3 or 4 minutes up to nearly half an hour after the pressure had become sensibly uniform, depend on a complication of circumstances, which appear to consist of (1) the change of cooling effect due to the instantaneous change of pressure ; (2) a heating or cooling effect produced instantaneously by compression or expansion in all the air flowing towards and entering the plug, and conveyed through the plug to the issuing stream ; and (3) heat or cold communicated by contact from the air on the high-pressure side, to the metals and boxwood, and conducted through them to the issuing stream.

The first of these causes may be expected to influence the issuing stream instantaneously on any change in the stop-cock; and after fluctuations from other sources have ceased it must leave a permanent effect in those cases in which the stop-cock is permanently changed. But after a certain interval the reverse agency of the second cause, much more considerable in amount, will begin to affect the issuing stream, will soon preponderate over the first, and (always on the supposition that this convection is uninfluenced by conduction of any of the materials) will affect it with all the variations, undiminished in amount, which the air entering the plug experiences, but behind time by a constant interval equal to the time occupied by as much air as is equal in thermal capacity to the cotton of the plug, in passing through the apparatus *. This, in the experiments with the stop-cock

* To prove this, we have only to investigate the convection of heat through a prismatic solid of porous material, when a fluid entering it with a varying temperature is forced through it in a continuous and uniform stream. Let A B be the porous body, of length a and transverse section S; and let a fluid be pressed continuously through it in the direction from

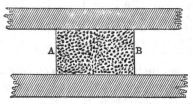

A to B, the temperature of this fluid as it enters at A being an arbitrary function $F(t)$ of the time. Then if v be the common temperature of the porous body and fluid passing through it, at a distance x from the end A, we have

$$\sigma \frac{dv}{dt} = k \frac{d^2v}{dx^2} - \frac{\theta}{S} \frac{dv}{dx}; \quad \ldots \ldots \quad (1)$$

if k be the conducting power of the porous solid for heat (the solid surrounding it being supposed to be an infinitely bad conductor, or the circumstances to be otherwise arranged, as is practicable in a variety of ways, so that there may be no lateral conduction of heat), σ the thermal capacity of unity of its bulk, and θ the thermal capacity of as much of the fluid as passes in the unit time. Now if, as is probably the case in the actual circumstances, conduction through the porous solid itself is

shut, would be very exactly a quarter of a minute; but it
appears to have averaged more nearly one third of a minute
in the varying circumstances of the actual experiments, since
our observations (as may be partially judged from the preceding charts) showed us with very remarkable sharpness, in
each case about twenty seconds after the shutting or opening
of the stop-cock, the commencement of the heating or cooling
effect on the issuing stream, due to the sudden compression
or rarefaction instantaneously produced in the air on the
other side of the plug.

The entering air will, very soon after its pressure ceases to
vary, be reduced to the temperature of the bath by the excellent conducting action of the spiral copper pipe through which
it passes; and, consequently, twenty seconds or so later, the
issuing stream can experience no further fluctuations in
temperature except by the agency depending on the third
cause.

That the third cause may produce very considerable effects
is obvious, when we think how great the variations of tempe-

---

insensible in its influence as compared with the convection of the fluid,
this equation will become approximately

$$\sigma \frac{dv}{dt} = -\frac{\theta}{S} \frac{dv}{dx}, \quad \cdots \cdots \cdots \quad (2)$$

which, in fact, expresses rigorously the effect of the second cause
mentioned in the text if alone operative.

If F denote any arbitrary function, and if θ be supposed to be constant,
the general integral of this equation is—

$$v = F\left(t - \frac{\sigma S}{\theta} x\right); \quad \cdots \cdots \cdots \quad (3)$$

and if the arbitrary function be chosen to express by F(t) the given
variation of temperature where the fluid enters the porous body, we have
the particular solution of the proposed problem. We infer from it that,
at any distance x in the porous body from the entrance, the temperature
will follow the same law and extent of variation as at the entrance, only
later in time by an interval equal to $\frac{\sigma S}{\theta} x$. We conclude that the
variations of temperature in the issuing stream due to the second cause
alone, in the actual circumstances, are equal and similar to those of the

rature must be to which the surfaces of the solid materials in the neighbourhood of the plug on the high-pressure side are subjected during the sudden changes of pressure : and that the heat consequently taken in or emitted by these bodies may influence the issuing stream perceptibly for a quarter or a half hour after the changes of pressure from which it originated have ceased, is quite intelligible on account of the slowness of conduction of heat through the wood and metals, when we take into account the actual dimensions of the parts of the apparatus round the plug. It is not easy, however, to explain all the fluctuations of temperature which have been observed after the pressure had become constant in the different cases. Those shown in the first set of diagrams are just such as might be expected from the alternate heating and cooling which the solids must have experienced at their surfaces on the high-pressure side, and which must be conducted through so as to affect the issuing stream after a considerable time; but the great elevations of temperature shown in the second set of diagrams, which correspond to cases when the pressure was temporarily or permanently

---

air entering the plug, but later in time by $\frac{\sigma Sa}{\theta}$. In this expression, the numerator σSa denotes simply the thermal capacity of the whole plug. The plug, in the actual experiments, having consisted of 382 grains of cotton, of which the thermal capacity is about 191 times that of a grain of water, and (when the stopcock was closed) the air having been pumped through at the rate, per second, of 50 grains, of which the capacity is twelve times that of a grain of water, the value of $\frac{\sigma Sa}{\theta}$ must have been $\frac{191}{12}$ seconds, or about a quarter of a minute. When the stopcock was open, an unknown quantity of air escaped through it, and therefore the value of $\frac{\sigma Sa}{\theta}$ must have been somewhat greater. The variation which the value of θ must have experienced when the stopcock was opened or closed in the course of an experiment, or even merely in consequence of the change of pressure following the initial opening or closing of the stopcock, makes the circumstances not such as in any of the cases to correspond rigorously to the preceding solution; which, notwithstanding, represents the general nature of the convective effect nearly enough for the explanation in the text.

diminished, are not, so far as we see, explained by the causes we have mentioned, and the circumstances of these cases require further examination.

When we had thus examined the causes of the fluctuations of temperature in the issuing air, the precautions to prevent their injurious effect upon the accuracy of the determinations of the cooling effect in the passage of air through the porous plug became evident. These were simply to render the action of the pump as uniform as possible, and to commence the record of observations only after one hour and a half or two hours had elapsed from the starting of the pump. The system then adopted was to observe the thermometers in the bath and stream of air, and the pressure-gauge every two minutes or minute and a half; the means of which observations are recorded in the columns of the Tables. In some instances the air previous to passing into the pump was transmitted through a cylinder which had been filled with quicklime. But since by previous use its power of absorbing water had been considerably deteriorated, a portion of the air was always transmitted through a Liebig tube containing asbestos moistened with sulphuric acid or chloride of zinc. The influence of a small quantity of moisture in the air is trifling, but will hereafter be examined. That of the carbonic acid contained by the atmosphere was, as will appear in the sequel, quite inappreciable. It will be proper to observe that the thermometers by which the temperature of the bath and issuing air was ascertained, were repeatedly compared together to avoid any error which might arise from the alteration of their fixed points from time to time.

In each, excepting the first of the seven experiments recorded, the air was passed through the quicklime cylinder.

TABLE I.—Experiments with a plug consisting of 191 grains of cotton wool.

1.	2.	3.	4.	5.	6.	7.	8.
Number of observations from which the results in Columns 4, 6, and 7 are obtained.	Cubic inches passed through the nozle per minute.	Water in 100 grains of air, in grains.	Pressure in lbs. on the square inch.	Atmospheric pressure.	Temperature of the bath.	Temperature of the issuing air.	Cooling effect in Cent. degrees.
20	10822	0·51	21·326	14·400	20·295	20·201	0·094
20	10998	0·30	21·239	14·252	16·740	16·615	0·125
10	Not observed.	0·56	20·446	14·609	17·738	17·622	0·116
10	10769	0·66	20·910	14·772	16·039	15·924	0·115
10	10769	0·66	20·934	14·775	16·065	15·967	0·098
10	10769	0·66	20·995	14·779	16·084	15·984	0·100
10	10769	0·66	20·933	14·782	16·081	15·974	0·107
Mean ...	....	0·57	20·969	14·624	17·006	16·898	0·108

In the next experiments the nozle was filled with 382 grains of cotton wool. The intermediate stop-cock was however partly opened, in order that by discharging a portion of the air before its arrival at the nozle, the pressure might not be widely different from that employed in the last series. In all excepting the last experiment recorded in the following Table, the cylinder of lime was dispensed with.

TABLE II.—Experiments with a smaller quantity of air passed through a plug consisting of 382 grs. of cotton wool.

1.	2.	3.	4.	5.	6.	7.	8.
Number of observations from which the results in Columns 4, 6, and 7 are obtained.	Cubic inches passed through the nozle per minute.	Water in 100 grains of air, in grains.	Pressure in lbs. on the square inch.	Atmospheric pressure.	Temperature of the bath.	Temperature of the issuing air.	Cooling effect in Cent. degrees.
20	3865	0·59	22·614	14·513	20·363	20·224	0·139
30	3960	0·73	22·818	14·514	19·853	19·769	0·084
20	Not observed.	0·56	22·818	14·604	20·481	20·407	0·074
45	3125	0·65	22·296	14·590	20·584	20·313	0·271
20	Not observed.	1·23	23·000	14·518	18·636	18·476	0·160
36	Not observed.	1·20	22·616	14·520	20·474	20·336	0·138
50	Not observed.	1·36	22·582	14·518	20·485	20·325	0·160
Mean	....	0·90	22·678	14·540	20·125	19·979	0·146

TABLE III.—Experiments in which the entire quantity of air propelled by the pump was passed through a plug consisting of 382 grains of cotton wool. The cylinder of lime was not employed.

1.	2.	3.	4.	5.	6.	7.	8.
Number of observations from which the results in Columns 4, 6, and 7 are obtained.	Cubic inches passed through the nozle per minute.	Water in 100 grains of air, in grains.	Pressure in lbs. on the square inch.	Atmospheric pressure,	Temperature of the bath.	Temperature of the issuing air.	Cooling effect in Cent. degrees.
7	11766	0·56	36·625	14·583	19·869	19·535	0·334
10	Not observed.	0·56	35·671	14·790	20·419	20·098	0·321
10	Not observed.	0·36	35·772	14·504	16·096	15·730	0·366
10	Not observed.	0·36	35·872	14·504	16·104	15·721	0·383
10	Not observed.	0·36	36·026	14·504	16·232	15·869	0·363
Mean	....	0·44	35·993	14·577	17·744	17·390	0·354

In the next series of experiments the air was passed through a plug of silk, formed by rolling a silk handkerchief into a cylindrical shape, and then screwing it into the nozle. The silk weighed 580 grains, and the small quantity of cotton wool placed on the side next the thermometer in order to equalize the stream of air more completely, weighed 15 grains. The stop-cock was partly opened as in the experiments of Table II., in order to reduce the pressure to that obtained by passing the full quantity of air propelled by the pump through a more porous plug. The cylinder of lime was employed.

TABLE IV.—Experiments in which a smaller quantity of air was passed through a plug consisting of 580 grains of silk.

1.	2.	3.	4.	5.	6.	7.	8.
Number of observations from which the results in Columns 4, 6, and 7 are obtained.	Cubic inches passed through the nozle per minute.	Water in 100 grains of air, in grains.	Pressure in lbs. on the square inch.	Atmospheric pressure.	Temperature of the bath.	Temperature of the issuing air.	Cooling effect in Cent. degrees.
10	3071	0·18	33·168	14·727	18·882	18·524	0·358
10	Not observed.	0·18	33·024	14·732	18·884	18·536	0·348
10	Not observed.	0·14	33·820	14·660	19·066	18·686	0·380
10	Not observed.	0·14	33·226	14·650	19·068	18·695	0·373
Mean	....	0·16	33·309	14·692	18·975	18·610	0·365

TABLE V.—Experiments in which the entire quantity of air propelled by the pump was passed through the silk plug. The cylinder of lime was employed in all excepting the first two experiments.

1.	2.	3.	4.	5.	6.	7.	8.
Number of observations from which the results in Columns 4, 6, and 7 are obtained.	Cubic inches passed through the nozle per minute.	Water in 100 grains of air, in grains.	Pressure in lbs. on the square inch.	Atmospheric pressure.	Temperature of the bath.	Temperature of the issuing air.	Cooling effect in Cent. degrees.
10	7594	0·40	53·722	14·580	17·585	16·903	0·682
10	Not observed.	0·40	53·530	14·580	17·628	16·954	0·674
10	Not observed.	0·32	53·317	14·563	17·993	17·318	0·675
10	Not observed.	0·32	53·317	14·563	18·027	17·357	0·670
10	7742	0·11	55·797	14·615	17·822	17·063	0·759
10	Not observed.	0·11	54·074	14·611	17·813	17·079	0·734
10	Not observed.	0·11	55·720	14·608	17·808	17·082	0·726
10	Not observed.	0·11	56·174	14·605	17·796	17·058	0·738
Mean	....	0·23	54·456	14·591	17·809	17·102	0·707

In order to obtain a greater pressure, a plug was formed of silk " waste " compressed very tightly into the nozle.

TABLE VI.—Experiments in which the air, after passing through the cylinder of lime, was forced through a plug consisting of 740 grains of silk.

1.	2.	3.	4.	5.	6.	7.	8.
Number of observations from which the results in Columns 4, 6, and 7 are obtained.	Cubic inches passed through the nozle per minute.	Water in 100 grains of air, in grains.	Pressure in lbs. on the square inch.	Atmospheric pressure.	Temperature of the bath.	Temperature of the issuing air.	Cooling effect in Cent. degrees.
10	Not observed.	0·19	79·852	14·777	17·050	15·884	1·166
10	Not observed.	0·19	80·133	14·782	17·066	15·913	1·153
10	Not observed.	0·19	79·870	14·787	17·079	15·945	1·134
10	5650	0·19	80·013	14·793	17·083	15·967	1·116
10	Not observed.	0·15	79·814	14·960	16·481	15·338	1·143
10	Not observed.	0·15	80·274	14·957	16·489	15·374	1·115
10	Not observed.	0·15	79·903	14·953	16·505	15·392	1·113
10	5378	0·15	77·867	14·950	16·521	15·428	1·093
10	Not observed.	0·14	78·214	14·638	12·851	11·770	1·081
10	Not observed.	0·14	78·245	14·638	12·877	11·800	1·077
10	Not observed.	0·14	78·180	14·638	12·885	11·824	1·061
10	Not observed.	0·14	78·633	14·638	12·905	11·839	1·066
Mean	....	0·16	79·250	14·793	15·483	14·373	1·110

In the foregoing experiments the pressure of the air on its exit from the plug was always exactly equal to the atmospheric pressure. To ascertain the effect of an alteration in the pressure of the exit air, we now enclosed a long siphon-barometer within the glass tube (fig. 14). The upper part of this tube was surmounted with a cap, furnished with a stopcock, by partially closing which the air at its exit could be brought to the required pressure. The influence of pressure in raising the mercury in the thermometer by compressing its bulb, was ascertained by plunging the instrument into a bottle of water within the glass tube, and noting the amount of the sudden rise or fall of the quicksilver on a sudden augmentation or reduction of pressure. It was found that the pressure equal to that of 17 inches of mercury, raised the indication by 0°·09; which quantity was therefore subtracted after the usual reduction of the thermometric scale.

TABLE VII.—Experiments with the plug consisting of 740 grains of silk. Pressure of the exit air increased. Cylinder of lime used.

1.	2.	3.	4.	5.	6.	7.	8.
Number of observations from which the results in Columns 4, 6, and 7 are obtained.	Cubic inches passed through the nozle per minute.	Water in 100 grains of air, in grains.	Pressure in lbs. on the square inch.	Pressure of the exit air.	Temperature of the bath.	Temperature of the issuing air.	Cooling effect in Cent. degrees.
10	Not observed.	0·14	82·982	23·093	12°·673	11°·612	1°·061
10	Not observed.	0·14	82·510	22·878	12·713	11·676	1·037
10	Not observed.	0·14	81·895	22·798	12·755	11·725	1·030
10	Not observed.	0·14	80·630	22·488	12·795	11·792	1·003
Mean	Estimated at 5400	0·14	82·004	22·814	12·734	11·701	1·033

With reference to the experiments in Table VII. it may be remarked, that the cooling effect must be the excess of that which would have been obtained had the air been only resisted by the atmospheric pressure in escaping from the plug, above the cooling effect that would be found in an experiment with the temperature of the bath and the pressure of the entering air the same as the temperature and pressure of the exit air in the actual experiment, and the air issuing at atmospheric pressure. Hence, since two or three degrees of difference of temperature in the bath would not sensibly alter the cooling effect in any of the experiments on air, the cooling effect in an experiment in which the pressure of the exit air is increased, must be sensibly equal to the difference of the cooling effects in two of the ordinary experiments, with the high pressures the same as those used for the entering and issuing air respectively, and the low pressure that of the atmosphere in each case; a conclusion which is verified by the actual results, as the comparison given below shows.

The results recorded in the foregoing Tables are laid down on Chart No. 3, in which the horizontal lines represent the excess of the pressure of the air in the receiver over that of the exit air as found by subtracting the fifth from the fourth columns of the Tables, and the vertical lines represent the cooling effect in tenths of a degree Centigrade. It will be remarked that the line drawn through the points of observation is nearly straight, indicating that the cooling effect is, approximately at least, proportional to the excess of pressure, being about ·018° per pound on the square inch of difference of pressure. Or we may arrive at the same conclusion by dividing the cooling effect (δ) by the difference of pressures ($P - P'$) in the different experiments. We thus find, from the means shown in the different tables,—

$$\text{Table (I.)} \quad \frac{\delta}{P - P'} = ·0170$$

(II.)	·0179
(III.)	·0165
(IV.)	·0196
(V.)	·0177
(VI.)	·0172
(VII.)	·0174
Mean.........	·0176

On the Cooling Effects experienced by Carbonic Acid in passing through a porous Plug.

The position of the apparatus gave us considerable practical facilities in experimenting with carbonic acid. A fermenting tun 10 feet deep and 8 feet square was filled with wort to a depth of 6 feet. After the fermentation had been carried on for about forty hours, the gas was found to be produced in sufficient quantity to supply the pump for the requisite time. The carbonic acid was conveyed by a gutta-percha pipe, and passed through two glass vessels surrounded by ice in order to condense the greater portion of vapours. In the succeeding experiment the total quantity of liquid so condensed was 300 grains, which having a specific gravity of 9965, was composed of 10 grains of alcohol and 290 grains of water. On analysing a portion of the gas during the experiment by passing it through a tube containing chloride of zinc, it was found to contain 0·733 gr. of water to 100 grs. of carbonic acid.

In Table IX., as well as in the next series, the carbonic acid contained 0·35 per cent. of water.

In the experiment of Table X., as well as in those of the adjoining Tables, the sudden diminution of pressure on connecting the pump with the receiver containing carbonic acid, is in perfect accordance with the discovery of Prof. Graham of the superior facility with which that gas may be transmitted through a porous body compared with an equal volume of atmospheric air.

TABLE VIII.—Carbonic acid forced through a plug of 382 grs. of cotton wool. Mean barometric pressure 29·45 inches, equivalent to 14·399 lbs. Gauge under atmospheric pressure 151. The pump was placed in connexion with the pipe immersed in carbonic acid at $10^h 55^m$.

1. Time of observation.	2. Volume percentage of carbonic acid.	3. Pressure-gauge; mean pressure in lbs. on the square inch.	4. Temperature of the bath, by indications of thermometer.	5. Temperature of the issuing gas, by indications of thermometer.	6. Cooling effect in Cent. degrees.
h m					
10 47	0	79·0	486·0	198·5	
49	0	79·0	486·0	198·5	
53	0	79·6	486·0	198·2	
57		85·2	486·0	195·0	
58		86·0	486·0	186·0	
59	95·51	85·0	486·0	186·6	
11 0		85·0	486·0	188·5	
2		86·4	486·0	187·6	
4		86·7	486·0	187·8	
6	95·51	86·6	486·0	188·9	
9		86·6	486·0	188·9	
13		84·0	486·0	188·65	
14		84·2	486·0	188·1	
15	95·51 94·89	84·4 84·906=32·989 lbs.	486·0 486·00=20·001	188·0 188·36=18·611	1·390
19		84·5	486·0	188·0	
22		84·1	486·0	188·1	
24		84·6	486·0	188·3	
25	93·03	84·2	486·0	188·5	
28		84·1	486·0	188·6	
32		83·2	486·0	188·9	

No.				
33	86·82	83·8	486·0	189·9
35		84·0	486·0	189·0
40	80·61	83·8	486·0	189·6
41		83·9	486·0	189·7
43		85·0	485·9	189·9
45		86·0	485·9	190·4
49	79·37	84·6	485·9	190·8
51		84·5	485·9	190·8
53		83·9	485·9	190·6
55	75·65	83·6	485·9	190·6
12 0		83·6	485·9	190·8
2		83·0	485·7	190·8
5	70·68	82·7	485·7	190·9
9		82·7	485·4	190·8
13	68·82	82·9	485·4	191·1
15		82·7	485·5	191·3
21	66·96	82·7	485·4	191·5
23		82·8	485·4	191·55
25	65·72	82·9	485·4	191·6
28		82·9	485·4	191·7
33		82·2	485·4	191·8
35	63·23	82·3	485·4	191·7
40		81·9	485·3	191·65
44		81·9	485·2	191·6
45	63·23	82·1	485·2	191·6
52	63·85	82·4	485·0	191·65
55	62·0	83·9	485·0	192·0
1 2	63·23	84·1	485·0	192·1
5		84·9	485·0	192·1
11	63·23	85·4	485·0	192·3
15	65·72	82·1	484·9	192·1

Summary (block 1): 84·245 = 33·286 ; 485·94 = 19·998 ; 190·1 = 18·787 ; 1·211

Summary (block 2): 82·783 = 33·960 ; 485·52 = 19·980 ; 191·07 = 18·884 ; 1·096

Summary (block 3): 82·986 = 33·864 ; 485·18 = 19·966 ; 191·82 = 18·959 ; 1·007

TABLE IX.—Carbonic acid forced through a plug consisting of 191 grs. of cotton wool. Mean barometric pressure 29·6 inches, equivalent to 14·472 lbs. Gauge under atmospheric pressure 150·6. Pump placed in connexion with the pipe immersed in carbonic acid at $10^h\ 38^m$.

1. Time of observation.	2. Volume percentage of carbonic acid.	3. Pressure-gauge, and pressure in lbs. on the square inch equivalent thereto.	4. Indication of thermometer. Temperature of the bath.	5. Indication of thermometer. Temperature of the issuing gas.	6. Cooling effect in Cent. degrees.
h m					
10 40		123·0	461·5	189·5	
42		123·1	461·6	187·6	
44		123·1	461·6	187·25	
50	95·51	123·0	461·75	187·5	
53		123·2	461·75	187·45	
55		123·0	461·75	187·55	
57	94·58	123·0 } 122·91 = 20·43 lbs.	461·75 } 461·78 = 18·962	187·55 } 187·49 = 18·522	0·44
59		122·9	461·8	187·55	
11 0		122·6	461·9	187·55	
1	93·65	122·6	461·95	187·6	
3		122·6	462·0	187·55	
5		122·5	462·0	188·1	
7		122·8	462·0	188·1	
9		122·1	462·0		
10		121·6	462·0	188·4	
11	81·86	121·7	462·2	188·4	
15		121·6 } 121·91 = 20·682	462·15 } 462·11 = 18·976	188·55 } 188·35 = 18·609	0·367
17	76·27	121·7	462·2	188·7	
19		121·3	462·2	188·65	
20	70·68	121·2	462·2	188·7	
21			462·2		
25			462·2		

TABLE X.—Experiment in which carbonic acid was forced through a plug consisting of 580 grs. of silk. Mean barometric pressure 29·56, equivalent to 14·452 lbs. Gauge under atmospheric pressure 150·8. Pump placed in connexion with the pipe immersed in carbonic acid at 12h 53m. Quantity of gas forced through the plug about 7170 cubic inches per minute.

1.	2.	3.	4.	5.	6.
Time of observation.	Volume percentage of carbonic acid.	Pressure-gauge, and pressure in lbs. on the square inch equivalent thereto.	Indication of thermometer. Temperature of the bath.	Indication of thermometer. Temperature of the issuing gas.	Cooling effect in Cent. degrees.
h m					
12 42	0	52·2	464·2	185·6	
44	0	52·2	464·35	185·5	
46	0	52·2	464·4	185·5	
49	0	52·2 } 52·2 = 55·454 lbs.	464·35 } 464·34 = 19·072	185·55 } 185·53 = 18·323	0·749
50	0	52·2	464·35	185·55	
52		52·2	464·4	185·5	
54		56·0	464·55	179·0	
57		55·7	464·65	166·3	
1 0	95·51	56·0	464·3	165·0	
5		56·0	464·55	165·0	
7	96·0 } 94·85	56·0 } 55·92 = 51·7	464·5 } 464·47 = 19·077	165·0 } 165·0 = 16·256	2·821
9		56·0	464·4	164·9	
10		55·8	464·6	164·9	
11		55·6	464·55	164·8	
13		56·0	464·5	165·0	
17	98·03	56·0	464·4	165·4	
20		55·5	464·6	166·0	
24		55·7	464·6	166·3	
25		56·1	464·6	166·8	
27	85·92	56·0 } 55·94 = 51·68	464·7 } 464·71 = 19·088	167·9 } 167·8 = 16·538	2·550
30		56·1	464·8	168·9	
35		53·1	464·8	169·1	
36		56·1	464·9	169·6	
38					

TABLE XI.—Experiment in which carbonic acid was forced through a plug consisting of 740 grs. of silk. Mean barometric pressure 30·065, equivalent to 14·723 lbs. on the inch. Gauge under atmospheric pressure 145·65. Pump placed in connexion with the pipe immersed in carbonic acid at 11^h 37^m. Percentage of moisture in the carbonic acid 0·15.

1.	2.	3.	4.	5.	6.
Time of observation.	Volume percentage of carbonic acid.	Pressure-gauge, and pressure in lbs. on the square inch equivalent thereto.	Indication of thermometer. Temperature of the bath.	Indication of thermometer. Temperature of the issuing gas.	Cooling effect in Cent. degrees.
h m					
11 28		35·5	318·9	117·9	
30		35·1	318·95	118·0	
32		35·6	318·95	118·0	
34		35·2	318·95	117·9	
36		35·2	318·95	117·73	
37		36·0	318·95	117·5	
38	95·51	36·2	318·95	112·0	
39	95·51	36·6		94·0	
43		36·9	319·03	83·95	
45	95·51	37·0		88·6	
47		37·1		83·0	
50	95·51	37·0	319·05	82·6	
53		37·0		82·4	
55	95·51	37·0	319·15	82·35	
57		37·0		82·3	
12 0	95·51	37·0		82·3	
2		37·0	319·3	82·7	
5	95·51	37·0		83·0	
				88·0	
	95·51	87·0 = 75·324 lbs.	319·17 = 12·844°	82·62 = 7·974°	4·87°

In order to ascertain the cooling effect due to pure carbonic acid, we may at present neglect the effect due to the small quantity of watery vapour contained by the gas; and as the cooling effects observed in the various mixtures of atmospheric air and carbonic acid appear nearly consistent with the hypothesis that the specific heats of the two elastic fluids are for equal volumes equal to one another, and that each fluid experiences in the mixture the same absolute thermo-dynamic effect as if the other were removed, we may for the present take the following estimate of the cooling effects due to pure carbonic acid, at the various temperatures and pressures employed, calculated by means of this hypothesis from the observations in which the percentage of carbonic acid was the greatest, and in fact so great, that a considerable error in the correction for the common air would scarcely affect the result to any sensible extent.

	Temperature of the bath.	Excess of pressure, P−P′.	Cooling effect, δ.	Cooling effect divided by excess of pressure.
From Table IX.....	18·962	5·958	0·459	·0770
From Table VIII.....	20·001	18·590	1·446	·0778
From Table X.....	19·077	37·248	2·938	·0789
From Table XI.....	12·844	60·601	5·049	·0833
	Mean 17·721		Mean	of first three.. ·0779
				Mean of all .. ·0793

We shall see immediately that the temperature of the bath makes a very considerable alteration in the cooling effect, and we therefore select the first three results, obtained at nearly the same temperature, in order to indicate the effect of pressure. On referring to Chart No. 3, it will be remarked that these three results range themselves almost accurately in a straight line. Or, by looking to the numbers in the last column, we arrive at the same conclusion.

Cooling Effect experienced by Hydrogen in passing through a porous Plug.

Not having been able as yet to arrange the large apparatus so as to avoid danger in using this gas in it, we have

TABLE XII.—Experiment in which—1st, air; 2nd, carbonic acid; 3rd, air dried by quicklime, was forced through a plug consisting of 740 grs. of silk. Mean barometric pressure 30·015, equivalent to 14·68 lbs. on the inch. Gauge under the atmospheric pressure 150. Percentage of moisture in the carbonic acid 0·31. Pump placed in connexion with the pipe immersed in carbonic acid at 11h 24m. Disconnected and attached to the quicklime cylinder at 12h 22m.

1.	2.	3.	4.	5.	6.
Time of observation.	Volume percentage of carbonic acid.	Pressure-gauge, and pressure in lbs. on the square inch equivalent thereto.	Indication of thermometer. Temperature of the bath.	Indication of thermometer. Temperature of the issuing gas.	Cooling effect in Cent. degrees.
h m					
11 5	0	31·6	646·35	479·1	
7	0	31·4	646·3	478·8	
9	0	31·7	646·1	478·05	
11	0	31·6 lbs.	646·05	478·1	
13	0	31·9 } 31·62=91·508	646·05 } 646·15=91·452	478·2	478·43=90·008 } 1·444
15	0	31·5	646·05	478·35	
17	0	31·8	646·2	478·7	
19	0	31·5	646·0	478·6	
21	0	32·0	646·0	478·7	
22	0	32·2 } 31·95=90·576	646·1 } 646·08=91·442	478·6 } 478·58=90·043	1·399
23	0	32·2	646·1	478·1	
24	0	32·0	646·1	478·8	
25	0	32·0	646·1	477·0	
26	0	32·1	646·4	471·6	
30	95·51	32·2	646·7	469·2	
32	95·51	32·2 } 32·23=89·799	646·5 } 646·59=91·516	469·5 } 469·63=88·044	3·472
33	95·51	32·0	646·45	469·6	
36	95·51	32·6	646·7	469·6	
38	95·51	32·2	646·6	469·9	
40	95·51	32·2	646·6	469·98	

43	98·03 } 91·81	32·1	} 32·1 = 90·162	646·6	} 647·08 = 91·579	470·05 } 470·57 = 88·255 3·324
46	90·60	32·1		647·0		470·3
48	80·82 } 77·37	32·1		647·1		470·9
50	75·65	32·1		647·4		471·05
53	75·65	32·05	} 32·16 = 90·006	647·2	} 647·5 = 91·647	471·2 } 472·29 = 88·638 3·009
55	65·72 } 62·46	32·0		647·2		471·75
58	60·83	32·0		647·2		472·05
12 0	60·83	32·6		647·7		472·6
4	0	32·25		647·9		472·9
6	0	32·8		647·8		473·25
9	0	32·4	} 32·54 = 88·971	647·95	} 647·94 = 91·711	473·95 } 474·64 = 89·162 2·549
11	0	32·2		647·9		474·1
15	0	32·4		647·95		474·8
20	0	32·9		647·95		475·15
22	0	32·0		647·95		475·2
27	0	32·0		647·85		477·0
29	0	31·6		647·8		480·1
31	0	32·0	} 32·3 = 89·618	647·5	} 647·02 = 91·578	480·6 } 480·97 = 90·528 1·050
33	0	32·1		647·3		480·6
35	0	32·2		647·1		480·8
37	0	32·0		647·0		480·83
39	0	32·1		647·1		480·9
41	0	32·1		647·03		481·03
43	0	32·4		647·1		480·9
45	0	32·6		647·03		481·02
47	0	32·8		647·05		481·04
49	0			646·98		480·98
51	0			646·85		480·9

contented ourselves for the present with obtaining a determination by the help of the smaller force-pump employed in our preliminary experiments. The hydrogen, after passing through a tube filled with fragments of caustic potash, was forced, at a pressure of 68·4 lbs. on the inch, through a piece of leather in contact with the bulb of a small thermometer, the latter being protected from the water of the bath by a piece of india-rubber tube. At a temperature of about 10° Cent., a slight cooling effect was observed, which was found by repeated trials to be 0°·076. The pressure of the atmosphere being 14·7 lbs., it would appear that the cooling effect experienced by this gas is only one-thirteenth of that observed with atmospheric air. We state this result with some reserve, on account of the imperfection of such experiments on a small scale, but there can be no doubt that the effect of hydrogen is vastly inferior to that of atmospheric air.

Influence of Temperature on the Cooling Effect.

By passing steam through pipes plunged into the water of the bath, we were able to maintain it at a high temperature without any considerable variation. The passage of hot air speedily raised the temperature of the stem of the thermometer, as well as of the glass tube in which it was enclosed; but nevertheless the precaution was taken of enclosing the whole in a tin vessel, by means of which water in constant circulation with the water of the bath was kept within one or two inches of the level of the mercury in the thermometer. The bath was completely covered with a wooden lid, and the water kept in constant and vigorous agitation by a proper stirrer.

Although hot air had been passed through the plug for half an hour before the readings in the preceding Table were obtained, it is probable that the numbers 1·444 and 1·399, representing the cooling effect of atmospheric air, are not so accurate as the value 1°·050. Taking this latter figure for the effect of an excess of pressure of 89·618−14·68=74·938 lbs., we find a considerable decrease of cooling effect owing to elevation of temperature, for that pressure, at the low temperatures previously employed, is able to produce a cooling effect of 1°·309.

In order to obtain the effect of carbonic acid unmixed with atmospheric air, we shall, in accordance with the principle already adhered to, consider the thermal capacities of the gases to be equal for equal volumes. Then the cooling effect

of the pure gas $=\dfrac{3 \cdot 472 \times 100 - 1 \cdot 052 \times 4 \cdot 49}{95 \cdot 51} = 3° \cdot 586$.

Collecting these results, we have,

Temperature of bath.	Excess of pressure.	Cooling effect.	Cooling effect reduced to 100 lbs. pressure.	Theoretical cooling effect for 100 lbs. pressure.
$1\overset{\circ}{2}\cdot 844$	60·601	$\overset{\circ}{5}\cdot 049$	$\overset{\circ}{8}\cdot 33$	8·27
19·077	37·248	2·938	7·89	8·07
91·516	74·938	3·586	4·78	4·96

Note.—The numbers shown in the last column of the Table are calculated by the general expression given in our former paper* for the cooling effect, from an empirical formula for the pressure of carbonic acid, recently communicated by Mr. Rankine in a letter, from which the following is extracted.

"Glasgow, May 9, 1854.

" Annexed I send you formulæ for carbonic acid, in which the coefficient *a* has been determined *solely* from Regnault's experiments on the increase of pressure at constant volume between 0° and 100° Cent. It gives most satisfactory results for expansion at constant pressure, compression at constant temperature, and also (I think) for cooling by free expansion " [*i. e.* the cooling effect in our experiments].

" Carbonic Acid Gas.

" P pressure in pounds per square foot.

" V volume of one pound in cubic feet.

" P_0 one atmosphere.

" V_0 *theoretical* volume, in the state of *perfect gas*, of one lb. at the pressure P_0 and the temperature of melting ice.

" $P_0 V_0$ for carbonic acid 17116 feet, log $P_0 V_0 = 4 \cdot 2334023$.

* 'Philosophical Transactions,' June 1853.

[*Note* 1854.—Since this paper has been printed the authors have been informed by Mr. Rankine, that the value of P_0V_0 in his formula for carbonic acid is more nearly 17264 than 17116, as quoted in p. 269, from his letter of May 9. This correction does not sensibly affect the results calculated from his formula for comparison with their experiments.]

" (P_0V_0 *actually*, at 0°, 17145.)

" K_p dynam. spec. heat at constant pressure 300·7 feet; log $K_p = 2·4781334$.

" C absolute temperature of melting ice, 274° Cent.

" The absolute zeros of gaseous tension and of heat are supposed sensibly to coincide, i. e. κ is supposed inappreciably small.

" *Formulæ* :

$$\frac{PV}{P_0V_0} = \frac{T+C}{C} - \frac{a}{T+C}\frac{V_0}{V} \quad \cdot \quad \cdot \quad \cdot \quad \cdot \quad \cdot \quad (1)$$

$$a = 1·9, \quad \log a = 0·2787536.$$

" Cooling by free expansion, supposing the perfect gas-thermometer to give the true scale of absolute temperatures

$$\delta T = \frac{P_0V_0}{K} \cdot \frac{3a}{T+C}\left\{\frac{V_0}{V_1} - \frac{V_0}{V_2}\right\} \quad \cdot \quad \cdot \quad \cdot \quad (2)*$$

$$\log \frac{3P_0V_0a}{K_r} = 2·5111438."$$

By substituting for $\dfrac{V_0}{V_1}$ and $\dfrac{V_0}{V_2}$ their approximate values $\dfrac{C}{T+C}\cdot\dfrac{P_1}{P_0}$ and $\dfrac{C}{T+C}\dfrac{P_2}{P_0}$, we reduce it to

$$\delta = \frac{3P_0V_0aC}{K_r(T+C)^2} \cdot \frac{P_1 - P_2}{P_0},$$

from which we have calculated the theoretical results for different temperatures shown above, which agree remarkably well with those we have obtained from observation.

The interpretation given above for the experimental results on mixtures of carbonic acid and air depends on the assump-

---

* Obtained by using Mr. Rankine's formula (1) in the general expression for the cooling effect given in our former paper, and repeated below as equation (15) of Section V.

tion (rendered probable as a very close approximation to the truth, by Dalton's law), that in a mixture each gas retains all its physical properties unchanged by the presence of the other. This assumption, however, may be only approximately true, perhaps similar in accuracy to Boyle's and Gay-Lussac's laws of compression and expansion by heat; and the theory of gases would be very much advanced by accurate comparative experiments on all the physical properties of mixtures and of their components separately. Towards this object we have experimented on the thermal effect of the mutual interpenetration of carbonic acid and air. In one experiment we found that when 7500 cubic inches of carbonic acid at the atmospheric pressure were mixed with 1000 cubic inches of common air, and a perfect mutual interpenetration had taken place, the temperature had fallen by about ·2° Cent. We intend to try more exact experiments on this subject.

THEORETICAL DEDUCTIONS.

SECTION I. *On the Relation between the Heat evolved and the Work spent in Compressing a Gas kept at Constant Temperature.*

This relation is not a relation of simple mechanical equivalence, as was supposed by Mayer * in his ' Bemerkungen ueber die Kräfte der Unbelebten Natur,' in which he founded on it an attempt to evaluate numerically the mechanical equivalent of the thermal unit. The heat evolved may be less than, equal to, or greater than the equivalent of the work spent, according as the work produces other effects in the fluid than heat, produces only heat, or is assisted by molecular forces in generating heat, and according to the quantity of heat, greater than, equal to, or less than that held by the fluid in its primitive condition, which it must hold to keep itself at the same temperature when compressed. The *a priori* assumption of equivalence, for the case of air, without some special reason from theory or experiment, is not less unwarrantable than for the case of any fluid whatever sub-

* Annalen of Wöhler and Liebig, May 1842.

jected to compression.   Yet it may be demonstrated * that
water below its temperature of maximum density (39°·1 Fahr.),
instead of evolving any heat at all when compressed, actually
absorbs heat, and at higher temperatures evolves heat in
greater or less, but probably always very small, proportion
to the equivalent of the work spent ; while air, as will be
shown presently, evolves always, at least when kept at any
temperature between 0° and 100° Cent., somewhat more
heat than the work spent in compressing it could alone create.
The first attempts to determine the relation in question, for
the case of air, established an approximate equivalence
without deciding how close it might be, or the direction of
the discrepance, if any.   Thus experiments "On the Changes
of Temperature produced by the Rarefaction and Condensation of Air,"† showed an approximate agreement between
the heat evolved by compressing air into a strong copper
vessel under water, and the heat generated by an equal
expenditure of work in stirring a liquid ; and again, conversely, an approximate compensation of the cold of expansion
when air in expanding spends all its work in *stirring* its own
mass by rushing through the narrow passage of a slightly
opened stopcock.   Again, theory ‡, without any doubtful
hypothesis, showed from Regnault's observations on the
pressure and latent heat of steam, that unless the density of
saturated steam differs very much from what it would be if
following the gaseous laws of expansion and compression,
the heat evolved by the compression of air must be sensibly
less than the equivalent of the work spent when the temperature is as low as 0° Cent., and very considerably greater
than that equivalent when the temperature is above 40° or
50°.   Mr. Rankine is, so far as we know, the only other

* Dynamical Theory of Heat, § 63, equation (*b*), Trans. Roy. Soc.
Edinb. vol. xvi. p. 290; or Phil. Mag. vol. iv. series 4, p. 425.

† Communicated to the Royal Society, June 20, 1844, and published
in the Philosophical Magazine, May 1845.

‡ Appendix to "Account of Carnot's Theory," Roy. Soc. Edinburgh,
April 30, 1849, Transactions, vol. xvi. p. 568; confirmed in the Dyna-
mical Theory, § 22, Transactions Roy. Soc. Edinb. March 17, 1851; and
Phil. Mag. vol. iv. series 4, p. 20.

writer who independently admitted the necessity of experiment on the subject, and he was probably not aware of the experiments which had been made in 1844, on the rarefaction and condensation of air, when he remarked *, that "the value of κ is unknown; and as yet no experimental data exist by which it can be determined" (κ denoting in his expressions a quantity the vanishing of which for any gas would involve the equivalence in question). In further observing that probably κ is small in comparison with the reciprocal of the coefficient of expansion, Mr. Rankine virtually adopted the equivalence as probably approximate; but in his article "On the Thermic Phenomena of Currents of Elastic Fluids,"† he took the first opportunity of testing it closely, afforded by our preliminary experiments on the thermal effects of air escaping through narrow passages.

We are now able to give much more precise answers to the question regarding the heat of compression, and to others which rise from it, than those preliminary experiments enabled us to do. Thus if K denote the specific heat under constant pressure, of air or any other gas, issuing from the plug in the experiments described above, the quantity of heat that would have to be supplied, per pound of the fluid passing, to make the issuing stream have the temperature of the bath, would be $K\delta$, or

$$Km\frac{(P-P')}{\Pi},$$

where m is equal to $\cdot 26°$ for air and $1°\cdot 15$ for carbonic acid, since we found that the cooling effect was simply proportional to the difference of pressure in each case, and was $\cdot 0176°$ per pound per square inch, or $\cdot 26°$ per atmosphere, for air, and about $4\frac{1}{2}$ times as much for carbonic acid. This shows precisely how much the heat of friction in the plug falls short of compensating the cold of expansion. But the heat of friction is the thermal equivalent of all the work done actually in the narrow passages by the air expanding as it

* Mechanical Action of Heat, Section II. (10), communicated to the Roy. Soc. Edinb. Feb. 4, 1850, Transactions, vol. xx. p. 166.

† Mechanical Action of Heat, Subsection 4, communicated to the Roy. Soc. Edinb. Jan. 4, 1853, Transactions, vol. xx. p. 580.

flows through. Now this, in the cases of air and carbonic acid, is really not as much as the whole work of expansion, on account of the deviation from Boyle's law to which these gases are subject; but it exceeds the whole work of expansion in the case of hydrogen, which presents a contrary deviation; since P'V', the work which a pound of air must do to escape against the atmospheric pressure, is, for the two former gases, rather greater, and for hydrogen rather less, than PV, which is the work done on it in pushing it through the spiral up to the plug.   In any case, w denoting the whole work of expansion, $w - (\text{P}'\text{V}' - \text{PV})$ will be the work actually spent in friction within the plug ; and

$$\frac{1}{\text{J}} \{ w - (\text{P}'\text{V}' - \text{PV}) \}$$

will be the quantity of heat into which it is converted, a quantity which, in the cases of air and carbonic acid, falls short by

$$\text{K}m\frac{\text{P} - \text{P}'}{\Pi}$$

of compensating the cold of expansion. If, therefore, H denote the quantity of heat that would exactly compensate the cold of expansion, or which amounts to the same, the quantity of heat that would be evolved by compressing a pound of the gas from the volume V' to the volume V, when kept at a constant temperature, we have

$$\frac{1}{\text{J}} \{ w - (\text{P}'\text{V}' - \text{PV}) \} = \text{H} - \text{K}m\frac{\text{P} - \text{P}'}{\Pi},$$

whence

$$\text{H} = \frac{w}{\text{J}} + \left\{ -\frac{1}{\text{J}} (\text{P}'\text{V}' - \text{PV}) + \text{K}m\frac{\text{P} - \text{P}'}{\Pi} \right\}.$$

Now, from the results derived by Regnault from his experiments on the compressibility of air, of carbonic acid, and of hydrogen, at three or four degrees above the freezing-point, we find, approximately,

$$\frac{\text{P}'\text{V}' - \text{PV}}{\text{PV}} = f\frac{\text{P} - \text{P}'}{\Pi},$$

where $\qquad$ $f=\ \cdot00082$ for air,

$\qquad\qquad$ $f=\ \cdot0064$ for carbonic acid,

and $\qquad\qquad$ $f=-\cdot00043$ for hydrogen.

No doubt the deviations from Boyle's law will be somewhat different at the higher temperature (about 15° or 16° Cent.) of the bath in our experiments, probably a little smaller for air and carbonic acid, and possibly greater for hydrogen; but the preceding formula may express them accurately enough for the rough estimate which we are now attempting.

We have, therefore, for air or carbonic acid,

$$H=\frac{w}{J}+\left(Km-\frac{PVf}{J}\right)\frac{P-P'}{\Pi}=\frac{w}{J}+\frac{PV}{J}\left(\frac{JKm}{PV}-f\right)\frac{P-P'}{\Pi}.$$

The values of JK and PV for the three gases in the circumstances of the experiments are as follow :—

$\qquad$ For atmospheric air JK$=1390\times\ \cdot238\ =331$;

$\qquad$ For carbonic acid $\quad$ JK$=1390\times\ \cdot217\ =301$;

$\qquad$ For hydrogen . . $\quad$ JK$=1390\times3\cdot4046=4732$;

and for atmospheric air,

$\qquad$ at 15° Cent. PV$=26224\ (1+15\times\cdot00366)=27663$;

$\qquad$ for carbonic acid,

$\qquad$ at 10° Cent. PV$=17154\ (1+10\times\cdot00366)=17782$;

$\qquad$ for hydrogen,

$\qquad$ at 10° Cent. PV$=378960\ (1+10\times\cdot00367)=393000$.

Hence we have, for air and carbonic acid,

$$H=\frac{w}{J}+\frac{PV}{J}\cdot\lambda\frac{P-P'}{\Pi},$$

where λ denotes $\cdot0024$ for air, and $\cdot013$ for carbonic acid; showing (since these values of λ are positive) that in the case of each of these gases, more heat is evolved in compressing it than the equivalent of the work spent (a conclusion that would hold for hydrogen even if no cooling effect, or a heating effect less than a certain limit, were observed for it in our form of experiment). To find the proportion which this excess bears to the whole heat evolved, or to the thermal equivalent of the work spent in the compression, we may use the expression

$$w=PV\log\frac{P}{P'},$$

as approximately equal to the mechanical value of either of those energies ; and we thus find for the proportionate excess,

$$\frac{H - \frac{1}{J}w}{\frac{1}{J}w} = \lambda \frac{P - P'}{\Pi \log \frac{P}{P'}} = \cdot 0024 \frac{P - P'}{\Pi \log \frac{P}{P'}} \text{ for air,}$$

or $\qquad\qquad = \cdot 013 \dfrac{P - P'}{\Pi \log \dfrac{P}{P'}}$ for carbonic acid.

This equation shows in what proportion the heat evolved exceeds the equivalent of the work spent in any particular case of compression of either gas. Thus for a very small compression from $P' = \Pi$, the atmospheric pressure, we have

$$\log \frac{P}{P'} = \log \left(1 + \frac{P - \Pi}{\Pi} \right) = \frac{P - \Pi}{\Pi} \text{ approximately;}$$

and therefore $\qquad \dfrac{H - \frac{1}{J}w}{\frac{1}{J}w} = \cdot 0024$ for air,

or $\qquad\qquad\qquad\qquad = \cdot 013$ for carbonic acid.

Therefore, when slightly compressed from the ordinary atmospheric pressure, and kept at a temperature of about 60° Fahr., common air evolves more heat by $\frac{1}{417}$, and carbonic acid more by $\frac{1}{77}$ than the amount mechanically equivalent to the work of compression. For considerable compressions from the atmospheric pressure, the proportionate excesses of the heat evolved are greater than these values, in the ratio of the Napierian logarithm of the number of times the pressure is increased, to this number diminished by 1. Thus, if either gas be compressed from the standard state to double density, the heat evolved exceeds the thermal equivalent of the work spent, by $\frac{1}{290}$ in the case of air, and by $\frac{1}{53}$ in the case of carbonic acid.

As regards these two gases, it appears that the observed cooling effect was chiefly due to an actual preponderance of the mechanical equivalent of the heat required to compensate the cold of expansion over the work of expansion, but that rather more than one fourth of it in the case of air, and about one third of it in the case of carbonic acid, depended on a portion of the work of expansion going to do the extra work spent by the gas in issuing against the atmospheric pressure above that gained by it in being sent into the plug. On the other hand, in the case of hydrogen, in such an experiment as we have performed, there would be a heating effect, if the work of expansion were precisely equal to the mechanical equivalent of the cold of expansion, since not only the whole work of expansion, but also the excess of the work done in forcing the gas in above that performed by it in escaping, is spent in friction in the plug. Since we have observed actually a cooling effect, it follows that the heat absorbed in expansion must exceed the equivalent of the work of expansion, enough to over-compensate the whole heat of friction mechanically equivalent, as this is, to the work of expansion together with the extra work of sending the gas into the plug above that which it does in escaping. In the actual experiment* we found a cooling effect of ·076°, with a difference of pressures, $P - P'$, equal to 53·7 lbs. per square inch, or 3·7 atmospheres. Now the mechanical value of the specific heat of a pound of hydrogen is, according to the result stated above, 4732 foot-pounds, and hence the mechanical value of the heat that would compensate the observed cooling effect per pound of hydrogen passing is 360 foot-pounds. But, according to Regnault's experiments on the compression of hydrogen, quoted above, we have

$$PV - P'V' = PV \times ·00043 \, \frac{P - P'}{\Pi} \text{ approximately;}$$

and as the temperature was about 10° in our experiment, we have, as stated above, $PV = 393000$.

---

* From the single experiment we have made on hydrogen we cannot conclude that at other pressures a cooling effect proportional to the difference of pressures would be observed, and therefore we confine the comparison of the three gases to the particular pressure used in the hydrogen experiment. It should be remarked too, that we feel little confidence in the value assigned to the thermal effect for the case observed in the experiment on hydrogen, and only consider it established that it is a cooling effect, and very small.

Hence, for the case of the experiment in which the difference of pressures was 3·7 atmospheres, or

$$\frac{P-P'}{\Pi}=3·7,$$

we have $$PV-P'V'=625;$$

that is, 625 foot-pounds more of work, per pound of hydrogen, is spent in sending the hydrogen into the plug at 4·7 atmospheres of pressure, than would be gained in allowing it to escape at the same temperature against the atmospheric pressure. Hence the heat required to compensate the cold of expansion, is generated by friction from (1) the actual work of expansion, together with (2) the extra work of 625 foot-pounds per pound of gas, and (3) the amount equivalent to 360 foot-pounds which would have to be communicated from without to do away with the residual cooling effect observed. Its mechanical equivalent therefore exceeds the work of expansion by 985 foot-pounds; which is $\frac{1}{630}$ of its own amount, since the work of expansion in the circumstances is approximately $393000 \times \log 4·7 = 608000$ foot-pounds. Conversely, the heat evolved by the compression of hydrogen at 10° Cent., from 1 to 4·7 atmospheres, exceeds by $\frac{1}{630}$ the work spent. The corresponding excess in the case of atmospheric air, according to the result obtained above, is $\frac{1}{174}$, and in the case of carbonic acid $\frac{1}{32}$.

It is important to observe how much less close is the compensation in carbonic acid than in either of the other gases, and it appears probable that the more a gas deviates from the gaseous laws, or the more it approaches the condition of a vapour at saturation, the wider will be the discrepancy. We hope, with a view to investigating further the physical properties of gases, to extend our method of experimenting to steam (which will probably present a large cooling effect), and perhaps to some other vapours.

In Mr. Joule's original experiment* to test the relation

* The second experiment mentioned in the abstract published in the Proceedings of the Royal Society, June 20, 1844, and described in the Philosophical Magazine, May 1845, p. 377.

between heat evolved and work spent in the compression of air, without an independent determination of the mechanical equivalent of the thermal unit, air was allowed to expand through the aperture of an open stopcock from one copper vessel into another previously exhausted by an air-pump; and the whole external thermal effect on the metal of the vessels, and a mass of water below which they are kept, was examined. We may now estimate the actual amount of that external thermal effect, which observation only showed to be insensibly small. In the first place it is to be remarked, that, however the equilibrium of pressure and temperature is established between the two air-vessels, provided only no appreciable amount of work is emitted in sound, the same quantity of heat must be absorbed from the copper and water to reduce them to their primitive temperature; and that this quantity, as was shown above, is equal to

$$\frac{PV}{J} \times \cdot0024 \times \frac{P-P'}{\Pi} = \frac{27000 \times \cdot0024}{1390} \times \frac{P-P'}{\Pi} = \cdot046\frac{P-P'}{\Pi}.$$

In the actual experiments the exhausted vessel was equal in capacity to the charged vessel, and the latter contained 13 of a pound of air under 21 atmospheres of pressure, at the commencement. Hence $P' = \frac{1}{2} P$, and

$$\frac{P-P'}{\Pi} = 10\cdot5;$$

and the quantity of heat required from without to compensate the total internal cooling effect must have been

$$\cdot046 \times 10\cdot5 \times \cdot13 = \cdot063.$$

This amount of heat, taken from $16\frac{1}{2}$ lbs. of water, 28 lbs. of copper, and 7 lbs. of tinned iron, as in the actual experiment, would produce a lowering of temperature of only $\cdot003°$ Cent. We need not therefore wonder that no sensible external thermal effect was the result of the experiment when the two copper vessels and the pipe connecting them were kept under water, stirred about through the whole space surrounding them, and that similar experiments, more recently made by M. Regnault, should have led only to the same negative conclusion.

If, on the other hand, the air were neither allowed to take in heat from nor to part with heat to the surrounding matter in any part of the apparatus, it would experience a resultant cooling effect (after arriving at a state of uniformity of temperature as well as pressure) to be calculated by dividing the preceding expression for the quantity of heat which would be required to compensate it, by ·17, the specific heat of air under constant pressure. The cooling effect on the air itself therefore amounts to

$$0°\!\cdot\!27 \times \frac{P-P'}{\Pi},*$$

which is equal to $2°\!\cdot\!8$, for air expanding, as in Mr. Joule's experiment, from 21 atmospheres to half that presure, and is 900 times as great as the thermometric effect when spread over the water and copper of the apparatus. Hence our present system, in which the thermometric effect on the air itself is directly observed, affords a test hundreds of times more sensitive than the method first adopted by Mr. Joule, and no doubt also than that recently practised by M. Regnault, in which the dimensions of the various parts of the apparatus (although not yet published) must have been on a corresponding scale, or in somewhat similar proportions, to those used formerly by Mr. Joule.

SECTION II.

On the Density of Saturated Steam.

The relation between the heat evolved and the work spent, approximately established by the air-experiments communi-

---

* It is worthy of remark that this, the expression for the cooling effect experienced by a mass of atmospheric air expanding from a bulk in which its pressure is P to a bulk in which, at the same (or very nearly the same) temperature its pressure is P′, and spending all its work of expansion in friction among its own particles, agrees very closely with the expression, $·26 \times \dfrac{P-P'}{\Pi}$, for the cooling effect in the somewhat different circumstances of our experiments.

cated to the Royal Society in 1844, was subjected to an independent indirect test by an application of Carnot's theory, with values of "Carnot's function" which had been calculated from Regnault's data as to the pressure and latent heat of steam, and the assumption (in want of experimental data) that the density varies according to the gaseous laws. The verification thus obtained was very striking, showing an exact agreement with the relation of equivalence at a temperature a little above that of observation, and an agreement with the actual experimental results quite within the limits of the errors of observation; but a very wide discrepancy from equivalence for other temperatures. The following table is extracted from the Appendix to the "Account of Carnot's Theory," in which the theoretical comparison was first made, to facilitate a comparison with what we now know to be the true circumstances of the case.

"Table of the Values of $\dfrac{\mu(1+Et)}{E}$" $= [W]$.

"Work requisite to produce a unit of heat by the compression of a gas $\dfrac{[\mu](1+Et)}{E} = [W]$.	"Temperature of the gas t.	"Work requisite to produce a unit of heat by the compression of a gas $\dfrac{[\mu](1+Et)}{E} = [W]$.	"Temperature of the gas t.
ft. lbs.	°	ft. lbs.	°
1357·1	0	1446·4	120
1368·7	10	1455·8	130
1379·0	20	1465·3	140
1388·0	30	1475·8	150
1395·7	40	1489·2	160
1401·8	50	1499·0	170
1406·7	60	1511·3	180
1412·0	70	1523·5	190
1417·6	80	1536·5	200
1424·0	90	1550·2	210
1430·6	100	1564·0	220
1438·2	110	1577·8	230 "

We now know, from the experiments described above in the present paper, that the numbers in the first column, and, we may conclude with almost equal certainty, that the numbers in the third also, ought to be each very nearly the mechanical equivalent of the thermal unit. This having been ascertained to be 1390 (for the thermal unit Centigrade) by the experiments on the friction of fluids and solids, communicated to the Royal Society in 1849, and the work having been found above to fall short of the equivalent of heat produced, by about $\frac{1}{417}$, at the temperature of the air-experiments at present communicated, and by somewhat less at such a higher temperature as 30°, we may infer that the agreement of the tabulated theoretical result with the fact is perfect at about 30° Cent. Or, neglecting the small discrepance by which the work truly required falls short of the equivalent of heat produced, we may conclude that the true value of $\frac{\mu(1+\mathrm{E}t)}{\mathrm{E}}$ for all temperatures is about 1390; and hence that if [W] denote the numbers shown for it in the preceding table, μ the true value of Carnot's function, and [μ] the value tabulated for any temperature in the "Account of Carnot's Theory," we must have, to a very close degree of approximation,

$$\mu = [\mu] \times \frac{1390}{[\mathrm{W}]}.$$

But if [σ] denote the formerly assumed specific gravity of saturated steam, p its pressure, and λ its latent heat per pound of matter, and if ρ be the mass (in pounds) of water in a cubic foot, the expression from which the tabulated values of [μ] were calculated is

$$[\mu] = \frac{1-[\sigma]}{\rho[\sigma]} \frac{1}{\lambda} \frac{dp}{dt};$$

while the true expression for Carnot's function in terms of properties of steam is

$$\mu = \frac{1-\sigma}{\rho\sigma} \cdot \frac{1}{\lambda} \frac{dp}{dt}.$$

Hence

$$\frac{\mu}{[\mu]} = \frac{[\sigma]}{\sigma} \cdot \frac{1-\sigma}{1-[\sigma]};$$

or, approximately, since σ and [σ] are small fractions,

$$\frac{\mu}{[\mu]} = \frac{[\sigma]}{\sigma}.$$

We have, therefore,

$$\frac{\sigma}{[\sigma]} = \frac{[W]}{1390};$$

and we infer that the densities of saturated steam in reality bear the same proportions to the densities assumed, according to the gaseous laws, as the numbers shown for different temperatures in the preceding Table bear to 1390. Thus we see that the assumed density must have been very nearly correct, about 30° Cent., but that the true density increases much more at the high temperatures and pressures than according to the gaseous laws, and consequently that steam appears to deviate from Boyle's law in the same direction as carbonic acid, but to a much greater amount, which in fact it must do unless its coefficient of expansion is very much less, instead of being, as it probably is, somewhat greater than for air. Also, we infer that the specific gravity of steam at 100° Cent., instead of being only $\frac{1}{1693 \cdot 5}$, as was assumed, or about $\frac{1}{1700}$, as it is generally supposed to be, must be as great as $\frac{1}{1645}$. Without using the preceding Table, we may determine the absolute density of saturated steam by means of a formula obtained as follows. Since we have seen the true value of W is nearly 1390, we must have, very approximately,

$$\mu = \frac{1390E}{1+Et},$$

and hence, according to the preceding expression for μ in terms of the properties of steam,

$$\rho\sigma = \frac{1-\sigma}{1390E}(1+Et)\frac{1}{\lambda}\frac{dp}{dt},$$

or, within the degree of approximation to which we are going (omitting as we do fractions such as $\frac{1}{400}$ of the quantity evaluated),

$$\rho\sigma = \frac{(1+Et)}{1390E.\lambda}\frac{dp}{dt},$$

an equation by which $\rho\sigma$, the mass of a cubic foot of steam in fraction of a pound, or τ, its specific gravity (the value of ρ being 63·887), may be calculated from observations such as those of Regnault on steam. Thus, using Mr. Rankine's empirical formula for the pressure which represents M. Regnault's observations correctly at all temperatures, and M. Regnault's own formula for the latent heat; and taking $E=\dfrac{1}{273}$, we have

$$\rho\sigma=\frac{273+t}{1390}\;\frac{p\left(\dfrac{\beta}{(274\cdot6+t)^2}+\dfrac{2\gamma}{(274\cdot6+t)^3}\right)\times\cdot4342945}{(606\cdot5+0\cdot305t)-(t+\cdot00002t^2+0000003t^3)},$$

with the following equations for calculating p and the terms involving β and γ:—

$$\log_{10}p=a-\frac{\beta}{t+274\cdot6}-\frac{\gamma}{(274\cdot6+t)^2},$$
$$a=4\cdot950433+\log_{10}2114=8\cdot275538,$$
$$\log_{10}\beta=3\cdot1851091,$$
$$\log_{10}\gamma=5\cdot0827176.$$

The densities of saturated steam calculated for any temperatures, either by means of this formula, or by the expression given above, with the assistance of the Table of values of [W], are the same as those which, in corresponding on the subject in 1848, we found would be required to reconcile Regnault's actual observations on steam with the results of air-experiments which we then contemplated undertaking, should they turn out, as we now find they do, to confirm the relation which the air-experiments of 1844 had approximately established. They should agree with results which Clausius* gave as a consequence of his extension of Carnot's principle to the dynamical theory of heat, and his assumption of Mayer's hypothesis.

Section III.

Evaluation of Carnot's Function.

The importance of this object, not only for calculating the efficiency of steam-engines and air-engines, but for advancing the theory of heat and thermo-electricity, was a

* Poggendortt's Annalen, April and May 1850.

principal reason inducing us to undertake the present investigation. Our preliminary experiments, demonstrating
that the cooling effect which we discovered in all of them
was very slight for a considerable variety of temperatures
(from about 0° to 77° Cent.), were sufficient to show, as we
have seen in §§ I. and II., that $\frac{\mu(1+Et)}{E}$ must be very
nearly equal to the mechanical equivalent of the thermal unit;
and therefore we have

$$\mu = \frac{J}{\frac{1}{E}+t} \text{ approximately,}$$

or, taking for E the standard coefficient of expansion of
atmospheric air, ·003665,

$$\mu = \frac{J}{272\cdot85+t}.$$

At the commencement of our first communication to the
Royal Society on the subject, we proposed to deduce more
precise values for this function by means of the equation

$$\frac{J}{\mu} = \frac{JK\delta-(P'V'-PV)+w}{\frac{dw}{dt}};$$

where

$$w = \int_{V}^{V'} p\,dv;$$

v, V, V' denote, with reference to air at the temperature of
the bath, respectively, the volumes occupied by a pound
under any pressure p, under a pressure P, equal to that
with which the air enters the plug, and under a pressure P',
with which the air escapes from the plug; and JKδ is the
mechanical equivalent of the amount of heat per pound of
air passing that would be required to compensate the observed
cooling effect δ. The direct use of this equation for determining $\frac{J}{\mu}$ requires, besides our own results, information as
to compressibility and expansion, which is as yet but very

insufficiently afforded by direct experiments, and is consequently very unsatisfactory, so much so that we shall only give an outline, without details, of two plans we have followed, and mention the results. First, it may be remarked that, approximately,

$$w = (1 + \mathrm{E}t)\,\mathrm{H}\log\frac{\mathrm{P}}{\mathrm{P}'}, \quad \text{and } \frac{dw}{dt} = \mathrm{EH}\log\frac{\mathrm{P}}{\mathrm{P}'};$$

H being the "height of the homogeneous atmosphere," or the product of the pressure into the volume of a pound of air, at 0° Cent.; of which the value is 26224 feet. Hence, if **E** denote a certain mean coefficient of expansion suitable to the circumstances of each individual experiment, it is easily seen that $\dfrac{w}{\frac{dw}{dt}}$ may be put under the form $\dfrac{1}{\mathbf{E}} + t$; and thus we have

$$\frac{\mathrm{J}}{\mu} = \frac{1}{\mathbf{E}} + t + \frac{\mathrm{JK}\delta - (\mathrm{P'V'} - \mathrm{PV})}{\mathrm{EH}\log\dfrac{\mathrm{P}}{\mathrm{P}'}},$$

since the numerator of the fraction constituting the last term is so small, that the approximate value may be used for the denominator. The first term of the second member may easily be determined analytically in general terms; but as it has reference to the rate of expansion at the particular temperature of the experiment, and not to the mean expansion from 0° to 100°, which alone has been investigated by Regnault and others who have made sufficiently accurate experiments, we have not data for determining its values for the particular cases of the experiments. We may, however, failing more precise data, consider the expansion of air as uniform from 0° to 100°, for any pressure within the limits of the experiments (four or five atmospheres); because it is so for air at the atmospheric density by the hypothesis of the air-thermometer, and Regnault's comparisons of air-thermometers in different conditions show for all, whether on the constant-volume or constant-pressure principle, with

density or pressure from one half to double the standard
density or pressure, a very close agreement with the standard
air-thermometer. On this assumption then, when we take
into account Regnault's observations regarding the effect of
variations of density on the coefficient of increase of pressure,
we find that a suitable mean coefficient **E** for the circumstances of the preceding formula for $\dfrac{J}{\mu}$ is expressed, to a
sufficient degree of approximation, by the equation

$$\mathbf{E} = \cdot0036534 + \frac{\cdot0000441}{3\cdot81}\, \frac{P - P'}{\Pi \log \dfrac{P}{P'}}\; .$$

Also, by using Regnault's experimental results on compressibility of air as if they had been made, not at $4^\circ\cdot75$, but
at 16° Cent., we have estimated $P'V' - PV$ for the numerator
of the last term of the preceding expression. We have thus
obtained estimates for the value of $\dfrac{J}{\mu}$, from eight of our experiments (not corresponding exactly to the arrangement in
seven series given above), which, with the various items of
the correction in the case of each experiment, are shown in
the following Table (p. 288).

In consequence of the approximate equality of $\dfrac{J}{\mu}$ to $\dfrac{1}{E} + t$,
its value must be, within a very minute fraction, less by
16 at 0° than at 16°; and, from the mean result of the
Table, we therefore deduce $273\cdot68$ as the value of $\dfrac{J}{\mu}$ at the
freezing-point. The correction thus obtained on the approxi-
mate estimate $\dfrac{1}{E} + t = 272\cdot85 + t$, for $\dfrac{J}{\mu}$, at temperatures not
much above the freezing-point, is an augmentation of $\cdot83$.

For calculating the unknown terms in the expression for
$\dfrac{J}{\mu}$, we have also used Mr. Rankine's formula for the pressure
of air, which is as follows :—

$$pv = H\frac{C+t}{C}\left\{1 - \frac{aC}{(C+t)^2}\left(\frac{1}{\rho v}\right)^{\frac{2}{3}} + \frac{hC}{C+t}\left(\frac{1}{\rho v}\right)^{\frac{1}{2}}\right\},$$

No. of experiment.	Pressure of air forced into the plug. P.	Barometric pressure. P'.	Excess. $P-P'$.	Cooling effect. δ.	Correction by cooling effect. $\dfrac{JK\delta}{EH\log\frac{P}{P'}}$	Correction by reciprocal coefficient of expansion. $\dfrac{1}{E}-\dfrac{1}{\bar{E}}$.	Correction by compressibility (subtracted). $\dfrac{P'V'-PV}{EH\log\frac{P}{P'}}$	Value of J divided by Carnot's function for 16° Cent. $\dfrac{J}{\mu_{16}}$.
I.	20·943	14·777	6·166	0·105	1·031	0·174	0·290	289·4
II.	21·282	14·326	6·956	0·109	0·942	0·168	0·291	289·3
III.	35·822	14·504	21·318	0·375	1·421	0·519	0·412	289·97
IV.	33·310	14·692	18·618	0·364	1·523	0·470	0·372	290·065
V.	55·441	14·610	40·831	0·740	1·892	0·923	0·480	289·705
VI.	53·471	14·571	38·900	0·676	1·814	0·883	0·475	289·59
VII.	79·464	14·955	64·509	1·116	2·272	1·379	0·592	289·69
VIII.	79·967	14·785	65·182	1·142	2·300	1·376	0·586	289·73
							Mean	289·68

where

$$C = 274\cdot6, \quad \log_{10}a = \cdot3176168, \quad \log_{10}h = \bar{3}\cdot8181546,$$
$$H = \frac{26224}{1-a+h};$$

and, v being the volume of a pound of air when at the temperature t and under the pressure p, ρ denotes the mass in pounds of a cubic foot at the standard atmospheric pressure of $29\cdot9218$ inches of mercury. The value of p according to this equation, when substituted in the general expression for $\frac{J}{\mu}$, gives

$$= C + t + \frac{\frac{JKC}{H}\delta + 3h\frac{C^{\frac{3}{2}}}{(C+t)^{\frac{1}{2}}}\left\{\left(\frac{P}{\Pi}\right)^{\frac{1}{4}} - \left(\frac{P'}{\Pi}\right)^{\frac{1}{3}}\right\} - \frac{13}{3}a\left(\frac{C}{C+t}\right)^{\frac{8}{5}}\left\{\left(\frac{P}{\Pi}\right)^{\frac{3}{5}} - \left(\frac{P'}{\Pi}\right)^{\frac{2}{5}}\right\}}{\log\frac{P}{P'}}.$$

From this we find, with the data of the eight experiments just quoted, the following values for $\frac{J}{\mu}$ at the temperature $16°$ Cent.,

$289\cdot044$, $289\cdot008$, $288\cdot849$, $289\cdot112$, $288\cdot787$, $288\cdot722$, $288\cdot505$, $288\cdot559$, the mean of which is $288\cdot82$,

giving a correction of only $\cdot03$ to be subtracted from the previous approximate estimate $\frac{1}{E} + t$.

It should be observed that Carnot's function varies only with the temperature; and therefore if such an expression as the preceding, derived from Mr. Rankine's formula, be correct, the cooling effect, δ, must vary with the pressure and temperature in such a way as to reduce the complex fraction, constituting the second term, to either a constant or a function of t. Now at the temperature of our experiments, δ is very approximately proportional simply to $P-P'$, and therefore all the terms involving the pressure in the numerator ought to be either linear or logarithmic; and the linear terms should balance one another so as to leave only terms which, when divided by $\log\frac{P}{P'}$, become independent

of the pressures.   This condition is not fulfilled by the
actual expression, but the calculated results agree with one
another as closely as could be expected from a formula obtained with such insufficient experimental data as Mr.
Rankine had for investigating the empirical forms which his
theory left undetermined.   We shall see in Section V. below
that simpler forms represent Regnault's data within their
limits of error of observation, and at the same time may be
reduced to consistency in the present application.

As yet we have no data regarding the cooling effect, of
sufficient accuracy for attempting an independent evaluation
of Carnot's function for other temperatures.   In the following section, however, we propose a new system of thermometry
the adoption of which will quite alter the form in which such
a problem as that of evaluating Carnot's function for any
temperature presents itself.

Section IV. *On an Absolute Thermometric Scale founded on
the Mechanical Action of Heat.*

In a communication to the Cambridge Philosophical
Society * six years ago, it was pointed out that any system
of thermometry, founded either on equal additions of heat, or
equal expansions, or equal augmentations of pressure, must
depend on the particular thermometric substance chosen, since
the specific heats, the expansions, and the elasticities of substances vary, and, so far as we know, not proportionally with
absolute rigour for any two substances.   Even the air-thermometer does not afford a *perfect standard,* unless the precise
constitution and physical state of the gas used (the density,
for a pressure-thermometer, or the pressure, for an expansion-
thermometer) be prescribed ; but the very close agreement
which Regnault found between different air- and gas-ther-
mometers removes, for all practical purposes, the incon-
venient narrowness of the restriction to atmospheric air kept
permanently at its standard density, imposed on the thermometric substance in laying down a rigorous definition of

* On an Absolute Thermometric Scale founded on Carnot's Theory of
the Motive Power of Heat, and calculated from Regnault's observations
on Steam," by Prof. W. Thomson ; Proceedings Camb. Phil. Soc. June 5,
1848, or Philosophical Magazine, Oct. 1848.

temperature. It appears then that the standard of practical thermometry consists essentially in the reference to a certain numerically expressible quality of a particular substance. In the communication alluded to, the question, "Is there any principle on which an absolute thermometric scale can be founded?" was answered by showing that Carnot's function (derivable from the properties of any substance whatever, but the same for all bodies at the same temperature), or any arbitrary function of Carnot's function, may be defined as temperature, and is therefore the foundation of an absolute system of thermometry. We may now adopt this suggestion with great advantage, since we have found that Carnot's function varies very nearly in the inverse ratio of what has been called "temperature from the zero of the air-thermometer," that is, Centigrade temperature by the air-thermometer increased by the reciprocal of the coefficient of expansion; and we may define temperature simply as the reciprocal of Carnot's function. When we take into account what has been proved regarding the mechanical action of heat *, and consider what is meant by Carnot's function, we see that the following explicit definition may be substituted:—

If any substance whatever, subjected to a perfectly reversible cycle of operations, takes in heat only in a locality kept at a uniform temperature, and emits heat only in another locality kept at a uniform temperature, the temperatures of these localities are proportional to the quantities of heat taken in or emitted at them in a complete cycle of the operations.

To fix on a unit or degree for the numerical measurement of temperature, we may either call some definite temperature such as that of melting ice, unity, or any number we please; or we may choose two definite temperatures, such as that of melting ice and that of saturated vapour of water under the pressure 29·9218 inches of mercury in the latitude 45°, and call the difference of these temperatures any number we please, 100 for instance. The latter assumption is the only one that can be made conveniently in the present state of science, on account of the necessity of retaining a connection with practical thermometry as hitherto practised; but the

---

* Dynamical Theory of Heat, §§ 42, 43.

former is far preferable in the abstract, and must be adopted
ultimately. In the meantime it becomes a question, What
is the temperature of melting ice, if the difference between
it and the standard boiling-point be called 100°? When
this question is answered within a tenth of a degree or so, it
may be convenient to alter the foundation on which the
degree is defined, by assuming the temperature of melting
ice to agree with that which has been found in terms of the
old degree; and then to make it an object of further experimental research, to determine by what minute fraction the
range from freezing to the present standard boiling-point
exceeds or falls short of 100. The experimental data at
present available do not enable us to assign the temperature
of melting ice, according to the new scale, to perfect certainty
within less than two or three tenths of a degree; but we
shall see that its value is probably about 273·7, agreeing
with the value of $\dfrac{J}{\mu}$ at 0° found by the first method in Section
III. From the very close approximation to equality between
$\dfrac{J}{\mu}$ and $\dfrac{1}{E} + t$, which our experiments have established, we
may be sure that temperature from the freezing-point by
the new system must agree to a very minute fraction of a
degree with Centigrade temperature between the two prescribed points of agreement, 0° and 100°, and we may consider it as highly probable that there will also be a very close
agreement through a wide range on each side of these limits.
It becomes of course an object of the greatest importance,
when the new system is adopted, to compare it with the old
standard; and this is in fact what is substituted for the problem, the evaluation of Carnot's function, now that it is
proposed to call the reciprocal of Carnot's function, temperature. In the next Section we shall see by what kind of
examination of the physical properties of air this is to be
done, and investigate an empirical formula expressing them
consistently with all the experimental data as yet to be had,
so far as we know. The following Table, showing the indications of the constant-volume and constant-pressure air-

thermometer in comparison for every twenty degrees of the
new scale, from the freezing-point to 300° above it, has been
calculated from the formulæ (9), (10), and (39) of Section V.
below.

Comparison of Air-thermometer with Absolute Scale.

Temperature by absolute scale in Cent. degrees from the freezing-point. $t-273{\cdot}7.$	Temperature Centigrade by constant-volume thermometer with air of specific gravity $\frac{\Phi}{v}$. $\theta=100\dfrac{p_t-p_{273{\cdot}7}}{p_{373{\cdot}7}-p_{273{\cdot}7}}.$	Temperature Centigrade by constant-pressure air-thermometer. $\vartheta=100\dfrac{v_t-v_{273{\cdot}7}}{v_{373{\cdot}7}-v_{273{\cdot}7}}.$
$\overset{\circ}{0}$	$\overset{\circ}{0}$	$\overset{\circ}{0}$
20	$20+{\cdot}0298\times\dfrac{\Phi}{v}$	$20+{\cdot}0404\times\dfrac{p}{\Pi}$
40	$40+{\cdot}0403$,,	$40+{\cdot}0477$,,
60	$60+{\cdot}0366$,,	$60+{\cdot}0467$,,
80	$80+{\cdot}0223$,,	$80+{\cdot}0277$,,
100	$100+{\cdot}0000$,,	$100+{\cdot}0000$,,
120	$120-{\cdot}0284$,,	$120-{\cdot}0339$,,
140	$140-{\cdot}0615$,,	$140-{\cdot}0721$,,
160	$160-{\cdot}0983$,,	$160-{\cdot}1134$,,
180	$180-{\cdot}1382$,,	$180-{\cdot}1571$,,
200	$200-{\cdot}1796$,,	$200-{\cdot}2018$,,
220	$220-{\cdot}2232$,,	$220-{\cdot}2478$,,
240	$240-{\cdot}2663$,,	$240-{\cdot}2932$,,
260	$260-{\cdot}3141$,,	$260-{\cdot}3420$,,
280	$280-{\cdot}3610$,,	$280-{\cdot}3897$,,
300	$300-{\cdot}4085$,,	$300-{\cdot}4377$,,

The standard defined by Regnault is that of the constant-
volume air-thermometer, with air at the density which it has
when at the freezing-point under the pressure of 760 mm. or
29·9218 inches of mercury, and its indications are shown in
comparison with the absolute scale by taking $\dfrac{\Phi}{v}=1$ in the
second column of the preceding Table. The greatest discrepance between 0° and 100° Cent. amounts to less than $\frac{1}{20}$ of
a degree, and the discrepance at 300° Cent. is only four tenths.
The discrepancies of the constant-pressure air-thermometer,
when the pressure is equal to the standard atmospheric
pressure, or $\dfrac{p}{\Pi}=1$, are somewhat greater, but still very
small.

294 ON THE THERMAL EFFECTS

SECTION V. *Physical Properties of Air expressed according to the absolute Thermo-dynamic scale of Temperature.*

All the physical properties of a fluid of given constitution are completely fixed when its density and temperature are specified; and as it is these qualities which we can most conveniently regard as being immediately adjustable in any arbitrary manner, we shall generally consider them as the independent variables in formulæ expressing the pressure, the specific heats, and other properties of the particular fluid in any physical condition.

Let v be the volume (in cubic feet) of a unit mass (one pound) of the fluid, and t its absolute temperature; and let p be its pressure in the condition defined by these elements.

Let also e be the "mechanical energy"* of the fluid, reckoned from some assumed standard or zero state, that is, the sum of the mechanical value of the heat communicated to it, and of the work spent on it, to raise it from that zero state to the condition defined by (v, t); and let N and K be its specific heats with constant volume, and with constant pressure, respectively. Then denoting, as before, the mechanical equivalent of the thermal unit by J, and the value of Carnot's function for the temperature t by μ, we have †

$$\frac{de}{dv} = \frac{J}{\mu}\frac{dp}{dt} - p, \quad \dots \dots \quad (1)$$

$$N = \frac{1}{J}\frac{de}{dt}, \quad \dots \dots \quad (2)$$

$$K = \frac{1}{J}\frac{de}{dt} + \frac{1}{J}\left(\frac{de}{dv} + p\right)\frac{\frac{dp}{dt}}{-\frac{dp}{dv}} \quad \dots \quad (3)$$

From these we deduce, by eliminating e,

$$K - N = \frac{1}{\mu}\frac{\left(\frac{dp}{dt}\right)^2}{-\frac{dp}{dv}} \quad \dots \dots \quad (4)$$

---

* Dynamical Theory of Heat, Part V.—On the Quantities of Mechanical Energy contained in a Fluid in different States as to Temperature and Density, § 82, Trans. Roy. Soc. Edin., Dec. 15, 1851.

† Ibid. §§ 89, 91.

and
$$\frac{dN}{dv} = \frac{d\left(\frac{1}{\mu}\frac{dp}{dt}\right)}{dt} - \frac{1}{J}\frac{dp}{dt}, \quad \ldots \quad (5)$$

equations which express two general theorems regarding the specific heats of any fluid whatever, first published * in the Transactions of the Royal Society of Edinburgh, March 1851. The former (4) is the extension of a theorem on the specific heats of gases originally given by Carnot †, while the latter (5) is inconsistent with one of his fundamental assumptions, and expresses in fact the opposed axiom of the Dynamical Theory. The use of the absolute thermo-dynamic system of thermometry proposed in Section IV., according to which the definition of temperature is

$$t = \frac{J}{\mu}, \quad \ldots \quad \ldots \quad \ldots \quad (6)$$

simplifies these equations, and they become

$$JK - JN = t\frac{\left(\frac{dp}{dt}\right)^2}{-\frac{dp}{dv}}, \quad \ldots \quad \ldots \quad (7)$$

$$\frac{d(JN)}{dv} = t\frac{d^2p}{dt^2}. \quad \ldots \quad \ldots \quad (8)$$

To compare with the absolute scale the indications of a thermometer in which the particular fluid (which may be any gas, or even liquid) referred to in the notation p, v, t, is used as the thermometric substance, let p_0 and p_{100} denote the pressures which it has when at the freezing- and boiling-points respectively, and kept in constant volume, v; and let v_0 and v_{100} denote the volumes which it occupies under the same pressure, p, at those temperatures. Then if θ and ϑ denote its thermometric indications when used as a constant-volume and as a constant-pressure thermometer respectively, we have

$$\theta = 100\frac{p - p_0}{p_{100} - p_0}, \quad \ldots \quad \ldots \quad (9)$$

$$\vartheta = 100\frac{v - v_0}{v_{100} - v_0}. \quad \ldots \quad \ldots \quad (10)$$

* Ibid. §§ 47, 48.

† See "Account of Carnot's Theory," Appendix III., Trans. Roy. Soc. Edin., April 30, 1849, p. 565.

Let also ϵ denote the " coefficient of increase of elasticity with temperature,"* and ε the coefficient of expansion at constant pressure, when the gas is in the state defined by (v, t); and let E and E denote the mean values of the same coefficients between 0° and 100° Cent. Then we have

$$\epsilon = \frac{dp}{p_0 dt}, \quad . \quad . \quad . \quad . \quad . \quad . \quad (11)$$

$$\varepsilon = \frac{\dfrac{dp}{dt}}{v_0 \times -\dfrac{dp}{dv}}, \quad . \quad . \quad . \quad . \quad . \quad (12)$$

$$E = \frac{p_{100} - p_0}{100 p_0}, \quad . \quad . \quad . \quad . \quad . \quad (13)$$

$$E = \frac{v_{100} - v_0}{100 v_0}. \quad . \quad . \quad . \quad . \quad . \quad (14)$$

Lastly, the general expression for $\dfrac{J}{\mu}$ quoted in Section II. from our paper of last year, leads to the following expression for the cooling effect on the fluid when forced through a porous plug as in our air experiments :—

$$\delta = \frac{1}{JK} \left\{ \int_{V}^{V'} \left(t \frac{dp}{dt} - p \right) dv + (P'V' - PV) \right\} . \quad . \quad . \quad (15)$$

(p, v), (P', V'), (P, V), as explained above, having reference to the fluid in different states of density, but always at the same temperature, t, as that with which it enters the plug.

From these equations, it appears that if p be fully given in terms of v and absolute values of t for any fluid, the various properties denoted by

$$JK - JN, \quad \frac{(dJN)}{dv}, \quad \theta, \quad \vartheta, \quad \epsilon, \quad \varepsilon, \quad E, \quad E, \quad \text{and } \delta,$$

may all be determined for it in every condition. Conversely, experimental investigations of these properties may be made to contribute, along with direct measurements of the pres-

---

* So called by Mr. Rankine. The same element is called by M. Regnault the coefficient of dilatation of a gas at constant volume.

sure for various particular conditions of the fluid, towards
completing the determination of the function which expresses
this element in terms of v and t. But it must be remarked
that even complete observations determining the pressure for
every given state of the fluid, could give no information as to
the values of t on the absolute scale, although they might
afford data enough for fully expressing p in terms of the
volume and the temperature with reference to some particular
substance used thermometrically. On the other hand, observations on the specific heats of the fluid, or on the thermal
effects it experiences in escaping through narrow passages,
may lead to a knowledge of the absolute temperature, t, of
the fluid when in some known condition, or to the expression
of p in terms of v, and absolute values of t; and accordingly
the formulæ (7), (8), and (15) contain t explicitly, each of
them in fact essentially involving Carnot's function. As for
actual observations on the specific heats of air, none which
have yet been published appear to do more than illustrate
the theory, by confirming (as Mr. Joule's, and the more
precise results more recently published by M. Regnault, do),
within the limits of their accuracy, the value for the specific
heat of air under constant pressure which we calculated*
from the *ratio of the specific heats*, determined according to
Laplace's theory by observations on the velocity of sound,
and the *difference of the specific heats* determined by Carnot's
theorem with the value of Carnot's function estimated from
Mr. Joule's original experiments on the changes of temperature produced by the rarefaction and condensation of air†,
and established to a closer degree of accuracy by our preliminary experiments on expansion through a resisting solid‡.
It ought also to be remarked, that the specific heats of air
can only be applied to the evaluation of absolute temperature
with a knowledge of the mechanical equivalent of the thermal
unit; and therefore it is probable that, even when sufficiently

* Philosopical Transactions, March 1852, p. 82.

† Royal Society Proceedings, June 20, 1844; or Phil. Mag., May
1845.

‡ Ibid. Dec. 1850.

accurate direct determinations of the specific heats are
obtained, they may be useful rather for a correction or
verification of the mechanical equivalent, than for the thermometric object.   On the other hand, a comparatively very
rough approximation to JK, the mechanical value of the
specific heat of a pound of the fluid, will be quite sufficient to render our experiments on the cooling effects
available for expressing with much accuracy, by means of
the formula (15), a thermo-dynamic relation between absolute
temperature and the mechanical properties of the fluid at
two different temperatures.

In the Notes to Mr. Joule's paper on the Air-Engine*, it
was shown that if Mayer's hypothesis be true we must have
approximately,

$$K = \cdot 2374 \text{ and } N = \cdot 1684,$$

because observations on the velocity of sound, with Laplace's
theory, demonstrate that

$$k = 1 \cdot 410$$

within $\frac{1}{700}$ of its own value.   Now the experiments at present
communicated to the Royal Society prove a very remarkable
approximation to the truth in that hypothesis (see above,
Section I.), and we may therefore use these values as very
close approximations to the specific heats of air.   The experiments on the friction of fluids and solids made for the purpose
of determining the mechanical value of heat†, give for J the
value 1390 ; and we therefore have $JN = 234 \cdot 1$ with sufficient
accuracy for use in calculating small terms.

Now according to Regnault we have, for dry air at the
freezing-point, in the latitude of Paris,

$$H = 26215 ;$$

and since the force of gravity at Paris, with reference to a
foot as the unit of space and a second as the unit of time, is
$32 \cdot 1813$, it follows that the velocity of sound in dry air at $0°$
Cent. would be, according to Newton's unmodified theory,

---

* Philosophical Transactions, March 1852, p. 82.
† Philosophical Transactions, 1849.

$$\sqrt{26215 \times 32 \cdot 1813} = 918 \cdot 49,$$

or in reality, according to Laplace's theory,

$$\sqrt{k}. \ \sqrt{26215 \times 32 \cdot 1813}.$$

But according to Bravais and Martins it is in reality

1090·5, which requires that $k = 1 \cdot 4096$,

or according to Moll and Van Beck

1090·1, which requires that $k = 1 \cdot 4086$.

The mean of these values of k is $1 \cdot 4091$.

[Note of Jan. 5th, 1882, by Sir W. THOMSON.—That portion of this Second Part of our researches which was devoted to working out an empirical formula for the thermo-elastic properties of air, and the calculation of specific heats from it, is not reproduced here, because at the conclusion of Part IV. we have derived a better and simpler empirical formula from more comprehensive experimental data.]

Abstract of the above paper " On the Thermal Effects of Fluids in Motion."—No. II. *By* J. P. JOULE, *Esq., F.R.S., and* Professor W. THOMSON, *F.R.S.*

[Proceedings of the Royal Society, vol. vii. p. 127.]

THE first experiments described in this paper show that the anomalies exhibited in the last table of experiments, in the paper preceding it *, are due to fluctuations of temperature in the issuing stream consequent on a change of the pressure with which the entering air is forced into the plug. It appears from these experiments, that when a considerable alteration is suddenly made in the pressure of the entering stream, the issuing stream experiences remarkable successions

* Proc. Roy. Soc., and Phil. Mag. Sept. 1853, p. 230.

of augmentations and diminutions of temperature, which are sometimes perceptible for half an hour after the pressure of the entering stream has ceased to vary.

Several series of experiments are next described in which air is forced (by means of the large pump and other apparatus described in the first paper) through a plug of cotton wool, or unspun silk pressed together, at pressures varying in their excess above the atmospheric pressure, from five or six up to fifty or sixty pounds on the square inch. ·By these it appears that the cooling effect which the air, as found in the authors' previous experiments, always experiences in passing through the porous plug, varies proportionally to the excess of the pressure of the air on entering the plug above that with which it is allowed to escape. Seven series of experiments, in each of which the air entered the plug at a temperature of about 16° Cent., gave a mean cooling effect of about 0°·0175 Cent., per pound on the square inch, or 0°·27 Cent. per atmosphere of difference of pressure. Experiments made at lower and at higher temperatures showed that the cooling effect is very sensibly less for high than for low temperatures, but have not yet led to sufficiently exact results at other temperatures than that stated (16° Cent.) to indicate the law according to which it varies with the temperature.

Experiments on carbonic acid at different temperatures are also described, which show that at about 16° Cent. this gas experiences $4\frac{1}{2}$ times as great a cooling effect as air. They agree well at all the different temperatures with a theoretical result, derived according to the general dynamical theory from empirical formulæ for the pressure of carbonic acid in terms of its temperature and density, which was kindly communicated by Mr. Rankine to the authors, having been investigated by him upon no other experimental data than those of Regnault on the expansion of the gas by heat, and its compressibility.

Experiments were also made on hydrogen gas, which, although not such as to lead to accurate determinations, appeared to indicate very decidedly a cooling effect amounting to a small fraction, perhaps about $\frac{1}{16}$, of that which air would experience in the same circumstances.

The following theoretical deductions from these experiments are made :—

I. The relations between the heat generated and the work spent in compressing carbonic acid, air, and hydrogen, are investigated from the experimental results. In each case the relation is nearly that of equivalence, but the heat developed exceeds the equivalent of the work spent, by a very small amount for hydrogen, considerably more for air, and still more for carbonic acid. For slight compressions with the gases kept about the temperature 16°, this excess amounts to about $\frac{1}{77}$ of the whole heat emitted in the case of carbonic acid and $\frac{1}{120}$ in the case of air.

II. It is shown in the general dynamical theory, that the air experiments, taken in connection with Regnault's experimental results on the latent heat and pressure of saturated steam, make it certain that the density of saturated steam increases very much more with the pressure than according to Boyle's and Gay-Lussac's gaseous laws, and numbers are given expressing the theoretical densities of saturated steam, at different temperatures, which it is desired should be verified by direct experiments.

III. Carnot's function in the "Theory of the Motive Power of Heat" is shown to be very nearly equal to the mechanical equivalent of the thermal unit divided by the temperature from the zero of the air-thermometer (that is, temperature Centigrade with a number equal to the reciprocal of the coefficient of expansion added), and corrections, depending on the amount of the observed cooling effects in the new air-experiments, and the deviations from the gaseous laws of expansion and compression determined by Regnault, are applied to give a more precise evaluation.

IV. An absolute scale of temperature—that is, a scale not founded on reference to any particular thermometric substance or to any special qualities of any class of bodies—is founded on the following definition :—

If a physical system be subjected to cycles of perfectly

reversible operations and be not allowed to take in or to emit heat except in localities, at two fixed temperatures, these temperatures are proportional to the whole quantities of heat taken in or emitted at them respectively during a complete cycle of the operations.

The principles upon which the unit or degree of temperature is to be chosen, so as to make the difference of temperatures on the absolute scale, agree with that on any other scale for a particular range of temperatures. If the difference of temperatures between the freezing- and the boiling-points of water be made 100° on the new scale, the absolute temperature of the freezing-point is shown to be about 273°·7; and it is demonstrated that the temperatures from the freezing-point on the new scale will agree very closely with Centigrade temperature by the standard air-thermometer; quite within the limits of the most accurate practical thermometry when the temperature is between 0° and 100° Cent., and very nearly, if not quite, within these limits for temperatures up to 300° Cent.

Expressions for the specific heat of any fluid in terms of the absolute temperature, the density, and the pressure, derived from the general dynamical theory, are worked out for the case of air according to the empirical formula; and tables of numerical results derived exclusively from these expressions and the ratio of the specific heats as determined by the theory of sound, are given. These tables show the mechanical values of the specific heats of air at different constant pressures and at constant densities. Taking 1390 as the mechanical equivalent of the thermal unit as determined by Mr. Joule's experiment on the friction of fluids, the authors find, as the mean specific heat of air under constant pressure,

0·2390, from 0° to 100° Cent.
0·2384, from 0° to 300° Cent.

On the Thermal Effects of Fluids in Motion. By Professor WILLIAM THOMSON, *F.R.S.*, and J. P. JOULE, *Esq.*, *F.R.S.*

[Proceedings of the Royal Society, vol. viii. p. 41.]

A VERY great depression of temperature has been remarked by some observers when steam of high pressure issues from a small orifice into the open air. After the experiments we have made on the rush of air in similar circumstances, it could not be doubted that a great elevation of temperature of the issuing steam might be observed as well as the great depression usually supposed to be the only result. The method to obtain the entire thermal effect is obviously that which we have already employed in our experiments on permanently elastic fluids, viz., to transmit the steam through a porous material and to ascertain its temperature as it enters into and issues from the resisting medium. We have made a preliminary experiment of this kind which may be sufficiently interesting to place on record before proceeding to obtain more exact numerical results.

A short pipe an inch and a half diameter was screwed into an elbow pipe inserted into the top of a high-pressure steam-boiler. A cotton plug placed in the short pipe had a fine wire of platina passed through it, the ends of which were connected with iron wires passing away to a sensitive galvanometer. The deflection due to a given difference of temperature of the same metallic junctions having been previously ascertained, we were able to estimate the difference of temperature of the steam at the opposite ends of the plug.

The result of several experiments showed that for each lb. of pressure by which the steam on the pressure side exceeded that of the atmosphere on the exit side there was a cooling effect of 0·2 Cent. The steam, therefore, issued at a temperature above 100° Cent., and, consequently, *dry;* showing

the correctness of the view which we brought forward some
years ago * as to the non-scalding property of steam issuing
from a high-pressure boiler.

On the Thermal Effects of Fluids - in Motion. By
J. P. JOULE, *Esq.*, *F.R.S.*, *and* Professor W.
THOMSON, *F.R.S.*

[Proceedings of the Royal Society, vol. viii. p. 178.]

On the Temperature of Solids exposed to Currents of Air.

IN examining the thermal effects experienced by air
rushing through narrow passages, we have found, in various
parts of the stream, very decided indications of a lowering
of temperature (see Phil. Trans. June 1853), but never
nearly so great as theoretical considerations at first led us to
expect, in air forced by its own pressure into so rapid motion
as it was in our experiments. The theoretical investigation
is simply as follows :— Let P and V denote the pressure and
volume of a pound of air moving very slowly up a wide pipe
towards the narrow passage. Let p and v denote the pressure
and volume per pound in any part of the narrow passage,
where the velocity is q. Let also $e - E$ denote the difference
of intrinsic energies of the air per pound in the two situations.
Then the equation of mechanical effect is

$$\frac{q^2}{2g} = (\text{PV} - pv) + (\text{E} - e),$$

since the first member is the mechanical value of the motion,
per pound of air; the first bracketed term of the second
member is the excess of work done in pushing it forward,
above the work spent by it in pushing forward the fluid
immediately in advance of it in the narrow passage ; and the
second bracketed term is the amount of intrinsic energy

---

* See letter from Mr. Thomson to Mr. Joule, published in the
Philosophical Magazine, November 1850.

given up by the fluid in passing from one situation to the other.

Now to the degree of accuracy to which air follows Boyle's and Gay-Lussac's laws, we have

$$pv = \frac{t}{\mathrm{T}}\,\mathrm{PV},$$

if t and T denote the temperatures of the air in the two positions reckoned from the absolute zero of the air-thermometer. Also, to about the same degree of accuracy, our experiments on the temperature of air escaping from a state of high pressure through a porous plug, establish Mayer's hypothesis as the thermo-dynamic law of expansion; and to this degree of accuracy we may assume the intrinsic energy of a mass of air to be independent of its density when its temperature remains unaltered. Lastly, Carnot's principle, as modified in the dynamical theory, shows that a fluid which fulfils those three laws must have its capacity for heat in constant volume constant for all temperatures and pressures, a result confirmed by Regnault's direct experiments to a corresponding degree of accuracy. Hence the variation of intrinsic energy in a mass of air is, according to those laws, simply the difference of temperatures multiplied by a constant, irrespectively of any expansion or condensation that may have been experienced. Hence, if N denote the capacity for heat of a pound of air in constant volume, and J the mechanical value of the thermal unit, we have

$$\mathrm{E} - e = \mathrm{JN}\,(\mathrm{T} - t).$$

Thus the preceding equation of mechanical effect becomes

$$\frac{q^2}{2g} = \mathrm{PV}\left(1 - \frac{t}{\mathrm{T}}\right) + \mathrm{JN}\,(\mathrm{T} - t).$$

Now (see "Notes on the Air-Engine," Phil. Trans. March 1852, p. 81, or "Thermal Effects of Fluids in Motion," Part 2, Phil. Trans. June 1854, p. 361) we have

$$\mathrm{JN} = \frac{1}{k-1}\,\frac{\mathrm{H}}{t^{\circ}} = \frac{1}{k-1}\,\frac{\mathrm{PV}}{\mathrm{T}},$$

where k denotes the ratio of specific heat of air under constant pressure to the specific heat of air in constant volume; H, the product of the pressure into the volume of a pound, or the " height of the homogeneous atmosphere " for air at the freezing-point (26,215 feet, according to Regnault's observations on the density of air), and $t°$ the absolute temperature of freezing (about 274° Cent.)

Hence we have

$$\frac{q^2}{2g} = \text{PV}\left(1 + \frac{1}{k-1}\right)\left(1 - \frac{t}{\text{T}}\right) = \frac{k\,\text{PV}}{k-1}\left(1 - \frac{t}{\text{T}}\right).$$

Now the velocity of sound in air at any temperature is equal to the product of $\sqrt{k}$ into the velocity a body would acquire in falling under the action of a constant force of gravity through half the height of the homogeneous atmosphere; and therefore if we denote by a the velocity of sound in air at the temperature T, we have

$$a^2 = kg\,\text{PV}.$$

Hence we derive from the preceding equation,

$$\frac{\text{T}-t}{\text{T}} = \frac{k-1}{2}\left(\frac{q}{a}\right)^2,$$

which expresses the lowering of temperature, in any part of the narrow channel, in terms of the ratio of the actual velocity of the air in that place to the velocity of sound in air at the temperature of the stream where it moves slowly up towards the rapids. It is to be observed, that the only hypothesis which has been made is, that in all the states of temperature and pressure through which it passes the air fulfils the three gaseous laws mentioned above; and that whatever frictional resistance, or irregular action from irregularities in the channel, the air may have experienced before coming to the part considered, provided only it has not been allowed either to give out heat or to take in heat from the matter around it, nor to lose any mechanical energy in sound, or in other motions not among its own particles, the preceding formulæ will give the lowering of temperature it experiences in acquiring the velocity q. It is to be observed that this is not the velocity the air would have in issuing in the same quantity at the density which it has in the slow stream approaching

the narrow passage. Were no fluid friction operative in the
circumstances, the density and pressure would be the same in
the slow stream flowing away from, and in the slow stream
approaching towards, the narrow passage; and each would
be got by considering the lowering of temperature from T to
t as simply due to expansion; so that we should have

$$\frac{t}{T}=\left(\frac{V}{v}\right)^{k-1}$$

by Poisson's formula. Hence if Q denote what we may call
the "reduced velocity" in any part of the narrow channel,
as distinguished from q, the actual or true velocity in the same
locality, we have

$$Q=\frac{V}{v}\,q=\left(\frac{t}{T}\right)^{\frac{1}{k-1}}q,$$

and the rate of flow of the air will be, in pounds per second,
$w\,QA$, if w denote the weight of the unit of volume, under
pressure P, and A the area of the section in the part of the
channel considered. The preceding equation, expressed in
terms of the "reduced velocity," then becomes

$$1-\frac{t}{T}=\frac{k-1}{2}\left(\frac{T}{t}\right)^{\frac{2}{k-1}}(a)^2,$$

and therefore we have

$$\frac{Q}{a}=\sqrt{\left\{\frac{2}{k-1}\left(\frac{t}{T}\right)^{\frac{2}{k-1}}\left(1-\frac{t}{T}\right)\right\}}.$$

The second member, which vanishes when $t=0$, and when
$t=T$, attains a maximum when

$$t=\cdot83T,$$

the maximum value being

$$\frac{Q}{a}=\cdot578.$$

Hence if there were no fluid friction, the "reduced velocity"
could never, in any part of a narrow channel, exceed ·578 of
the velocity of sound in air of the temperature which the air

has in the wide parts of the channel, where it is moving
slowly. If this temperature be 13° Cent. above the freezing-
point, or 287° absolute temperature (being 55° Fahr., an
ordinary atmospheric condition), the velocity of sound would
be 1115 feet per second, and the maximum reduced velocity
of the stream would be 644 feet per second. The cooling
effect that air must, in such circumstances, experience in
acquiring such a velocity would be from 287° to 268° absolute
temperature, or 19° Cent.

The effects of fluid friction in different parts of the stream
would require to be known in order to estimate the reduced
velocity in any narrow part, according to either the density
on the high-pressure side or the density on the low-pressure
side. We have not as yet made any sufficient investigation
to allow us to give even a conjectural estimate of what these
effects may be in any case. But it appears improbable that
the "reduced velocity," according to the density on the high-
pressure side, could ever with friction exceed the greatest
amount it could possibly have without friction. It therefore
seems improbable that the "reduced velocity" in terms of
the density on the high-pressure side can ever in the narrowest part of the channel exceed 644 feet per second, if the
temperature of the high-pressure air moving slowly be about
the atmospheric temperature of 13° Cent. used in the preceding estimate.

Experiments in which we have forced air through apertures
of $\frac{29}{1000}$, $\frac{53}{1000}$, and $\frac{84}{1000}$ of an inch in diameter drilled in
thin plates of copper, have given us a maximum velocity
reduced to the density of the high-pressure side equal to 550
feet per second. But there can be little doubt that the
stream of air, after issuing from an orifice in a thin plate,
contracts as that of water does under similar circumstances.
If the velocity were calculated from the area of this contracted
part of the stream it is highly probable that the maximum
velocity reduced to the density on the high-pressure side
would be found as near 644 feet as the degree of accuracy of
the experiments warrants us to expect.

As an example of the results we have obtained on examining the temperature of the rushing stream by a thermo-electric junction placed $\frac{1}{8}$th of an inch above the orifice, we cite an experiment in which the total pressure of the air in the receiver being 98 inches of mercury, we found the velocity in the orifice equal to 535 and 1780 feet respectively as reduced to the density on the high-pressure and on the atmospheric side. The actual velocity in the small aperture must have been greater than either of these, perhaps not much greater than 1780, the velocity reduced to atmospheric density. If it had been only this, the cooling effect would have been exactly $T \dfrac{k-1}{2} \left(\dfrac{1780}{1115}\right)^2$, that is, a lowering of tem-perature amounting to 150° Cent. But the amount of cooling effect observed in the experiment was only 13° Cent.; nor have we ever succeeded in observing (whether with thermometers held in various positions in the stream, or with a thermo-electric arrangement constituted by a narrow tube through which the air flows, or by a straight wire of two different metals in the axis of the stream, with the junction in the place of most rapid motion, and in other positions on each side of it) a greater cooling effect than 20° Cent. We therefore infer *that a body round which air is flowing rapidly acquires a higher temperature than the average temperature of the air close to it all round*. The explanation of this conclusion probably is, that the surface of contact between the air and the solid is the locality of the most intense frictional generation of heat that takes place, and that consequently a stratum of air round the body has a higher average temperature than the air further off; but whatever the explanation may be, it appears certainly demonstrated that the air does not give its own temperature even to a tube through which it flows or to a wire or thermometer-bulb completely surrounded by it.

Having been convinced of this conclusion by experiments on rapid motion of air through small passages, we inferred of course that the same phenomenon must take place universally whenever air flows against a solid or a solid is carried through air. If the velocity of 1780 feet per second in the foregoing experiment gave 137° Cent. difference of temperature between the air and the solid, how probable is it that meteors moving at from six to thirty miles per second, even through a rarified atmosphere, really acquire, in accordance with the same law, all the heat which they manifest! On the other hand, it seemed worth while to look for the same kind of effect on a much smaller scale in bodies moving at moderate velocities through the ordinary atmosphere. Accordingly, although it has been a practice in general undoubtingly followed, to whirl a thermometer through the air for the purpose of finding the atmospheric temperature, we have tried and found, with thermometers of different sizes and variously shaped bulbs, whirled through the air at the end of a string, with velocities of from 80 to 120 feet per second, temperatures always higher than when the same thermometers are whirled in exactly the same circumstances at smaller velocities. By alternately whirling the same thermometers for half a minute or so fast, and then for a similar time slow, we have found differences of temperature sometimes little if at all short of a Fahrenheit degree. By whirling a thermo-electric junction alternately fast and slow, the same phenomenon is most satisfactorily and strikingly exhibited by a galvanometer. This last experiment we have performed at night, under a cloudy sky, with the galvanometer within doors, and the testing thermo-electric apparatus whirled in the middle of a field; and thus, with as little as can be conceived of disturbing circumstances, we confirmed the result we had previously found by whirling thermometers.

Velocity of Air escaping through Narrow Apertures.*

In the foregoing part of this communication, referring to the circumstances of certain experiments, we have stated our opinion that the velocity of atmospheric air impelled through narrow orifices was, in the narrowest part of the stream, greater than the reduced velocity corresponding to the atmospheric pressure; in other words, that the density of the air,

* Received June 19, 1856.

kept at a constant temperature, was, in the narrowest part, less than the atmospheric density. In order to avoid misconception, we now add that this holds true only when the difference of pressures on the two sides is small, and the friction plays but a small part in bringing down the velocity of the exit stream. If there is a great difference between the pressures on the two sides, the reduced velocity will, on the contrary, be *less* than that corresponding with the atmospheric pressure ; and even if the pressure in the most rapid part falls short of the atmospheric pressure, the density may, on account of the cooling experienced, exceed the atmospheric density.

We stated that at 57° Fahr., the greatest velocity of air passing through a small orifice is 550 feet per second, if reduced to the density on the high-pressure side. The experiments from which we obtained this result enable us also to say that this maximum occurs, with the above temperature and a barometric pressure of 30·14 inches, when the pressure of the air is equal to about 50 inches of mercury above the atmospheric pressure. At a higher or lower pressure a smaller volume of the compressed air escapes in a given time.

Surface-Condenser.—A three-horse power high-pressure steam-engine was procured for our experiments. Wishing to give it equal power with a lower pressure, we caused the steam from the eduction port to pass downwards through a perpendicular iron gas-pipe, ten feet long and an inch and a half in diameter, placed within a larger pipe through which water was made to ascend. The lower end of the gas-pipe was connected with the feed-pump of the boiler, a small orifice being contrived in the pump-cover in order to allow the escape of air before it could pass, along with the condensed water, into the boiler. This simple arrangement constituted a "surface-condenser" of a very efficient kind, giving a vacuum of 23 inches, although considerable leakage of air took place, and the apparatus generally was not so perfect as subsequent experience would have enabled us to make it.

Besides the ordinary well-known advantages of the "sur-

face-condenser," such as the prevention of incrustation of
the boiler, there is one which may be especially remarked as
appertaining to the system we have adopted, of causing the
current of steam to move in an opposite direction to that of
the water employed to condense it. The refrigerating water
may thus be made to pass out of the condenser at a high
temperature, while the vacuum is that due to a low temperature*, and hence the quantity of water used for the purpose
of condensation may be materially reduced. We find that
our system does not require an amount of surface so great
as to involve a cumbrousness or cost which would prevent
its general adoption, and have no doubt that it will shortly
supersede that at the present time almost universally used.

---

*On the Thermal Effects of Fluids in Motion. Tem-
perature of a Body moving through Air. By* Prof.
WILLIAM THOMSON, *F.R.S., and* JAMES P. JOULE,
Esq., F.R.S.

[Proceedings of the Royal Society, vol. viii. p. 556.]

THE motion of air in the neighbourhood of a body moving
very slowly through it, may be approximately determined
by treating the problem as if air were an incompressible
fluid. The ordinary hydro-dynamical equations, so applied,
give the velocity and the pressure of the fluid at any point;
and the variations of density and temperature actually experienced by the air are approximately determined by using
the approximate evaluation of the pressure thus obtained.
Now, if a solid of any shape be carried uniformly through a
perfect liquid † it experiences fluid-pressure at different parts
of its surface, expressed by the following formula,

$$p = \Pi + \tfrac{1}{2}\rho \left(V^2 - q^2\right),$$

* This was not confirmed by subsequent experiments: see Phil. Trans.
1861, p. 157, where it is stated to be a matter of indifference in which
direction the water is transmitted. [Note, 1886.]

† That is, as we shall call it for brevity, an ideal fluid, perfectly incompressible and perfectly free from mutual friction among its parts.

where Π denotes the fluid-pressure at considerable distances from the solid, ρ the mass of unity of volume of the fluid, V the velocity of translation of the solid, and q the velocity of the fluid relatively to the solid, at the point of its surface in question. The effect of this pressure on the whole is, no resultant force; and only a resultant couple which vanishes in certain cases, including all in which the solid is symmetrical with reference to the direction of motion. If the surface of the body be everywhere convex, there will be an augmentation of pressure in the fore and after parts of it, and a diminution of pressure round a medium zone. There are clearly in every such case just two points in the surface of the solid, one in the fore part, and the other in the after part, at which the velocity of the fluid relatively to it is zero, and which we may call the fore and after pole respectively. The middle region round the body in which the relative velocity exceeds V, and where consequently the fluid-pressure is diminished by the motion, may be called the equatorial zone ; and where there is a definite middle line, or line of maximum relative velocity, this line will be called the equator.

If the fluid be air instead of the ideal " perfect liquid," and if the motion be slow enough to admit of the approximation referred to above, there will be a heating effect on the fore and after parts of the body, and a cooling effect on the equatorial zone. If the dimensions and the thermal conductivity of the body be such that there is no sensible loss on these cooling and heating effects by conduction, the temperature maintained at any point of the surface by the air flowing against it will be given by the equation

$$t = \Theta \left(\frac{\rho}{\Pi} \right)^{\frac{\cdot 41}{1 \cdot 41}},$$

where Θ denotes the temperature of the air as uninfluenced by the motion, and ρ and Π denote the same as before.*

* The temperatures are reckoned according to the absolute thermodynamic scale which we have proposed, and may, to a degree of accuracy correspondent with that of the ordinary " gaseous laws," be taken as temperature Centigrade by the air-thermometer, with 273°·7 added in each case. See the Authors' previous paper, " On the Thermal Effects of Fluids in Motion," Part II., Phil. Trans. 1854, part 2, p. 353.

Hence using for ρ its value by the preceding equation, we have

$$t = \Theta \left\{ 1 + \frac{\rho}{2\Pi} (V^2 - q^2) \right\}^{\frac{\cdot 41}{1 \cdot 41}}$$

But if H denote the length of a column of homogeneous atmosphere, of which the weight is equal to the pressure on its perpendicular section, and if g denote the dynamical measure of the force of gravity (32·2 feet per second of velocity generated per second), we have

$$g\rho \, \mathrm{H} = \Pi;$$

and if we denote by a the velocity of sound in air, which is equal to $\sqrt{1 \cdot 41 \times g\mathrm{H}}$, the expression for the temperature becomes

$$t = \Theta \left\{ 1 + \frac{1 \cdot 41}{2} \cdot \frac{V^2 - q^2}{a^2} \right\}^{\frac{\cdot 41}{1 \cdot 41}}.$$

According to the supposition on which our approximation depends, that the velocity of the motion is small—that is, as we now see, a small fraction of the velocity of sound—this expression becomes

$$t = \Theta \left\{ 1 + \cdot 41 \times \frac{V^2 - q^2}{2a^2} \right\}.$$

At either the fore or after pole, or generally at every point where the velocity of the air relatively to the solid vanishes (at a re-entrant angle for instance, if there is such), we have $q = 0$, and therefore an elevation of temperature amounting to

$$\cdot 41 \times \frac{V^2}{2a^2} \, \Theta.$$

If, for instance, the absolute temperature, Θ, of the air at a distance from the solid be 287° (that is, 55° on the Fahr. scale), for which the velocity of sound is 1115 feet per second, the elevation of temperature at a pole, or at any point of no relative motion, will be, in degrees Centigrade,

$$58°{\cdot}8 \times \left(\frac{V}{a}\right)^2, \text{ or } 58°{\cdot}8 \times \left(\frac{V}{1115}\right)^2,$$

the velocity V being reckoned in feet per second. If, for instance, the velocity of the body through the air be 88 feet per second (60 miles an hour), the elevation of temperature at the points of no relative motion is ·36°, or rather more than ⅓ of a degree Centigrade.

To find the greatest depression of temperature in any case, it is necessary to take the form of the body into account. If this be spherical, the absolute velocity of the fluid backwards across the equator will be half the velocity of the ball forwards; or the relative velocity (q) of the fluid across the equator will be ³⁄₂ of the velocity of the solid. Hence the depression of temperature at the equator of a sphere moving slowly through the air will be just ⁵⁄₄ of the elevation of temperature at each pole. It is obvious from this that a spheroid of revolution, moving in the direction of its axis, would experience at its equator a depression of temperature, greater if it be an oblate spheroid, or less if it be a prolate spheroid, than ⁵⁄₄ of the elevation of temperature at each pole.

It must be borne in mind that, besides the limitation to velocities of the body small in comparison with the velocity of sound, these conclusions involve the supposition that the relative motions of the different parts of the air are unresisted by mutual friction, a supposition which is not even approximately true in ·most cases that can come under observation. Even in the case of a ball-pendulum vibrating in air, Professor Stokes * finds that the motion is seriously influenced by fluid friction. Hence with velocities which could give any effect sensible in the, most delicate of the ether thermometers yet made (330 divisions to a degree), it is not to be expected that anything like a complete verification or even illustration of the preceding theory, involving the assumption of no friction, can be had. It is probable that the forward polar region of heating effect will, in consequence of fluid friction, become gradually larger as the velocity is increased, until it spreads over the whole equatorial region, and does away with all cooling effects.

* "On the Effect of the Internal Friction of Fluids on the Motion of Pendulums," read to the Cambridge Philosophical Society, Dec. 9, 1850, and published in vol. ix. pt. 2 of their Transactions.

Our experimental inquiry has hitherto been chiefly directed to ascertain the law of the thermal effect upon a thermometer rapidly whirled in the air. We have also made some experiments on the modifying effects of resisting envelopes, and on the temperatures of different parts of the surface of a whirled globe. The whirling apparatus consisted of a wheel worked by hand, communicating rapid rotation to an axle, at the extremity of which an arm, carrying the thermometer with its bulb outwards, was fixed. The distance between the centre of the axle and the thermometer-bulb was in all the experiments 39 inches. The thermometers made use of were filled with ether or chloroform, and had, the smaller 275, and the larger 330 divisions to the degree Centigrade. The lengths of the cylindrical bulbs were $\frac{9}{10}$ and $1\frac{4}{10}$ inch, their diameters ·26 and ·48 of an inch respectively.

TABLE I.—Small-bulb Thermometer.

Velocity in feet per second.	Rise of temperature, in divisions of the scale.	Rise divided by square of velocity.
46·9	27½	·0125
51·5	32	·0121
68·1	46½	·0100
72·7	57½	·0109
78·7	67½	·0109
84·8	74	·0103
104 5	91	·0083
130·2	151	·0089
133·2	172	·0097
145·4	191	·0090
	Mean..	·01026

The above Table shows an increase of temperature nearly proportional to the square of the velocity.

$$V = \sqrt{\frac{275}{\cdot01026}} = 163\cdot7 = \text{the velocity in feet per second}$$

which would have raised the temperature 1° Centigrade.

TABLE II.—Larger-bulb Thermometer.

Velocity in feet per second.	Rise of temperature, in divisions of the scale.	Rise divided by square of velocity.
36·3	18	·0125
66·6	42	·0095
84·8	57	·0071
125·6	146	·0093

Mean.. ·0098

In this instance $V = \sqrt{\dfrac{330}{·0098}} = 183·5$ feet per second for 1° Centigrade. It is, however, possible that the full thermal effect was not so completely attained in three minutes (the time occupied by each whirling) as with the smaller bulb. On the whole it did not appear to us that the experiments justified the conclusion, that an increase of the dimensions of the bulb was accompanied by an alteration of the thermal effect.

TABLE III.—Larger-bulb Thermometer covered with five folds of writing paper.

Velocity in feet per second.	Rise of temperature, in divisions of the scale.	Rise divided by square of velocity.
36·3	20	·0152
51·5	43	·0162
72·6	53	·0101
118	132	·0095

The increased thermal effect at comparatively slow velocities, exhibited in the above Table, appeared to be owing to the friction of the air against the paper surface being greater than against the polished glass surface.

One quarter of the enveloping paper was now removed, and the bulb whirled with its bared part in the rear. The results were as follows:—

TABLE IV.—Paper removed from posterior side.

Velocity in feet per second.	Rise of temperature, in divisions of the scale.	Rise divided by square of velocity.
75·6	60	·0105
96·8	87	·0093

On whirling in the contrary direction, so that the naked part of the bulb went first, we got,—

TABLE V.—Paper removed from the anterior side.

Velocity in feet per second.	Rise of temperature, in divisions of the scale.	Rise divided by square of velocity.
81·7	56	·0084
93·8	72	·0082

On rotating with the bare part, posterior and anterior in turns, at the constant velocity of 90 feet per second, the mean result did not appear to indicate any decided difference of thermal effect.

Another quarter of paper was now removed from the opposite side. Then on whirling so that the bared parts were anterior and posterior, we obtained a rise of 83 divisions with a velocity of 93·8. Bnt on turning the thermometer on its axis one quarter round, so that the bared parts were on each side, we found the somewhat smaller rise of 62 divisions for a velocity of 90·8 feet per second.

The effect of surface friction having been exhibited at slow velocities with the papered bulb, we were induced to try the effect of increasing it by wrapping iron wire round the bulb.

TABLE VI.—Larger-bulb Thermometer wrapped with iron wire.

Velocity in feet per second.	Rise of temperature, in divisions of the scale.	Rise divided by square of velocity.
15·36	10·25	·0434
23·04	33	·0623
30·71 *	49·25	·0522
46·08	68·75	·0324
69·12	98	·0206
111·34	185	·0149
126·72	207	·0129
153·55	above 280	above ·0118

* The whirring sound began at this velocity. According to its intensity the thermal effect must necessarily suffer diminution; unless, indeed, it gives rise to increased resistance.

On inspecting the above Table it will be seen that the thermal effect produced at slow velocities was five times as great as with the bare bulb. This increase is evidently due to friction. In fact, as one layer of wire was employed, and the coils were not so close as to prevent the access of air between them, the surface must have been about four times as great as that of the uncovered bulb. At high velocities it is probable that a cushion of air which has not time to escape past resisting obstacles, makes the actual friction almost independent of variations of surface, which leave the magnitude of the body unaltered. In conformity with this observation, it will be seen that at high velocities the thermal effect was nearly reduced to the quantity observed with the uncovered bulb. Similar remarks apply to the following result obtained after wrapping round the bulb a fine spiral of thin brass wire.

TABLE VII.—Bulb wrapped with a spiral of
thin brass wire.

Velocity in feet per second.	Rise of temperature, in divisions of the scale.	Rise divided by square of velocity.
7·68	2·5	·0424
15·36	13·5	·0572
23·04	36·5	·0687
30·71	48	·0509
46·08	64·5	·0304
76·8	103·5	·0175
115·18	224·5	·0169
148·78	264	·0119

The thermal effects on different sides of a sphere moving through air have been investigated by us experimentally by whirling a thin glass globe of 3·58 inches diameter, along with the smaller thermometer, the bulb of which was placed successively in three positions, viz. in front, at one side, and in the rear. In each situation it was placed as near the glass globe as possible without actually touching it.

TABLE VIII.—Smaller Thermometer whirled along with glass globe.

Velocity in feet per second.	Rise in divisions of the scale.		
	Therm. in front.	Therm. at side.	Therm. in rear.
3·84	·66	10	4
7·68	2·66	40	10·5
15·36	41·9	78	51
23·04	71·2	90	71·7
38·4	78·4	90	68
57·5	99·9	112	76
70·92			107

The effects of fluid friction are strikingly evident in the above results, particularly at the slow velocities of 3 and 7 feet per second. It is clear from these, that the air, after coming in contact with the front of the globe, traverses with friction the equatorial parts, giving out an accumulating thermal effect, a part of which is carried round to the after pole. At higher velocities the effects of friction seem rapidly to diminish, so that at the velocities between 23 and 38 feet per second, the mean indication of thermometers placed all round the globe would be nearly constant. Our anticipation (written before these latter experiments were made), that a complete verification of the theory propounded at the commencement was impossible with our present means, is thus completely justified.

It may be proper to observe, that in the form of experiment hitherto adopted by us, the results are probably, to a trifling extent, influenced by the vortex of air occasioned by the circular motion.

———————

We have on several occasions noticed the effect of sudden changes in the force of wind on the temperature of a thermometer held in it. Sometimes the thermometer was observed to rise, at other times to fall, when a gust came suddenly on. When a rise occurred, it was seldom equivalent to the effect, as ascertained by the foregoing experiments, due to the increased velocity of the air. Hence we draw the conclusion, that the actual temperature of a gust of wind is lower than that of the subsequent lull. This is probably owing to the air in the latter case having had its *vis viva* converted into heat by collision with material objects. In fact we find that in sheltered situations, such for instance as one or two inches above a wall opposite to the wind, the thermometer indicates a higher temperature than it does when exposed to the blast. The question, which is one of great interest for meteorological science, has hitherto been only partially discussed by us, and for its complete solution will require a careful estimate of the temperature of the earth's surface, of the effects of radiation, &c., and also a knowledge of the causes of gusts in different winds.

On the Thermal Effects of Fluids in Motion. By J. P. JOULE, *LL.D., F.R.S., and* Professor W. THOMSON, *LL.D., F.R.S.*

[Proceedings of the Royal Society, vol. x. p. 502.]

IN our paper published in the 'Philosophical Transactions' for 1854, we explained the object of our experiments to ascertain the difference of temperature between the high- and low-pressure sides of a porous plug through which elastic fluids were forced. Our experiments were then limited to air and carbonic acid. With new apparatus, obtained by an allotment from the Government grant, we have been able to determine the thermal effect with various other elastic fluids. The following is a brief summary of our principal results at a low temperature (about 7° Centigrade).

Elastic Fluid.	Thermal effect per 100 lbs. pressure on the square inch, in degrees Centigrade.
Atmospheric Air	1·6 Cold.
3·9 Air+96·1 Hydrogen	0·116 Heat.
7·9 Air+92·1 Nitrogen	1·772 Cold.
5·1 Air+94·9 Oxygen	1·936 Cold.
3·5 Air+96·5 Carbonic acid	8·19 Cold.
58·3 Air+41·7 Hydrogen	0·7 Cold.
62·5 Air+37·5 Carbonic acid	3·486 Cold.
54·6 Nitrogen+45·4 Oxygen	1·696 Cold.
4·23 Air { +46·47 Hydrogen ...... } { +49·3 Carbonic acid .... }	2·848 Cold.

Further experiments are being made at high temperatures, which show, in the gases in which a cooling effect is found, a decrease of this effect, and an increase of the heating effect in hydrogen. The results at present arrived at indicate invariably that a mixture of gases gives a smaller cooling effect than that deduced from the mean of the effects of the pure gases.

On the Thermal Effects of Fluids in Motion.— Part III. *On the Changes of Temperature experienced by Bodies moving through Air.* By Professor W. Thomson, *A.M., LL.D., F.R.S., &c.,* and J. P. Joule, *LL.D., F.R.S., &c.*

[Phil. Trans. 1860, p. 325.]

This interesting branch of our researches has been prosecuted by us from time to time since 1856. In the spring of that year we commenced our experiments by trying the effect of whirling thermometers in the air. This process had been confidently recommended as a means of obtaining the temperature of the atmosphere, but we were sure that the plan was not absolutely correct, and one of us had *, as early as 1847, explained the phenomena of "shooting stars" by the heat developed by bodies rushing into our atmosphere. In our early experiments we whirled a thermometer by means of a string, alternately quickly and slowly, and it was found that the thermometer was invariably higher after quick than after slow whirling, in some cases the difference amounting to as much as a degree Fahrenheit. We also succeeded in exhibiting the same phenomenon by whirling a thermo-electric junction. In 1857 we resumed the subject, using an apparatus consisting of a wheel worked by hand, communicating rapid rotation to an axle, at the extremity of which an arm carrying a thermometer, with its bulb outwards, was fixed. The distance between the centre of the axle and the thermometer-bulb was 39 inches. The thermometers made use of were filled with ether or chloroform, and had, the smaller 275, and the larger 330 divisions to the degree C. The lengths of the cylindrical bulbs were $\frac{9}{10}$ and $1\frac{4}{10}$ inch, their diameters ·26 and ·48 of an inch respectively. The method of experimenting was to revolve the thermometer-bulb at a certain velocity until we knew by experience that it had obtained the full thermal effect, then to stop it as suddenly as possible and observe the temperature.

Alternately with these observations others were made to ascertain the temperature after a slow velocity, the effect due to which was calculated from the other observations, on the hypothesis that it varied with the square of the velocity. In all cases the results in the Tables are means of several experiments.

* See Joule, " On Shooting Stars," Phil. Mag. 1848.

Series I.—Bulb ·26 inch diameter.

Velocities in the alternate experiments, in feet per second.	Difference of thermal effect.	Estimated effect of low velocity.	Thermal effect of high velocity.	Velocity due to 1° C.
46·9 and 24	0·082	0·018	0·1	148·8
51·5 and 24	0·098	0·018	0·116	151·2
68·1 and 24	0·151	0·018	0·169	165·6
72·7 and 24	0·191	0·018	0·209	159
78·7 and 24	0·228	0·018	0·246	158·6
84·8 and 24	0·251	0·018	0·269	163·5
103·7 and 24	0·333	0·018	0·351	175
130·2 and 24	0·531	0·018	0·549	175·7
133·2 and 24	0·607	0·018	0·625	168·5
145·4 and 24	0·676	0·018	0·695	174·6

Series II.—Bulb ·48 inch diameter.

Velocities in the alternate experiments, in feet per second.	Difference of thermal effect.	Estimated effect of low velocity.	Thermal effect of high velocity.	Velocity due to 1° C.
36·3 and 18	0·039	0·015	0·054	156·2
66·6 and 18	0·112	0·015	0·127	186·9
84·8 and 18	0·158	0·015	0·173	203·9
125·6 and 18	0·427	0·015	0·442	189

In the following experiments, made in the spring of 1859, thermo-electric junctions of copper and iron wire were whirled, and the effect measured by a Thomson's reflecting galvanometer. The arrangement will be understood from the adjoining sketch, where a is the axle of the whirling apparatus; b a block of wood placed on the end of the axle; to this is attached $c\,c'$, a copper tube, $\frac{3}{16}$ths of an inch in diameter, with a hole in its side. $d\,e$ is a copper wire, which, entering the hole, passes along the axis of the tube, from which it is insulated by non-conducting material. $d\,f$ is an iron wire soldered at d to the copper wire. $g\,g$ are

thick copper wires, communicating at their remote ends with
the galvanometer. They apply to the tube and wire with a
springing force, perfect contact being maintained by keeping
the touching surfaces clean, and lubricated with oil. A thin
piece of wood, not drawn in the sketch, was attached to the
block of wood. It was made to extend to within 1, 2, or 3
feet off *d*, according as the velocity was to be slow or quick.

Fig. 19.

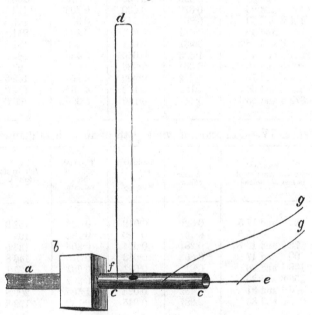

The wires being tied to it, were prevented from twisting out
of their proper position. The distance of *d* from the axis
of revolution was generally 44 inches. The thermal value
of the indications of the galvanometer was repeatedly ascertained by direct observations of the effect of heating the
junctions.

The following Tables comprise the results of
those experiments in which the junction was
placed at right angles to the direction of its
motion.

Series III.—Junction of wires $\frac{1}{100}$th of an inch in diameter.

Velocities in the alternate experiments, in feet per second.	Difference of thermal effect.	Estimated effect of low velocity.	Thermal effect of high velocity.	Velocity due to 1° C.
53·6 and 21	0·21	0·037	0·247	108
77·8 and 17·5	0·258	0·025	0·283	146·3
105 and 18	0·373	0·026	0·399	166·2
126·4 and 44·8	0·320	0·058	0·378	205·6
146 and 25	0·673	0·030	0·703	174·1
159·3 and 34	0·866	0·041	0·907	167·3
180 and 48	0·671	0·051	0·722	211
186·6 and 48·3	0·967	0·071	1·038	181·3
221 and 47	1·393	0·050	1·443	184
300·5 and 105·8	2·364	0·333	2·697	183
315·6 and 73	3·572	0·202	3·774	162·5
326·5 and 66·2	4·133	0·172	4·305	157·3
372·5 and 66·4	5·21	0·170	5·380	160·6

Series IV.—Junction of wires $\frac{1}{40}$th of an inch in diameter.

Velocities in the alternate experiments, in feet per second.	Difference of thermal effect.	Estimated effect of low velocity.	Thermal effect of high velocity.	Velocity due to 1° C.
29 and 17·5	0·022	0·012	0·034	157·2
46·2 and 15·5	0·157	0·026	0·183	108
73·7 and 16·75	0·281	0·024	0·305	133·5
90 and 17	0·363	0·013	0·376	146·8
139·1 and 26	0·61	0·021	0·631	175·1
155·6 and 26	0·878	0·024	0·902	163·9
246·4 and 31	1·482	0·023	1·505	200·8
262·6 and 35	2·087	0·045	2·132	179·8

Series V.—Junction of wires $\frac{1}{17·5}$th of an inch in diameter.

Velocities in the alternate experiments, in feet per second.	Difference of thermal effect.	Estimated effect of low velocity.	Thermal effect of high velocity.	Velocity due to 1° C.
64·03 and 33·96	0·127	0·049	0·176	152·6
91·48 and 41·17	0·204	0·073	0·277	173·8
134·9 and 51	0·67	0·112	0·782	152·5
160·73 and 47·34	0·685	0·097	0·782	181·8
177·9 and 48·25	0·863	0·100	0·963	181·3
208 and 44	1·469	0·071	1·540	168

SERIES VI.—Junction of wires $\frac{1}{8\cdot7}$th of an inch in diameter.

Velocities in the alternate experiments, in feet per second.	Difference of thermal effect.	Estimated effect of low velocity.	Thermal effect of high velocity.	Velocity due to 1° C.
93·1 and 39	0·198	0·042	0·240	190
109·6 and 52·4	0·239	0·072	0·311	196·5
133·96 and 58·3	0·432	0·100	0·532	183
163·7 and 55·2	0·654	0·084	0·738	190·5

From the above Tables it is manifest that the thermal effect increases nearly with the square of the velocity; it is, however, a little greater at low velocities than accords with this law. Taking, therefore, the means of the foregoing results, and rejecting all those obtained from a velocity under 100 feet per second, we obtain the following summary :—

Material of the whirled cylinder.	Diameter.	Velocity due to 1° Cent.
Glass	0·26	173·45
Glass	0·48	189
Copper-iron . . .	0·01	177·54
Copper-iron . . .	0·025	179·9
Copper-iron . . .	0·057	170·9
Copper-iron . . .	0·115	190
Mean		180·13

It may be inferred from the above that the thermal effect is independent of the kind of material whirled, provided its surface is smooth; and that it is likewise independent of the diameter of the cylinder moving in a direction perpendicular to its length.

In the next experiments we whirled the junctions parallel to the direction of motion.

Series VII.

Diameter of wire.	Velocities in the alternate experiments, in feet per second.	Difference of thermal effect.	Estimated effect of low velocity.	Thermal effect of high velocity.	Velocity due to 1° C.	Means.
·01	123·4 and 45·2 186·7 and 54·2	0̊·415 1·160	0̊·064 0·107	0̊·479 1·267	178·3 165·9	172·1
·057	126·5 and 39·6 206·6 and 44	0·59 1·518	0·058 0·072	0·648 1·59	157·1 164	160·55
·115	100·5 and 44·8 182·7 and 50	0·215 1·053	0·053 0·085	0·268 1·138	194·1 171·3	182·7

The general mean of the velocities due to 1° Cent. is therefore 171·78, which is not notably different from the result obtained when the wire was placed at right angles to the direction of motion. The absence of any considerable effect arising from the shape of the body whirled, was also shown by the following results obtained with a junction of flattened wires a quarter of an inch broad and one thirtieth of an inch thick.

Series VIII.

Position of junction.	Velocities in the alternate experiments, in feet per second.	Difference of thermal effect.	Estimated effect of low velocity.	Thermal effect of high velocity.	Velocity due to 1° C.	Mean.
Flat side against the air	85·8 and 40 156·5 and 52 164·4 and 48	0̊·165 0·683 0·736	0̊·046 0·085 0·074	0̊·211 0·768 0·810	186·6 178·6 182·6	180·6
Thin edge against the air	182·4 and 54·3	0·811	0·074	0·885	193·9	

The general mean of all the foregoing results is 179·15 feet per 1° Cent. The phenomena hitherto observed seemed to point to the effect of stopping air as a cause, since 145 feet per second is the velocity of air equivalent to the quantity of heat required to raise its substance, under constant pressure, by 1° Cent. temperature; and it was reasonable to infer

SERIES IX.

Position of junction.	Velocities in the alternate experiments, in feet per second.	Difference of thermal effect.	Estimated effect of low velocity.	Thermal effect of high velocity.	Velocity due to 1° C.	Mean.
Cotton-wool closely tied about the junction of fine wires	91·3 and 34	°0·25	°0·04	°0·29	169·6	
	112 and 26	0·447	0·025	0·472	162·8	
	135·8 and 40·6	1·37	0·14	1·51	110·5	148·76
	140·3 and 24	0·823	0·024	0·847	152·4	
	167·6 and 43·5	1·188	0·085	1·273	148·5	
Junction of fine wires placed in a small wicker basket filled with cotton-wool or tow	94·1 and 35	0·494	0·079	0·573	124	
	95·7 and 25·3	0·426	0·032	0·458	141·1	
	105·5 and 27·6	0·33	0·026	0·356	168·3	144·58
	113·5 and 33	0·409	0·038	0·447	169·5	
	116·8 and 33	0·639	0·055	0·694	140·1	
	116·8 and 47·4	0·739	0·145	0·884	124·2	

that a portion of the effect was lost by radiation. The following experiments (Series IX.), made with a junction of fine wires covered loosely with cotton-wool or tow, enabled us to eliminate all effects but those due to stopped air. Their results will be found to agree closely with theory.

When the junction was placed in the basket, without any cotton-wool or tow, a velocity of 160·1 ft. per second was required to give 1°. N.B. The basket was so open that its orifices amounted to half the entire area.

In several of our experiments with very slow velocities there appeared to be a greater evolution of heat than could be due to the stopping of air. This circumstance induced us to try various modifications of the surface of the whirled body. In the first instance we covered the bulb of the thermometer used in the second series of experiments with five folds of writing-paper, and then obtained the following results :—

Series X.

Velocities in the alternate experiments, in feet per second.	Difference of thermal effect.	Estimated effect of low velocity.	Thermal effect of high velocity.	Velocity due to 1° C., or $V_1°$, on the hypothesis that $V_1° = \dfrac{v}{\sqrt{t}}$.
36·3 and 18	0·045	0·015	0·060	148·2
51·5 and 18	0·115	0·015	0·130	142·8
72·6 and 18	0·146	0·015	0·161	180·9
118 and 18	0·385	0·015	0·400	186·6

It will be seen from the last column that the effect at slow velocities was greater than that which might have been anticipated. We were thus led to try the effect of a further

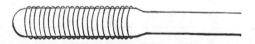

increase of what we may call "fluid friction." In the next series the bulb was wrapped with fine iron wire.

Series XI.

Velocities in the alternate experiments, in feet per second.	Difference of thermal effect.	Estimated effect of low velocity.	Thermal effect of high velocity.	Velocity due to 1° C., or $V_1°$, on the hypothesis that $V_1° = \dfrac{v}{\sqrt{t}}$.
15·36 and 7·68	0̊·022	0̊·008	0̊·030	88·8
23·04 and 15·36	0·069	0·03	0·099	73·2
30·71 and 15·36	0·118	0·03	0·148	79·8
46·08 and 15·36	0·177	0·03	0·207	101·3
69·12 and 15·36	0·267	0·03	0·297	126·8
111·34 and 15·36	0·530	0·03	0·560	148·8
126·72 and 15·36	0·598	0·03	0·628	160
153·55 and 15·36	0·850+	0·03	0·880+	163·4

In the next experiments the bulb was wrapped with a spiral of fine brass wire.

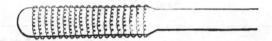

Series XII.

Velocities in the alternate experiments, in feet per second.	Difference of thermal effect.	Estimated effect of low velocity.	Thermal effect of high velocity.	Velocity due to 1° C., or $V_1°$, on the hypothesis that $V_1° = \dfrac{v}{\sqrt{t}}$.
7·68 and 1·92	0̊·006	0̊·002	0̊·008	86·3
15·36 and 7·68	0·033	0·008	0·041	75·8
23·04 and 15·36	0·070	0·041	0·111	69·1
30·71 and 15·36	0·105	0·041	0·146	80·3
46·08 and 19·2	0·120	0·075	0·195	104·4
76·8 and 23·04	0·203	0·111	0·314	137·1
115·18 and 23·04	0·570	0·111	0·681	139·5
148·78 and 76·8	0·488	0·314	0·802	166·2

The last columns of the above Tables clearly indicate that
at slow velocities a source of heat exists besides that from
stopped air. It is also evident that, as the velocity increases,
this thermal cause decreases; for at a velocity of 150 feet
per second the thermal effect is such as would be due to the
influence of stopped air alone.

In prosecuting still further this part of our subject we
made the following arrangement. A disk of mill-board, 32
inches in diameter, was fixed to the end of the axis of the
whirling apparatus. An ether thermometer, whose bulb was
one fourth of an inch in diameter, was tied by its stem to
the face of the disk, so that the bulb was 15 inches distant
from the axis of revolution, and 1 inch from the margin of
the disk. In the following Table the first five experiments
were made with the above arrangement, but in the last two
a thermo-electric junction of thin copper and iron wires,
tied closely to the mill-board, was substituted for the ether
thermometer.

SERIES XIII.

Velocities in the alternate experiments, in feet per second.	Difference of thermal effect.	Estimated effect of low velocity.	Thermal effect of high velocity.	Velocity due to 1° C., or $V_1°$, on the hypothesis that $V_1° = \frac{v}{\sqrt{t}}$.
3·15 and ·5	$\overset{\circ}{0}·029$	$\overset{\circ}{0}·005$	$\overset{\circ}{0}·034$	17·1
7·85 and 3·15	0·027	0·034	0·061	31·7
15·7 and 7·85	0·052	0·061	0·113	46·6
31·4 and 15·7	0·022	0·113	0·135	85·5
Thermo-electric junction { 63·3 and 27·4	0·106	0·120	0·226	133·3
90·2 and 25	0·286	0·116	0·402	142·3

The surface of the mill-board disk being rather rough, it
was judged desirable to make similar experiments with a disk
of sheet zinc. This was perfectly smooth, 36½ inches in
diameter. The thermometer-bulb was fixed at 17·1 inches
distance from the axis.

SERIES XIV.

Velocities in the alternate experiments, in feet per second.	Difference of thermal effect.	Estimated effect of low velocity.	Thermal effect of high velocity.	Velocity due to 1° C., or $V_1°$ on the hypothesis that $V_1° = \dfrac{v}{\sqrt{t}}$.
1·71 and ·57	$\overset{\circ}{0}$·024	$\overset{\circ}{0}$·010	$\overset{\circ}{0}$·034	9·2
3·42 and 1·71	0·017	0·034	0·051	15·1
8·55 and 3·42	0·027	0·051	0·078	30·7
17·1 and 8·55	0·023	0·078	0·101	53·8
34·2 and 17·1	0·046	0·102	0·148	88·8
57·28 and 17·1	0·070	0·102	0·172	138·3

The last columns of the two foregoing Tables clearly show the inapplicability of the law of the increase of temperature with the square of the velocity, at low velocities. The thermal effect appears even to increase at a slower rate than simply with the velocity. This phenomenon may, we think, be ascribed to the internal fluid friction of the particles of air among themselves, which Professor Stokes has proved to exist, by his researches on the motion of pendulums. We may easily apprehend that in such experiments as our last, the entire face of the disk is covered with a film of air which revolves along with it at very slow velocities. As the velocity increases there will still be a film of air adhering to the disk, but with the difference that it will be constantly replaced by fresh stopped air, the thermal effect of which will ultimately be the only recognizable phenomenon.

A very interesting and important branch of our subject was to inquire into the thermal phenomena which take place at the surface of a sphere passing rapidly through air. Some of our experiments on this subject have been made by blowing air from a large bellows against a ball; others by whirling a ball or sphere in the air by means of the apparatus already described. We shall commence by describing the latter, in some of which a thermo-electric junction was employed, and in others an ether thermometer.

Series XV.—Wooden ball 2 inches in diameter, with a thermo-electric junction of fine copper and iron wires made even with the surface.

Position of the junction in respect to the direction of motion.	Velocities in alternate experiments.	Difference of thermal effect.
In front, or anterior........	75·6 and 23·1	$\overset{\circ}{0}$·269
	118·4 and 23·1	0·517
	141·5 and 39·5	0·745
At the side, or equatorial ..	74 and 28·5	−0·146
	115 and 26·3	0·283
	120 and 40	0·020
In the rear, or posterior	71·5 and 25	0·093
	112·4 and 19·3	0·414
	113·7 and 42	0·280

In the above experiments differential results for the several pairs of velocities are alone given, so that, although one of the quantities has a negative sign, there is no proof of actual cooling effect.   In the next experiments (Series XVI.) we whirled a thin glass globe, 3·58 inches in diameter, placed at a distance of 38 inches from the axle of the apparatus.   The small bulb of an ether thermometer was kept in contact with the glass.

In the experiments of Series XVII., a 12-inch globe, such as is used in schools, was fixed at a distance of 3 feet from the axis of the revolving apparatus.   The ether thermometer was generally employed, as in the last series, but for the highest velocity a thermo-electric junction of thin wires placed close to the globe registered the thermal effect.

SERIES XVI.

Position of the bulb of the thermometer in respect to the direction of motion.	Velocities in the alternate experiments, in feet per second.	Difference of thermal effect.	Thermal effect due to low velocity.	Thermal effect due to high velocity.
In front of the globe	3·84 and 1·92	°0·002	°0·001	°0·003
	7·68 and 3·84	0·007	0·003	0·010
	15·36 and 7·68	0·143	0·010	0·153
	23·04 and 15·36	0·106	0·153	0·259
	38·4 and 15·36	0·133	0·153	0·286
	57·5 and 15·36	0·211	0·153	0·364
At the side of the globe	3·84 and 1·92	0·029	0·007	0·036
	7·68 and 3·84	0·109	0·036	0·145
	15·36 and 7·68	0·138	0·145	0·283
	23·04 and 7·68	0·181	0·145	0·326
	38·4 and 23·04	0·000	0·326	0·326
	57·5 and 23·04	0·087	0·326	0·413
In the rear of the globe	3·84 and 1·92	0·011	0·004	0·015
	7·68 and 3·84	0·024	0·015	0·039
	15·36 and 7·68	0·147	0·039	0·186
	23·04 and 15·36	0·076	0·186	0·262
	38·4 and 15·36	0·062	0·186	0·248
	57·5 and 15·36	0·091	0·186	0·277
	70·92 and 15·36	0·204	0·186	0·390

Series XVII.

Measurer of heat, and its position in respect to the direction of the motion.	Velocities in the alternate experiments, in feet per second.	Difference of thermal effect.	Thermal effect due to low velocity.	Thermal effect due to high velocity.
Anterior { Ether thermometer	3·72 and 1·24	0·019	0·009 estimated.	0·028
	7·44 and 3·72	0·008	0·028	0·036
Thermo-electric junction	14·88 and 7·44	0·028	0·036	0·064
	39·68 and 7·44	0·200	0·036	0·236
Equatorial. { Ether thermometer	3·72 and 1·24	0·007	0·003 estimated.	0·010
	7·44 and 3·72	0·013	0·010	0·023
Thermo-electric junction	14·88 and 7·44	0·024	0·023	0·047
	39·37 and 7·44	0·170	0·023	0·193
Posterior. { Ether thermometer	3·72 and 1·24	0·024	0·012 estimated.	0·036
	7·44 and 3·72	0·022	0·036	0·058
Thermo-electric junction	14·88 and 7·44	0·046	0·058	0·104
	37·2 and 7·44	0·140	0·058	0·198

In the experiments in which air was blown against a sphere, we made use of a large organ-bellows, from which a constant stream of air could be kept up at velocities dependent upon the weights laid on. In our first trials, the air issued from a circular aperture 2½ inches in diameter, and the ball was placed, at half an inch distance, in front of the aperture. We shall, as before, call that point of the ball which was nearest the wind, the Anterior Pole; the most sheltered point, the Posterior Pole; and the intermediate part, the Equator. The balls were furnished with thermo-electric junctions of thin copper and iron wires, made flat with the surface, the junctions being in each case 90° apart from one another.

SERIES XVIII.—½-inch Wooden Ball.

Velocity
of air.

68 ft. per sec. .. Equator $\overset{\circ}{0}$·114 colder   Posterior Pole $\overset{\circ}{0}$·067 colder than
                 than Anterior Pole.   Equator.

SERIES XIX.—1-inch Wooden Ball.

Velocity
of air.

1·2 .. Equator $\overset{\circ}{0}$·088 warmer than
         Anterior Pole.

3·6 .. Equator 0·129 warmer than   Posterior Pole $\overset{\circ}{0}$·03 warmer than
         Anterior Pole.   Equator.

7·2 .. Equator 0·160 warmer than   Posterior Pole 0·022 warmer than
         Anterior Pole.   Equator.

14·4 .. Equator 0·120 warmer than Posterior Pole 0·018 colder than
         Anterior Pole.   Equator.

28·8 .. Equator 0·056 warmer than Posterior Pole 0·018 colder than
         Anterior Pole.   Equator.

36 .. Equator 0·008 colder than
         Anterior Pole.

48 .. Equator 0·035 colder than
         Anterior Pole.

57·6 .. Equator 0·056 colder than   Posterior Pole 0·090 colder than
         Anterior Pole.   Equator.

73 .. Equator 0·245 colder than
         Anterior Pole.

105 .. Equator 0·380 colder than Posterior Pole 0·232 colder than
         Anterior Pole.   Equator.

In our next series, one junction was placed within the bellows, and the other in contact with the different parts of the 1-inch ball. All the results will be seen to indicate, as might have been anticipated, that the junction within the bellows was warmer than any part of the ball.

SERIES XX.

Velocity of air.	Pressure of air in the bellows, in inches of water.	Cold of Anterior Pole, in respect to the inner junction.	Cold of Equator, in respect to the inner junction.	Cold of Posterior Pole, in respect to the inner junction.
2·4	0·003 estimated	0·098	0·065	0·028
3·6	0·006 estimated	0·094	0·065	0·028
7·2	0·025 estimated	0·083	0·103	0·060
14·4	0·105 estimated	0·110	0·089	0·109
28·8	0·42 estimated	0·102	....	0·112
73	2·7 estimated	0·188	0·309	0·300
105	5·6 measured	0·195	....	0·360

A further modification of the experiments was made by placing a glass tube 3 feet long and of 1½-inch interior diameter, within the aperture, so that two thirds of the tube was inside, and one third outside of the bellows. A ball furnished with junctions 90° distant from each other was placed within the tube.

The general result is that at slow velocities of air there is a gradual increase of temperature from the anterior to the posterior pole, but the reverse at high velocities. We observed a great effect for slow velocities at the commencement of an experiment, which gradually declined on continued blowing. This phenomenon was apparently owing to circumstances in connexion with the temperature of the orifice and of the bellows.

SERIES XXI.—Wooden Ball, 1 inch diameter.

Velocity of air.

1·8	Equator 0°045 warmer than Anterior Pole	Posterior Pole 0°052 warmer than Equator.
2·7	Equator 0·056 warmer than Anterior Pole	Posterior Pole 0·052 warmer than Equator.
5·4	Equator 0·074 warmer than Anterior Pole	Posterior Pole 0·035 warmer than Equator.
10·8	Equator 0·052 warmer than Anterior Pole	Posterior Pole 0·017 warmer than Equator.
21·6	Equator 0·037 warmer than Anterior Pole	Posterior Pole 0·008 colder than Equator.
43·2	Equator 0·011 warmer than Anterior Pole	Posterior Pole 0·013 colder than Equator.
54	Equator 0·019 colder than Anterior Pole	Posterior Pole 0·014 colder than Equator.
62	......	Posterior Pole 0·023 colder than Equator.
83·8	......	Posterior Pole 0·041 colder than Equator.
108	Equator 0·091 colder than Anterior Pole	Posterior Pole 0·086 colder than Equator.

SERIES XXII.—Wooden Ball, ½ inch diameter.

Velocity of air.

1·8	Equator 0°048 warmer than Anterior Pole	Posterior Pole 0°050 warmer than Equator.
5·4	Equator 0·030 warmer than Anterior Pole	Posterior Pole 0·047 warmer than Equator.
10·8	Equator 0·023 warmer than Anterior Pole	Posterior Pole 0·031 warmer than Equator.
21·6	Equator 0·008 warmer than Anterior Pole	Posterior Pole 0·012 warmer than Equator.
43·2	Equator 0·006 colder than Anterior Pole	Posterior Pole 0·009 warmer than Equator.
54	Equator 0·019 colder than Anterior Pole	Posterior Pole 0·006 colder than Equator.
62	Equator 0·026 colder than Anterior Pole	Posterior Pole 0·014 colder than Equator.
83·8	Equator 0·040 colder than Anterior Pole	Posterior Pole 0·031 colder than Equator.
108	Equator 0·068 colder than Anterior Pole	Posterior Pole 0·050 colder than Equator.

The causes of the thermal effects on the surface of balls slowly passing through air are very complicated, as they arise from the effects of stopped air, fluid friction, and varied pressures. In order, if possible to throw some light on these, we made the following observations :—

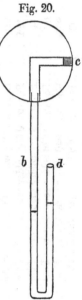

Fig. 20.

1st. That when the 12–inch globe passed through the air at the velocity of about 12 feet per second or under, the air at the equatorial part moved in the reverse direction. We did not observe the velocity, if it existed, at which this phenomenon ceased to take place*.

2nd. An ivory ball, 1·7 inch diameter, had holes drilled from points of the surface 90° asunder, which holes met at the centre of the ball. Into the lower one (see adjoining figure) a bent glass tube, partly filled with water, was cemented; in the other, at c, a porous wooden plug was placed. It was then found that when c was made the anterior pole in a blast from the bellows, a pressure was experienced able to produce a difference of level in the tubes b and d equal to 2·5 inches. When c was put in the equatorial position, there was, on the contrary, a suction equal to 1·2 inch. When c was made the posterior pole there was also a suction, equal, however, to only 0·1 inch. Having tied a thick fold of silk over the orifice d, we tried the same thing in a strong breeze of wind, when we found that on making c the anterior pole, we had a pressure amounting to 0·6 of an inch; on making c equatorial, a suction of 0·3; and on making c the posterior pole, a suction of 0·05 inch.

We have not hitherto been able to detect any change in the thermal effect owing to the whistling sound of wire or other bodies rapidly whirled. We think it possible that this vibratory action decreases the resistance and the evolution of

* See 'Proceedings of the Royal Society,' June 18, 1857, p. 558.

heat. Some of the sounds produced are interesting and worthy of further investigation. When a small piece of paper was attached to the revolving wire, we obtained a continuous succession of loud cracks similar to those of a whip.

But although this and other parts of our subject remain to be cleared up, we believe that it will be found that at all high velocities the thermal effect arises entirely from stopped air, and thus is independent of the shape and mass of the body, and of the temperature and density of the atmosphere. From some experiments described in the 'Proceedings of the Royal Society' of June 19, 1856, p. 183, we inferred that a body placed in a stream of air moving with a velocity of 1780 feet per second, was raised 137° C. above the temperature of that stream. This gives 152 feet per second as the velocity due to 1°, while our direct results, given in the present paper, indicate 179.

It must be obvious that a thermometer placed in the wind registers the temperature of the air, plus the greater portion, but not the whole, of the temperature due to the *vis viva* of its motion. In a place perfectly sheltered from the wind, the temperature of a thermometer immersed in the air will be that of the wind, plus the whole temperature due to the *vis viva* of the moving air. In accordance with this we have found that a thermometer placed in a sheltered situation, such as on the top of a wall opposite the wind, indicates a higher temperature than when it is exposed to the blast. A minute examination of these phenomena cannot fail to interest the meteorologist.

On the Thermal Effects of Fluids in Motion.—
Temperature of Bodies Moving in Air. By J. P.
JOULE, *LL.D., F.R.S., and* Professor W. THOMSON,
LL.D., F.R.S.

[' Proceedings of the Royal Society,' vol. x. p. 519.]

AN abstract of a great part of the present paper has appeared
in the ' Proceedings,' vol. viii. p. 556. To the experiments
then adduced a large number have since been added, which
have been made by whirling thermometers and thermo-
electric junctions in air. The result shows that at high
velocities the thermal effect is proportional to the square of
the velocity, the rise of temperature of the whirled body
being evidently that due to the communication of the velocity
to a constantly renewed film of air. With very small
velocities of bodies of large surface, the thermal effect was
very greatly increased by that kind of fluid friction the effect
of which on the motion of pendulums has been investigated
by Profesor Stokes.

On the Thermal Effects of Fluids in Motion.—Part
IV.—*By* J. P. JOULE, *LL.D., F.R.S., &c., and*
Professor W. THOMSON, *A.M., LL.D., F.R.S., &c.*

[Phil. Trans. 1862, p. 579.]

[PLATE III.]

IN the Second Part of these researches we have given the
results of our experiments on the difference between the
temperatures of an elastic fluid on the high- and low-
pressure sides of a porous plug through which it was
transmitted. The gases employed were atmospheric air and
carbonic acid. With the former, 0°·0176 of cooling effect
was observed for each pound per square inch of difference of
pressure, the temperature on the high-pressure side being
17°·125. With the latter gas, 0°·0833 of cooling effect was

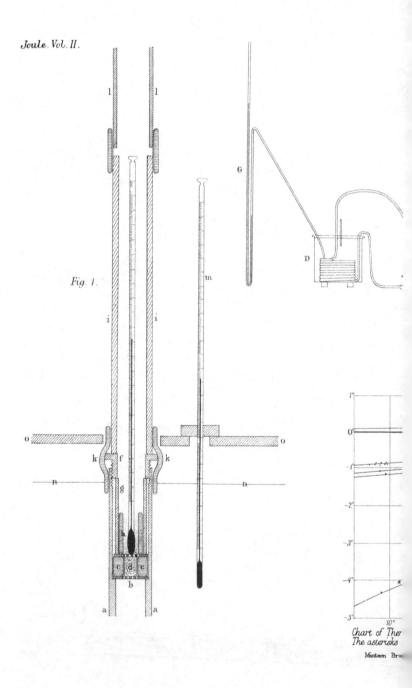

Fig. 1.

Chart of Ther
The asterisks

Mintern Bro

Plate III.

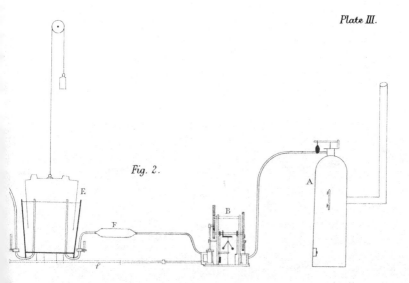

Fig. 2.

E

F

f

B

A

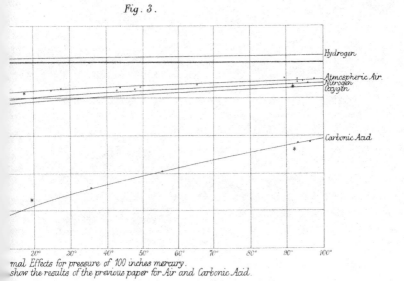

Fig. 3.

Hydrogen

Atmospheric Air.
Nitrogen
Oxygen

Carbonic Acid

20°   30°   40°   50°   60°   70°   80°   90°   100°

mal Effects for pressure of 100 inches mercury.
show the results of the previous paper for Air and Carbonic Acid.

s. lith.

produced per lb. of difference of pressure, the temperature on the high-pressure side being 12°·844.

It was also shown that in each of the above gases the difference of the temperatures on the opposite sides of the porous plug is sensibly proportional to the difference of the pressures.

An attempt was also made to ascertain the cooling effect when elastic fluids of high temperature were employed; and it was satisfactorily shown that in this case a considerable diminution of the effect took place. Thus, in air at 91°·58, the effect was only 0°·014; and in carbonic acid at 91°·52, it was 0°·0474.

In the experiments at high temperatures there appeared to be some grounds for suspecting that the apparent cooling effect was too high; for the quantity of transmitted air was very considerable, and its temperature possibly had not arrived accurately at that of the bath by the time it reached the porous plug.

The obvious way to get rid of all uncertainty on this head was to increase the length of the coil of pipes. Hence in the following experiments the total length of 2-inch copper pipe immersed in the bath was 60 feet instead of 35, as in the former series. The volume of air transmitted in a given time was also considerably less. There could therefore be no doubt that the temperature of the air on its arrival at the plug was sensibly the same as that of the bath.

The nozle employed in the former series of experiments was of box-wood,—the space occupied by cotton-wool, or other porous material, being 2·72 inches long and an inch and a half in diameter. The box-wood was protected from the water of the bath by being enveloped by a tin can filled with cotton-wool. This was unquestionably in most respects the best arrangement for obtaining accurate results; but it was found necessary to make each experiment last one hour or more before we could confidently depend on the thermal effect. The oscillations of temperature which took place during the first part of the time were traced to various causes, one of the principal being the length of time which, on account of the large capacity for heat and the small conductivity of the box-wood nozle, elapsed before the first large thermal effects consequent on the getting up of the pressure were dissipated. No doubt the results we arrived at were very accurate with the elastic fluids employed, viz. atmospheric air and carbonic acid; but we possessed an unlimited supply of the former and a supply of the latter equal to 120 cubic feet, which was sufficient to last for more than half an hour without being exhausted. In extending the inquiry to gases not so readily procured in large quantities, it was therefore desirable to use a porous plug of smaller dimensions enclosed in a nozle of less capacity for heat, so as to arrive rapidly at the normal effect.

Various alterations of the apparatus were made in order to meet the new requirements of our experiments. A small high-pressure engine of about one horse-power was placed in gear with a double-acting compressing air-pump, which had a cylinder $4\frac{1}{2}$ inches in diameter, with a length of stroke of 9 inches. The engine was able to work the piston of the pump sixty complete strokes in the minute. The quantity of air which it ought to have discharged at low pressure was therefore upwards of 16,000 cubic inches per minute. But much loss, of course, occurred from leakage past the metallic piston, and in consequence of the necessary clearance at the top and bottom of the cylinder when the pressure increased by a few atmospheres ; so that in practice we never pumped more than 8000 cubic inches per minute.

The nozle we employed will be understood by inspecting Plate III. fig. 1, where $a\,a$ is the upright end of the coil of copper pipes. On a shoulder within the pipe a perforated metallic disk (b) rests. Over this is a short piece of india-rubber tube ($c\,c$) enclosing a silk plug (d), which is kept in a compressed state by the upper perforated metallic plate (e). This upper plate is pressed down with any required force by the operation of the screw f on the metallic tube $g\,g$. A tube of cork ($h\,h$) is placed within the metallic tube, in order to protect the bulb of the thermometer from the effects of a too rapid conduction of heat from the bath. Cotton-wool is loosely packed round the bulb, so as to distribute the flowing air as evenly as possible. The glass tube (*i i*) is attached to the nozle by means of a piece of strong india-rubber tubing, and through it the indications of the thermometer are read. The top of the glass tube is attached to the metallic tube *l l*, for the purpose of conveying the gas to the meter.

The thermometer (*m*) for registering the temperature of the bath is placed with its bulb near the nozle. The level of the water is shown by *n n*; and *o o* represents the wooden cover of the bath.

When a high temperature was employed, it was maintained by introducing steam into the bath by means of a pipe led from the boiler. The water of the bath was in every case constantly and thoroughly stirred, especially when high temperatures were used.

The general disposition of the apparatus will be understood from fig. 2, in which A represents the boiler, B the steam-engine geared to the condensing air-pump C. From this pump the compressed air passes through a train of pipes 60 feet long and 2 inches in diameter, and then enters the coil of pipes in the bath D. Thence, after issuing from the porous plug, it passes through the gasometer E, and ultimately arrives again at the pump C. This complete circulation is of great importance, inasmuch as it permits the gas which has been collected in the meter to be used for a much longer period than would otherwise have been possible. A glass vessel full of chloride of calcium is placed in the circuit at F, and chloride of calcium is also placed in the pipe at *f*. A small tube leading from the coil is carried to the shorter leg of the glass siphon gauge G, of which the longer leg is 17 feet, and the shorter 12 feet long.

The thermometers employed were all carefully calibrated, and had about ten divisions to the degree Centigrade. We took the precaution of verifying the air- and bath-thermometers from time to time, especially when high temperatures were used, in which latter case a comparison between the thermometers at high temperature was made immediately after each experiment.

Atmospheric Air (Table I.).

In the experiments described in the present paper, the air was not deprived of its carbonic acid. It was simply dried by transmitting it in the first place, before it entered the pump, through a cylinder 18 inches long and 12 inches in diameter filled with chloride of calcium, and afterwards, in its compressed state, through a pipe 12 feet long and 2 inches in diameter filled with the same substance. The experiments were principally carried on in the winter season; so that the chloride kept dry for a long time. From its condition after some weeks' use, it was evident that the water was removed, almost as much as chloride of calcium can remove it, after the air had traversed three inches of the chloride contained by the first vessel.

Oxygen Gas (Table II.).

This elastic fluid was procured by cautiously heating chlorate of potash mixed with a small quantity of peroxide of manganese. In its way to the meter it passed through a tube containing caustic potash, in order to deprive it of any carbonic acid it might contain. The same drying-apparatus was employed as in the case of atmospheric air.

Nitrogen Gas (Table III.).

In preparing this gas the meter was first filled with air, and then a long shallow tin vessel was floated under it, containing sticks of phosphorus so disposed as to burn in succession. Some hours were allowed to elapse after the combustion had terminated, in order to allow of the deposition of the phosphoric acid formed.

Carbonic Acid (Table IV.).

This gas was formed by adding sulphuric acid to a solution of carbonate of soda. It was dried in the same manner as all the other gases.

TABLE I.

1 No. of experiment.	2 Cubical inches of air transmitted per minute.	3 Pressure over that of the atmosphere, in inches of mercury.	4 Temperature of the bath.	5 Thermal effect.	6 Correction on account of conduction of heat.	7 Corrected thermal effect.	8 Thermal effect reduced to the pressure of 100 inches of mercury.	9 Time occupied by experiment, in minutes.	10 Number of observations comprised in each mean.	11 Extreme range of the temperature of the bath.	12 Extreme range of the temperature of the air.	13 Extreme range of the pressure.
1	3000	83·96	4·499	−0·711	−0·044	−0·755	−0·900	14	5	0·020	0·015	2·25
2	3600	136·19	6·112	−1·11	−0·058	−1·168	−0·858	24	7	0·017	0·055	1·7
3	2600	156·59	6·082	−1·307	−0·094	−1·401	−0·895	15	5	0·009	0·065	8·0
4	1750	139·58	7·471	−1·137	−0·122	−1·259	−0·902	24	15	0·006	0·19	5·3
5	2250	153·9	7·640	−1·231	−0·103	−1·334	−0·867	12·5	20	0·008	0·028	3·6
6	2300	159·3	8·546	−1·252	−0·102	−1·354	−0·850	18	10	0·017	0·105	3·0
7	2060	165·73	8·2	−1·329	−0·121	−1·450	−0·875	14	20	0·034	0·128	4·8
8	1500	129·73	8·72	−1·019	−0·127	−1·146	−0·883	8	15	0·008	0·135	2·9
9	1500	128·9	24·92	−0·983	−0·037	−1·020	−0·791	12	9	0·015	0·09	7·0
10	5000	128·8	27·81	−0·947	−0·036	−0·983	−0·764	26	8	0·029	0·064	7·0
11	4600	122·8	42·64	−0·874	−0·036	−0·910	−0·741	8	15	0·127	0·122	2·0
12	5000	123·5	43·54	−0·937	−0·037	−0·974	−0·789	6	4	0·058	0·09	0
13	4800	137	47·92	−0·943	−0·037	−0·980	−0·715	6	3	0·02	0·02	0
14	5000	147	49·96	−0·969	−0·049	−1·018	−0·692	35	30	0·05	0·08	0
15	3700	146	53·375	−0·860	−0·028	−0·888	−0·608	28	30	0·14	0·26	0
16	5600	146	64·9	−0·870	−0·029	−0·899	−0·616	24	20	0·18	0·23	0
17	5700	112·43	89·901	−0·469	−0·033	−0·502	−0·446	20	10	0·202	0·273	8·6
18	2700	147	90·353	−0·821	−0·091	−0·912	−0·620	4	3	0·022	0·085	0
19	1700	153·16	92·486	−0·756	−0·083	−0·839	−0·547	19	10	0·112	0·23	0
20	1700	156·5	92·603	−0·674	−0·040	−0·714	−0·456	12	20	0·18	0·08	7·0
21	3150	146	93·78	−0·700	−0·036	−0·736	−0·504	24	10	0·236	0·255	3·5
22	3800	158·5	97·528	−0·722	−0·029	−0·751	−0·474	20	16	0·112	0·115	0

TABLE II.

		1	2	3	4	5	6
1	No. of experiment.	1	2	3	4	5	6
2	Cubical inches of elastic fluid transmitted per minute.	2000	2000	1700	3150	3150	4500
3	Composition of the elastic fluid.	{5·095 N / 94·905 O}	{54·62 N / 45·38 O}	{3·64 N / 96·36 O}	{22·37 N / 77·63 O}	{51·03 N / 48·97 O}	{4 N / 96 O}
4	Pressure over that of the atmosphere, in inches of mercury.	159·28	161·81	151	159·77	154·1	152
5	Temperature of the bath.	8·682	87·5	89·466	90·8	92·792	95·453
6	Thermal effect.	-1·547	-1·373	-1·069	-0·840	-0·734	-0·795
7	Correction on account of conduction of heat.	-0·145	-0·129	-0·118	-0·050	-0·043	-0·083
8	Corrected thermal effect.	-1·692	-1·502	-1·187	-0·890	-0·777	-0·828
9	Thermal effect reduced to the pressure of 100 inches of mercury.	-1·061	-0·928	-0·786	-0·557	-0·504	-0·544
10	Ditto, calculated for pure oxygen.	-1·075	-1·074	-0·800	-0·580	-0·527	-0·570
11	Time occupied by experiment, in minutes.	9	11	14	12	12	11
12	Number of observations comprised in each mean.	10	10	10	10	10	8
13	Extreme range of the temperature of the bath.	0·007	0·017	0·45	0·326	0·18	0·135
14	Extreme range of the temperature of the elastic fluid.	0·35	0·046	0·43	0·336	0·19	0·158
15	Extreme range of the pressure.	7·8	0·9	6·2	8·0	4·0	0

TABLE III.

		1	2	3
1	No. of experiment.	1	2	3
2	Cubical inches of elastic fluid transmitted per minute.	2050	2500	2500
3	Composition of the elastic fluid.	{7·9 O / 92·1 N}	{2·2 O / 97·8 N}	{12·5 O / 87·5 N}
4	Pressure over that of the atmosphere, in inches of mercury.	163·38	162·65	164·61
5	Temperature of the bath.	7·204°	91·415	91·965
6	Thermal effect.	−1·448°	−0·857	−0·869
7	Correction on account of conduction of heat.	−0·133°	−0·064	−0·065
8	Corrected thermal effect.	−1·581°	−0·921	−0·934
9	Thermal effect reduced to the pressure of 100 inches of mercury.	−0·967°	−0·587	−0·567
10	Ditto, calculated for pure nitrogen.	−1·034°	−0·576	−0·691
11	Time occupied by experiment, in minutes.	7	13	12
12	Number of observations comprised in each mean.	8	10	9
13	Extreme range of the temperature of the bath.	0·008°	0·036	0·337
14	Extreme range of the temperature of the elastic fluid.	0·25°	0·48	0·378
15	Extreme range of the pressure.	6·2	4·5	3·0

TABLE IV.

No. of experiment.	Cubical inches of elastic fluid transmitted per minute.	Composition of the elastic fluid.	Pressure over that of the atmosphere, in inches of mercury.	Temperature of the bath.	Thermal effect.	Correction on account of conduction of heat.	Corrected thermal effect.	Thermal effect reduced to the pressure of 100 inches of mercury.	Ditto, calculated for pure carbonic acid, calling its sp. heat for equal vol. 1·39.	Time occupied by experiment, in minutes.	Number of observations comprised in each mean.	Extreme range of the temperature of the bath.	Extreme range of the temperature of the elastic fluid.	Extreme range of the pressure.
1	2450	Air 68·42, CO₂ 31·58	163·7	7·362	−2·699	−0·190	−2·889	−1·765	−3·166	12	10	0	0·16	3·2
2	2350	Air 89·16, CO₂ 10·84	148·82	7·360	−1·621	−0·125	−1·746	−1·173	−2·990	14	10	0·004	0·282	9·2
3	3100	Air 96·48, CO₂ 3·52	164·07	7·384	−6·719	−0·299	−7·018	−4·277	−4·367	6·5	6	0·008	0·021	1·4
4	2500	Air 62·5, CO₂ 37·5	162·925	7·407	−2·839	−0·191	−3·030	−1·860	−3·052	8	8	0·007	0·11	5·8
5	2300	Air 88·13, CO₂ 11·87	158·08	7·433	−1·682	−0·132	−1·814	−1·147	−2·648	10	10	0·005	0·107	5·2
6	2260	Air 97·46, CO₂ 2·54	163·52	7·608	−1·407	−0·116	−1·523	−0·931	−2·753	8	8	0·007	0·064	2·0
7	3800	Air 4·0, H 5·286, CO₂ 90·714	161·97	7·960	−6·131	−0·262	−6·393	−3·947	−4·215	6	8	0	0·18	4·8

[illegible]	[illegible]	[illegible] H 49·3 CO_2	[illegible]	8·020	−2·189	−0·117	−2·306	−1·500	−2·631	5	5	0	0·19	1·6
9	1500	7·09 Air 67·05 H 25·86 CO_2	97·56	8·296	−0·543	−0·063	−0·606	−0·622	−1·940	15	15	0·012	0·146	5·4
10	2925	2·11 Air 97·89 CO_2	167·25	93·523	−3·418	−0·160	−3·578	−2·139	−2·164	10	10	0·382	0·49	4·0
11	2925	56·78 Air 43·22 CO_2	167·4	91·26	−1·746	−0·099	−1·845	−1·102	−1·674	30	20	0·292	0·49	11·0
12	2925	77·77 Air 22·23 CO_2	146·83	91·642	−1·292	−0·077	−1·369	−0·938	−2·053	9	6	0·045	0·245	3·5
13	5500	0·83 Air 99·17 CO_2	146	54·0	−4·184	−0·104	−4·288	−2·937	−2·951	24	16	0·24	0·46	0
14	5300	67·7 Air 32·3 CO_2	147	49·703	−1·832	−0·059	−1·891	−1·286	−2·225	24	16	0·025	0·17	0
15	5600	87·77 Air 12·23 CO_2	145	49·764	−1·250	−0·032	−1·282	−0·884	−2·025	20	16	0·01	0·11	0
16	5100	1·83 Air 98·17 CO_2	127·5	35·604	−4·186	−0·112	−4·298	−3·371	−3·407	18	15	0·03	0·095	0
17	5000	1·66 Air 98·34 CO_2	151	97·553	−3·11	−0·084	−3·194	−2·115	−2·135	20	16	0·292	0·272	0
1	2	3	4	5	6	7	8	9	10	11	12	13	14	15

TABLE V.

No. of experiment.	Cubical inches of elastic fluid transmitted per minute.	Composition of the elastic fluid.	Pressure over that of the atmosphere, in inches of mercury.	Temperature of the bath.	Thermal effect.	Correction on account of conduction of heat.	Corrected thermal effect.	Thermal effect reduced to the pressure of 100 inches of mercury.	Ditto, calculated for pure hydrogen.	Time occupied by experiment, in minutes.	Number of observations comprised in each mean.	Extreme range of the temperature of the bath.	Extreme range of the temperature of the elastic fluid.	Extreme range of the pressure.
1	3000	17·635 Air 82·365 H	64·1	6·34	−0·144	−0·009	−0·153	−0·239	−0·104	3	3	0·0	0·0	0
2	3000	75·16 Air 24·84 H	99·86	6·355	−0·564	−0·035	−0·599	−0·600	+0·226	10	4	0	0·15	1·5
3	3900	4·866 Air 95·134 H	49·91	6·132	+0·033	+0·002	+0·035	+0·070	+0·118	12	6	0·002	0·06	1·2
4	2900	78·295 Air 21·705 H	99·657	5·808	−0·535	−0·034	−0·569	−0·571	+0·525	34	12	0·03	0·11	1·85
5	2800	90·8 Air 9·2 H	86·885	7·244	+0·041	+0·008	+0·044	+0·05	+0·143	27	10	0·034	0·033	1·75
6	3300	1·798 Air 98·202 H	79·84	7·572	+0·043	+0·003	+0·046	+0·058	+0·075	28	8	0·008	0·023	2·85
7	2950	4·795 Air 95·205 H	74·08	6·654	+0·054	+0·004	+0·058	+0·078	+0·126	17	10	0·016	0·11	6·6

1	2	3	4	5	6	7	8	9	10	11	12	13	14	15
8	2650	67·75 Air / 32·25 H	130·97	6·717	−0·571	−0·040	−0·611	−0·466	+0·383	12	6	0·01	0·07	2·6
9	3800	4·07 Air / 95·93 H	100·72	6·781	+0·039	+0·002	+0·041	+0·041	+0·08	10	10	0·012	0·078	2·6
10	2700	58·29 Air / 41·71 H	144·02	6·846	−0·504	−0·035	−0·539	−0·375	+0·317	8·5	8	0·011	0·07	3·6
11	1900	91·81 Air / 8·19 H	152·67	7·406	−1·002	−0·099	−1·101	−0·721	+0·904	9	8	0	0·225	9·0
12	1760	97·56 Air / 2·44 H	138·55	7·474	−1·032	−0·11	−1·142	−0·825	+0·814	13	8	0·001	0·053	8·2
13	3100	4·375 Air / 95·625 H	88·66	88·66	+0·178	+0·011	+0·189	+0·215	+0·248	14	6	0·08	0·17	4·6
14	3300	6·08 Air / 93·92 H	91·52	92·951	+0·081	+0·005	+0·086	+0·094	+0·132	18	8	0·157	0·07	3·2
15	3000	5·043 Air / 94·957 H	73·99	90·353	+0·072	+0·005	+0·077	+0·104	+0·136	20	10	0·18	0·11	1·65
16	3000	2·99 Air / 97·01 H	85·15	89·242	+0·111	+0·007	+0·118	+0·139	+0·159	42	15	0·472	0·44	3·2
17	2900	4·13 Air / 95·87 H	104·72	89·858	+0·073	+0·004	+0·077	+0·073	+0·098	15·5	10	0·09	0·035	6·2

Hydrogen (Table V.).

Our method in procuring this elastic fluid was to pour sulphuric acid, prepared from sulphur, into a carboy nearly filled with water and containing fragments of sheet zinc. The gas was passed through a tube filled with rags steeped in a solution of sulphate of copper, and then through a tube filled with sticks of caustic potash. The rags became speedily browned, and we therefore adopted the plan of pouring a small quantity of solution of sulphate of copper from time to time into the carboy itself. This succeeded perfectly; the rags retained their blue colour, and the gas was rendered perfectly inodorous, whilst at the same time its evolution became much more free and regular.

Remarks on the Tables.

The correction for conduction of heat through the plug, inserted in column 6 of Table I., and in column 7 of the rest of the Tables, was obtained from data furnished by experiments in which the difference between the temperature of the bath and the air was purposely made very great. It was considered as directly proportional to the difference of temperature, and inversely to the quantity of elastic fluid transmitted in a given time.

The 10th column of Tables II., III., IV., and V. is calculated on the hypothesis that, in mixtures with other gases, atmospheric air retains its thermal qualities without change. This hypothesis is almost certainly incorrect, since it is reasonable to expect that the effect of mixture on the physical character is experienced by each of the constituent gases. The column is given as one method of showing the effect of mixture.

Effect of Mixture on the Constituent Gases.—Although the experiments on nitrogen given in Table III. are not so numerous as might be desired, we may infer from them, and the results in Table II., that common air and all other mixtures of oxygen and nitrogen behave more like a perfect gas, *i. e.* give less cooling effect than either one or the other gas

alone. We might expect the mixture to be something intermediate between the two. But this does not appear to be
the case. The two are very nearly equal in their deviations
from the condition of a perfect gas. Nitrogen deviates less
than oxygen, but oxygen mixed with nitrogen differs less
than nitrogen!

In the case of carbonic acid, which at low temperatures
(7°) deviates five times as much as atmospheric air, we
might expect that a mixture of CO_2 and air would deviate
more than air and less than CO_2. This is the case (see
Table IV.). Further, we might expect the two to contribute
each its proportion of cooling effect according to its own
amount, and its specific heat volume for volume. But do
the mixtures exhibit such a result? No! See column 10,
Table IV., in which also note, under experiments 8 and 9,
the great diminution produced by the admixture of hydrogen.

If, instead of attributing to air and carbonic acid moments
in proportion to their specific heats, or 1 : 1·39, as we have
done in column 10, we use 1: ·7, we obtain more consistent
results.

Let δ denote the cooling effect experienced by air per 100
inches of mercury, δ' that by carbonic acid, and Δ that by a
mixture of volume V of air, and V' of carbonic acid; then
we may take

$$\Delta = \frac{mV\delta + m'V'\delta'}{mV + m'V'}$$

to represent the cooling effect for the mixture, where m and
m' are numbers which we may call the moments (or impor-
tances) of the two in determining the cooling effect for the
mixture. The ratio of m to m' is the proper result of each
experiment on a mixture, if we knew with perfect accuracy
the cooling effect for each gas with none of the other mixed.
Now for common air we have direct experiments (Table I.),
and know the cooling effect for it better than from any
inferences from mixtures. But for pure CO_2 we know
the effect, for the most part, only inferentially. Hence,
having tried making $m : m' :: 1 : 1·39$ without obtaining

2 A 2

consistent results, we tried other proportions; and, after various attempts, found that $m : m' :: 1 : \cdot 7$, for all temperatures and pressures within the limits of our experiments, gives results as consistent with one another as the probable errors of the experiments justify us in expecting. Thus, using the formula

$$\Delta = \frac{V\delta + V'\delta' \times \cdot 7}{V + V' \times \cdot 7},$$

we have, for calculating the effect for CO_2 from any experiment on a mixture, the following formula,

$$\delta' = \frac{(V + V' \times \cdot 7)\Delta - V\delta}{V' \times \cdot 7}.$$

Hence, using the numbers in columns 3 and 9 of Table IV. which relate to mixtures of air and carbonic acid alone, we find :—

TABLE VI.

No. of experiment.	Proportions of mixtures.		Temperature of bath.	Thermal effect for air.	Deduced thermal effect for pure CO_2.
	Air.	CO_2.			
1.	68·42	31·58	7·36	−·88	−4·51
2.	89·16	10·84	7·36	−·88	−4·61
3.	3·52	96·48	7·38	−·88	−4·46
4.	62·5	37·5	7·41	−·88	−4·19
5.	88·13	11·87	7·43	−·88	−3·98
6.	97·46	2·54	7·61	−·88	−3·89
16.	1·83	98·17	35·6	−·75	−3·44
14.	67·7	32·3	49·7	−·70	−3·04
15.	87·77	12·23	49·76	−·70	−2·77
13.	0·83	99·17	54	−·66	−2·96
10.	2·11	97·89	93·52	−·51	−2·19
11.	56·78	43·22	91·26	−·51	−2·21
12.	77·77	22·23	91·64	−·51	−3·08
17.	1·66	98·34	97·55	−·49	−2·16
1	2		3	4	5

The agreement for each set of results at temperatures nearly agreeing (with one exception, No. 12), shows that the assumption $m : m' :: 1 : \cdot 7$ cannot be far wrong within our limits of temperature.

[Received subsequently to the reading of the Paper.]

Application of the preceding Results to Deduce approximately the Equation of Elasticity for the Gases experimented on.

The " equation of elasticity " for any fluid is the most appropriate name for the equation expressing the relation between the pressure and the volume of any portion of the fluid. As this relation depends on the temperature, the equation expressing it involves essentially three variables, which, as in our previous communications on this subject, we shall denote by p, v, t. Of these, p is the pressure in units of force per unit of area, v the volume of a unit mass of the fluid, and t the temperature according to the absolute thermo-dynamic system of thermometry * which we have proposed. As before, we shall still adopt a degree, or thermometric unit, agreeing approximately with the degree Centigrade of the air-thermometer; accordiug to which, as we have demonstrated by experiment †, the value of t for the freezing-point is within a few tenths of a degree of 273·7 (its value at the standard boiling-point being, by definition of the Centigrade scale, 100° more than at the freezing-point).

Instead of, as in our previous communications, taking v and t as independent variables, we shall now take p and t; and we shall accordingly consider the object of the equation of elasticity as being to express v explicitly as a function of p and t. Whatever may be the relation between these elements, the thermal effect, $d\vartheta$ (reckoned as positive when it is a rise in temperature), produced by forcing the fluid in a continuous stream through a narrow passage or porous plug by an infinitely small difference of pressures, dp, will be given by the formula

$$\frac{d\vartheta}{dp} = -\frac{1}{\mathrm{JK}}\left(t\frac{dv}{dt}-v\right),$$

where K denotes the thermal capacity, under constant

* Philosophical Transactions, 1854, p. 350.
† Ibid. p. 352.

pressure, of unit of mass of the fluid.   This formula may be derived from equation (15) of our previous communication already referred to, by substituting p, v, and $-\vartheta$ for P, V, and δ in that equation, changing to p and t, instead of v and t, as independent variables, and differentiating with reference to p.   It is scarcely necessary to remark that a direct demonstration of our present formula, founded on elementary thermodynamic principles, may be readily obtained.

Each experiment, of the several series recorded above, gives a value for $\dfrac{d\vartheta}{dp}$, which is found by multiplying the " corrected thermal effect " by $\dfrac{\cdot 299218}{2114}$, to reduce from the amounts per 100 inches of mercury to the amounts per pound per square foot.   Now by examining carefully the series of results for different temperatures, in the cases of atmospheric air and of carbonic acid, we find that they follow very closely the law of varying inversely as the square of the absolute temperature (or temperature Centigrade with 273·7 added). Thus for air the formula

$$^{\circ}\!\cdot 92 \times \left(\frac{273\cdot 7}{t}\right)^2,$$

and for carbonic acid

$$4^{\circ}\!\cdot 64 \times \left(\frac{273\cdot 7}{t}\right)^2,$$

express, the former almost accurately, the latter with a deviation which we shall hereafter investigate, the results through the whole range of temperature for which the investigation has been carried out.

Air.

Temperature.	Actual cooling effect.	Theoretical cooling effect.
$\overset{\circ}{0}$	$\overset{\circ}{\cdot 92}$	$\overset{\circ}{\cdot 92}$
7·1	·88	·87
39·5	·75	·70
92·8	·51	·51

Carbonic acid.

Temperature.	Actual cooling effect.	Theoretical cooling effect.
$\overset{\circ}{0}$	$\overset{\circ}{4\cdot 64}$	$\overset{\circ}{4\cdot 64}$
7·4	4·37	4·4
35·6	3·41	3·63
54·0	2·95	3·23
93·5	2·16	2·57
97·5	2·14	2·52

We have not experiments enough to establish the law of variation with temperature of the thermal effect for the pure gases oxygen and nitrogen, or for any stated mixture of them other than common air; but there can be no doubt, from the general character of the results, that the same law will be about as approximately followed by them as it is by air.

Hence we may presume that in all these cases the cooling effect is very well represented by the formula

$$-\frac{d\vartheta}{dp} = A\left(\frac{273\cdot 7}{t}\right)^2.$$

Comparing this with the general formula given above, we find

$$t\frac{dv}{dt} - v = AJK\left(\frac{273\cdot 7}{t}\right)^2.$$

The general integral of this differential equation, for v in terms of t, is

$$v = Pt - \tfrac{1}{3}AJK\left(\frac{273\cdot 7}{t}\right)^2,$$

P denoting an arbitrary constant with reference to t, which, so far as this integration is concerned, may be an arbitrary function of p. To determine its form, we remark in the first place, in consequence of Boyle's law, that it must be approximately $\dfrac{C}{p}$, C being independent of both pressure and temperature; and thus, if we omit the second term, we have two gaseous laws expressed by the approximate equation

$$v = \frac{Ct}{p}.$$

Now it is generally believed that at higher and higher temperatures the gases approximate more and more nearly to the rigorous fulfilment of Boyle's law. If this is true, the complete expression for P must be of the form $\frac{C}{p}$, since any other would simply show deviation from Boyle's law at very high temperatures, when the second term of our general integral disappears. Assuming then that no such deviation exists, we have, as the complete solution,

$$v = \frac{Ct}{p} - \tfrac{1}{3}\text{AJK}\left(\frac{273\cdot7}{t}\right)^2 p.$$

This is an expression of exactly the same form as that which Professor Rankine found applicable to carbonic acid, in the first place to express its deviations from the laws of Boyle and Gay-Lussac, as shown by Regnault's experiments, and which he afterwards proved to give correctly the law and the absolute amount of the cooling effect demonstrated by our first experiments on that gas *.

That more complicated formulæ were found for the law of elasticity for common air both by Mr. Rankine and by ourselves, now seems to be owing to an irreconcilability among the data we had from observation. The whole amounts of the deviations from the gaseous laws are so small for common air, that very small absolute errors in observations of so heterogeneous a character as those of Regnault on the law of compression and on the changes produced by pressure in the coefficients of expansion, and our own on the thermodynamic property on which we have experimented, may readily present us with results either absolutely inconsistent with one another, or only reconcilable by very strained assumptions. It is satisfactory now to find, when we have succeeded in extending our observations through a considerable range of temperature, that they lead to so simple a law; and it is probable that the formula we have

* Philosophical Transactions, 1854, Part II. p. 336.

been led to by these observations alone will give the deviations from Boyle's law, and the changes produced by pressure in the coefficients of expansion, with more accuracy than has hitherto been attained in attempts to determine these deviations by direct observation. We must, however, reserve for a future communication the comparison between such results of our theory and experiments and Regnault's direct observations. In the mean time we conclude by putting the integral equation of elasticity into a more convenient form, by taking $C = \dfrac{\mathfrak{H}}{t_o}$, where $\mathfrak{H}$ denotes the "height of the homogeneous atmosphere" for the gas under any excessively small pressure, at any temperature t, and taking t_o to denote the absolute temperature of freezing water, in which case we shall have, as nearly as observations hitherto made allow us to determine,

$$t_o = 273°{\cdot}7.$$

Then, in terms of this notation, and of that above explained in which t, p, v denote absolute temperature, pressure in pounds weight per square foot, and volume in cubic feet of one pound of air, the equation of elasticity investigated above becomes

$$v = \frac{\mathfrak{H}t}{pt_o} - \tfrac{1}{3}\mathrm{AJK}\left(\frac{t_o}{t}\right)^2,$$

where A denotes the amount of the thermal effect per pound per square foot, determined by our observations, reckoned positive when it is a depression of temperature.

On the Thermal Effects of Fluids in Motion.—
Part IV.   *By* J. P. JOULE, *LL.D.*, *F.R.S.*, *and*
Professor W. THOMSON, *A.M.*, *F.R.S.*

[Abstract.—Proceedings of the Royal Society, vol. xii. p. 202.]

A BRIEF notice of some of the experiments contained in
this paper has already appeared in the 'Proceedings.' Their
object was to ascertain with accuracy the lowering of temperature, in atmospheric air and other gases, which takes place
on passing them through a porous plug from a state of high
to one of low pressure. Various pressures were employed,
with the result (indicated by the Authors in their Part II.)
that the thermal effect is approximately proportional to the
difference of pressure on the two sides of the plug. The
experiments were also tried at various temperatures, ranging
from 5° to 98° Cent.; and have shown that the thermal effect,
if one of cooling, is approximately proportional to the inverse
square of the absolute temperature. Thus, for example, the
refrigeration at the freezing temperature is about twice that
at 100° Cent. In the case of hydrogen, the reverse phenomenon of a rise of temperature on passing through the plug
was observed, the rise being doubled in quantity when the
temperature of the gas was raised to 100°. This result is
conformable with the experiments of Regnault, who found
that hydrogen, unlike other gases, has its elasticity increased
more rapidly than in the inverse ratio of the volume. The
Authors have also made numerous experiments with mixtures
of gases, the remarkable result being that the thermal effect
(cooling) of the compound gas is less than it would be if
the gases after mixture retained in integrity the physical
characters they possessed while in a pure state.

PUBLICATIONS FROM WHICH THE PAPERS HAVE BEEN EXTRACTED.

ACADÉMIE DES SCIENCES: *Comptes Rendus*, 1847, Aug. 23.

Expériences sur l'Identité entre le Calorique et la Force mécanique. Détermination de l'équivalent par la Chaleur dégagée pendant la friction du Mercure, i. 283.

ANNALS OF ELECTRICITY.

Vol. ii. p. 122. Description of an Electro-magnetic Engine, i. 1.

Vol. iii. p. 437. Description of an Electro-magnetic Engine, with Experiments, i. 4.

Vol. iv. p. 58. On the use of Electro-magnets made of Iron Wire for the Electro-magnetic Engine, i. 6.

Vol. iv. p. 131. Investigations in Magnetism and Electro-magnetism, i. 10.

Vol. iv. p. 135. Investigations in Magnetism and Electro-magnetism, i. 15.

Vol. iv. p. 203. Description of an Electro-magnetic Engine, i. 16.

Vol. iv. p. 474. On Electro-magnetic Forces, i. 19.

Vol. v. p. 170. On Electro-magnetic Forces, i. 40.

Vol. v. p. 187. On Electro-magnetic Forces, i. 27.

Vol. vi. p. 431. Description of a new Electro-magnet, i. 42.

Vol. viii. p. 219. On a new Class of Magnetic Forces, i. 46.

BRITISH ASSOCIATION. *Reports.*

1845. Chemical Section, p. 31. On the Mechanical Equivalent of Heat, i. 202.

1847. Sections, p. 5. On the Mechanical Equivalent of Heat, as determined from the Heat evolved by the Agitation of Liquids, i. 276.

1848. Sections, p. 21. On the Mechanical Equivalent of Heat, and on the Constitution of Elastic Fluids, i. 288.

1850. Sections, p. 55. On some Amalgams, i. 331.

BRITISH ASSOCIATION. *Reports* (continued).

1855. An account of some Experiments with a large Electro-magnet, i. 368.

1859. Sections, p. 12. Notice of Experiments on the Heat developed by Friction in Air, i. 399.

1867. (Committee on Standards of Electrical Resistance.) Determination of the Dynamical Equivalent of Heat from the Thermal Effects of Electric Currents, i. 542.

1871. Notice of, and Observations with, a New Dip-Circle, i. 584.

CHEMICAL SOCIETY. *Memoirs.*

Vol. ii. p. 401. On Atomic Volume and Specific Gravity. [Series I. of the Joint Papers with Sir Lyon Playfair.] ii. 11.

Vol. iii. p. 57. Researches on Atomic Volume and Specific Gravity. Series II. On the Relation in Volumes between Simple Bodies, their Oxides and Sulphurets, and on the Differences exhibited by Polymorphous and Allotropic Substances. [Joint Paper with Sir Lyon Playfair.] ii. 117.

Vol. iii. p. 199. Researches on Atomic Volume and Specific Gravity. Series III. On the Maximum Density of Water. [Joint Paper with Sir Lyon Playfair.] ii. 173.

CHEMICAL SOCIETY. *Journal.*

Vol. i. p. 121. Researches on Atomic Volume and Specific Gravity. Series IV. Expansion by Heat of Salts in the Solid State. [Joint Paper with Sir Lyon Playfair.] ii. 180.

Vol. i. p. 139. Researches on Atomic Volume and Specific Gravity. Series V. On the Disappearance of the Volume of the Acid, and in some cases of the Volume of the Base, in the Crystals of Highly Hydrated Salts. [Joint Paper with Sir Lyon Playfair.] ii. 203.

COMPTES RENDUS : *See* Académie.

INSTITUTION OF ENGINEERS IN SCOTLAND. *Transactions.*

Vol. viii. p. 56. On a Self-Acting Apparatus for Steering Ships, i. 570.

LONDON ELECTRICAL SOCIETY. *Proceedings.* On Voltaic Apparatus, i. 53.

MANCHESTER COURIER, May 5 and 12, 1847. On Matter, Living Force, and Heat, i. 265.

MANCHESTER LITERARY AND PHILOSOPHICAL SOCIETY. *Memoirs.*

2nd Series, Vol. vii. p. 87. On the Heat evolved during the Electrolysis of Water, i. 109.

Vol. vii. p. 559. On a new Method for ascertaining the Specific Heat of Bodies, i. 192.

Vol. viii. p. 375. Note on the Employment of Electrical Currents for ascertaining the Specific Heat of Bodies, i. 201.

Vol. ix. p. 107. Some Remarks on Heat and the Constitution of Elastic Fluids, i. 290.

Vol. x. p. 173. On the Economical Production of Mechanical Effect from Chemical Forces, i. 363.

Vol. xiv. p. 49. On the Fusion of Metals by Voltaic Electricity, i. 381.

Vol. xv. p. 143. Note on Dalton's Determination of the Expansion of Air by Heat, i. 384.

Vol. xv. p. 146. On the Utilization of the Sewage of London and other large Towns, i. 386.

3rd Series, Vol. i. p. 97. On a Method of Testing the Strength of Steam-Boilers, i. 480.

Vol. i. p. 99. Experiments on the Total Heat of Steam, i. 482.

Vol. i. p. 102. Experiments on the Passage of Air through Pipes and Apertures in thin Plates, i. 485.

Vol. ii. p. 115. On some Amalgams, i. 490.

Vol. iii. p. 292. Observations on the Alteration of the Freezing-point in Thermometers, i. 558.

MANCHESTER LITERARY AND PHILOSOPHICAL SOCIETY. *Proceedings.*

Vol. ii. p. 218. On the Probable Cause of Electrical Storms, i. 500.

Vol. iii. p. 5. Notice of a Compressing Air-Pump, i. 531.

Vol. iii. p. 39. Note on a Mirage at Douglas, i. 532.

Vol. iii. p. 47. On a Sensitive Barometer, i. 534.

Vol. iii. p. 73. On a Sensitive Thermometer, i. 534.

Vol. iii. p. 203. Note on the Meteor of February 6, 1818, i. 536.

Vol. iv. p. 28. On a Method of Hardening Steel Wires for Magnetic Needles, i. 540.

PHILOSOPHICAL MAGAZINE (*continued*).

    Vol. xiv. p. 211. Some Remarks on Heat and the Constitution of Elastic Fluids, i. 290.

    Vol. xv. April 1858. On the Intensity of Light during the recent Solar Eclipse, i. 402.

    Vol. xv. p. 432. On an Improved Galvanometer, i. 404.

PHILOSOPHICAL TRANSACTIONS. *See* Royal Society.

POLE (W.). *Life of Sir W. Fairbairn.* On some Physical Properties of Beeswax, i. 605.

ROYAL SOCIETY. *Philosophical Transactions.*

    1850, vol. cxl. p. 61. On the Mechanical Equivalent of Heat, i. 298.

    1852, vol. cxlii. p. 65. On the Air-Engine, i. 331.

    1853, vol. cxliii. p. 357. On the Thermal Effects of Fluids in Motion. [Part I. of the Joint Papers with Sir W. Thomson.] ii. 231.

    1854, vol. cxliv. p. 321. On the Thermal Effects of Fluids in Motion. Part II. [Joint Paper with Sir. W. Thomson.] ii. 247.

    1856, vol. cxlvi. p. 287. Introductory Research on the Induction of Magnetism by Electrical Currents, i. 369.

    1859, vol. cxlix. p. 91. On some Thermo-dynamic Properties of Solids, i. 413.

    1859, vol. cxlix. p. 133. On the Thermal Effects of Compressing Fluids, i. 474.

    1860, vol. cl. p. 325. On the Thermal Effects of Fluids in Motion. Part III. On the Changes of Temperature experienced by Bodies Moving through Air. [Joint Paper with Sir W. Thomson.] ii. 322.

    1861, vol. cli. p. 133. On the Surface-condensation of Steam, i. 532.

    1862, vol. clii. p. 579. On the Thermal Effects of Fluids in Motion. Part IV. [Joint Paper with Sir W. Thomson.] ii. 342.

    1878, vol. clxix. p. 365. New Determination of the Mechanical Equivalent of Heat, i. 632.

Royal Society. *Proceedings.*

INDEX.

Kane on ammoniacal salts, discussed, ii. 104 *et seq.*

Keir first observed the passive state of iron, i. 162 *n.*

Kite, Common, utilization of, i. 628, 629.
  defects of, i. 628.
  improved form of, i. 629.

Kopp, on atomic volumes, alluded to, ii. 12.

Kupffer on the influence of temperature on the elasticity of metals, referred to, i. 408 *n.*

Laplace, confirmation of his theory of the velocity of sound, i. 282.

Latent heat, defined, i. 121.

Lead, thermal effect of longitudinal compression on, i. 463.
  vol. and sp. gr. of, ii. 122, 128.
  vol. and sp. gr. of when melted, ii. 137.
  vol. and sp. gr. of the Chromate, ii. 91.
    of the Sesquibasic Chromate, ii. 91.
    of the Subchromate, ii. 91.
    of the Peroxide, ii. 149.
    of the Protoxide, ii. 149.
  sp. gr. and expansion of do., ii. 189.
  vol. and sp. gr. of the Suboxide, ii. 149.
    of the Nitrate in solution, ii. 59.
  sp. gr. and expansion of do., ii. 191.
  vol. and sp. gr. of the Basic Nitrate in solution, ii. 62.
    of the Sulphuret, ii. 155.
  sp. gr. and expansion of the Sulphuret, ii. 188.
  vol. and sp. gr. of the Sesqui-sulphuret, ii. 155.
    of Minium, ii. 149.

Leather, thermal effects of tension on, i. 455.

Lenz, on the heat evolved by voltaic electricity, mentioned, i. 211.

Liebig on the value of animal excrements as manure, quoted, i. 393.

Lifting power of magnets, i. 13, 14, 32, 33.
  remains after electric force is cut off, i. 37.

Light, its intensity during an Eclipse determined, i. 402, 403.
  is evolved in combustion with small increase of heat, i. 106, 107.

Lightning, remarkable appearance of, i. 329, 330.

Lime, vol. and sp. gr. of, ii. 151.

Lime, Carbonate of, specific gravities of its two forms compared, ii. 161.

Liquids, method of determining their specific heat, i. 196.

Living force. See *Vis Viva.*

Local action in voltaic batteries, i. 53–59.

Lubbock, Sir J. W., his theory of shooting-stars referred to, i. 286.

Magnesia, vol. and sp. gr. of, ii. 151.
  of the Nitrate insolution, ii. 61.
  of the Sulphate, ii. 34.
  of the Sulphate, anhydrous, ii. 49.
  sp. gr. and expansion of the Sulphate, ii. 190.
  peculiar behaviour of the Sulphate when dissolved in cold water, ii. 34.
  ——— and Ammonia, vol. and sp. gr. of the Sulphate, ii. 44.
  of the Sulphate, anhydrous, ii. 51.
  sp. gr. and expansion of the Sulphate, ii. 194.
  ——— and Potash, vol. and sp. gr. of the Sulphate, ii. 44.
  of the Sulphate, anhydrous, ii. 51.
  sp. gr. and expansion of do., ii. 195.

Magnesian metals, peculiarities of, ii. 36, 39.
  their parallelism to hydrogen, ii. 55, 101.
  their relation to the potassium group, ii. 102.
  proposed increase in their equivalents, ii. 152.

Magnesium, sp. gr. from specimen prepared by Prof. Liebig, ii. 135.
  vol. and sp. gr. of Chloride in sol., ii. 23.
  of Chloride, anhydrous, ii. 53.

Magnetism, its effect upon the dimensions of iron and steel bars, i. 235–264.
  apparatus used, i. 236.
  does not influence their bulk, i. 236, 263.
  law of elongation, i. 49, 242, 243, 245.
  elongation not affected by pressure, i. 252.
  shortening produced by tension, i. 256, 260, 262.

Magnetism, dimensions change at right angles to polarity, i. 263, 264.

Printed by TAYLOR AND FRANCIS, Red Lion Court, Fleet Street.

Printed in the United States
By Bookmasters